Transport, Mobility, and the Production of Urban Space

The contemporary urban experience is defined by flow and structured by circulating people, objects, and energy. Geographers have long provided key insights into transportation systems. But today, concerns for social justice and sustainability motivate new, critical approaches to mobilities. Reimagining the city prompts an important question: How best to rethink urban geographies of transport and mobility? This original book explores connections—in theory and practice—between transport geographies and "new mobilities" in the production of urban space. It provides a broad introduction to intersecting perspectives of urban geography, transport geography, and mobilities studies on urban "places of flows." Diverse, international, and leading-edge contributions reinterpret everyday intersections as nodes, urban corridors as links, cities and regions as networks, and the discourses and imaginaries that frame the politics and experiences of mobility. The chapters illuminate nearly all aspects of urban transport—street regulation and roadway planning, intended and "subversive" practices of car and truck drivers, planning and promotion of mass transit investments, and the restructuring of freight and logistics networks. Together these offer a unique and important contribution for social scientists, planners, and others interested in the politics of the city on the move.

Julie Cidell is Associate Professor of Geography and GIS at the University of Illinois at Urbana-Champaign.

David Prytherch is Associate Professor of Geography at Miami University, Oxford, Ohio.

Routledge Studies in Human Geography

This series provides a forum for innovative, vibrant, and critical debate within human geography. Titles will reflect the wealth of research that is taking place in this diverse and ever expanding field. Contributions will be drawn from the main subdisciplines and from innovative areas of work that have no particular subdisciplinary allegiances.

For a full list of titles in this series, please visit www.routledge.com

27 **Whose Urban Renaissance?**
An international comparison of urban regeneration policies
Edited by Libby Porter and Katie Shaw

28 **Rethinking Maps**
Edited by Martin Dodge, Rob Kitchin and Chris Perkins

29 **Rural–Urban Dynamics**
Livelihoods, mobility and markets in African and Asian frontiers
Edited by Jytte Agergaard, Niels Fold and Katherine V. Gough

30 **Spaces of Vernacular Creativity**
Rethinking the cultural economy
Edited by Tim Edensor, Deborah Leslie, Steve Millington and Norma Rantisi

31 **Critical Reflections on Regional Competitiveness**
Gillian Bristow

32 **Governance and Planning of Mega-City Regions**
An international comparative perspective
Edited by Jiang Xu and Anthony G.O. Yeh

33 **Design Economies and the Changing World Economy**
Innovation, production and competitiveness
John Bryson and Grete Rustin

34 **Globalization of Advertising**
Agencies, cities and spaces of creativity
James R. Faulconbridge, Peter J. Taylor, Jonathan V. Beaverstock and Corinne Nativel

35 **Cities and Low Carbon Transitions**
Edited by Harriet Bulkeley, Vanesa Castán Broto, Mike Hodson and Simon Marvin

36 **Globalization, Modernity and the City**
John Rennie Short

37 **Climate Change and the Crisis of Capitalism**
A chance to reclaim self, society and nature
Edited by Mark Pelling, David Manual Navarette and Michael Redclift

38 **New Economic Spaces in Asian Cities**
From industrial restructuring to the cultural turn
Edited by Peter W. Daniels, Kong Chong Ho and Thomas A. Hutton

39 **Landscape and the Ideology of Nature in Exurbia**
Green sprawl
Edited by Kirsten Valentine Cadieux and Laura Taylor

40 **Cities, Regions and Flows**
Edited by Peter V. Hall and Markus Hesse

41 **The Politics of Urban Cultural Policy**
Global perspectives
Edited by Carl Grodach and Daniel Silver

42 **Ecologies and Politics of Health**
Edited by Brian King and Kelley Crews

43 **Producer Services in China**
Economic and urban development
Edited by Anthony G.O. Yeh and Fiona F. Yang

44 **Locating Right to the City in the Global South**
Tony Roshan Samara, Shenjing He and Guo Chen

45 **Spatial-Economic Metamorphosis of a Nebula City**
Schiphol and the Schiphol region during the 20th century
Abderrahman El Makhloufi

46 **Learning Transnational Learning**
Edited by Åge Mariussen and Seija Virkkala

47 **Cultural Production in and Beyond the Recording Studio**
Allan Watson

48 **Global Perspectives on Gender and Space**
Edited by Ann M. Oberhauser and Ibipo Johnston-Anumonwo

49 **Fieldwork in the Global South**
Ethical challenges and dilemmas
Edited by Jenny Lunn

50 **Intergenerational Space**
Edited by Robert Vanderbeck and Nancy Worth

51 **Performativity, Politics, and the Production of Social Space**
Edited by Michael R. Glass and Reuben Rose-Redwood

52 **Knowledge and the City**
Concepts, applications and trends of knowledge-based urban development
Francisco Javier Carrillo, Tan Yigitcanlar, Blanca García and Antti Lönnqvist

53 **Migration, Risk and Uncertainty**
Allan M. Williams and Vladimír Baláž

54 **Transport, Mobility, and the Production of Urban Space**
Edited by Julie Cidell and David Prytherch

Transport, Mobility, and the Production of Urban Space

Edited by Julie Cidell and
David Prytherch

LONDON AND NEW YORK

First published 2015 by Routledge

2 Park Square, Milton Park, Abingdon, Oxfordshire OX14 4RN
711 Third Avenue, New York, NY 10017

*Routledge is an imprint of the Taylor & Francis Group,
an informa business*

First issued in paperback 2018

Copyright © 2015 Taylor & Francis

The right of the editors to be identified as the authors of the editorial material, and of the authors for their individual chapters, has been asserted in accordance with sections 77 and 78 of the Copyright, Designs and Patents Act 1988.

All rights reserved. No part of this book may be reprinted or reproduced or utilised in any form or by any electronic, mechanical, or other means, now known or hereafter invented, including photocopying and recording, or in any information storage or retrieval system, without permission in writing from the publishers.

Notice:
Product or corporate names may be trademarks or registered trademarks, and are used only for identification and explanation without intent to infringe.

Library of Congress Cataloging-in-Publication Data

Transport, mobility, and the production of urban space / edited by Julie Cidell and David Prytherch.
    pages cm. — (Routledge studies in human geography ; 54)
    Includes bibliographical references and index.
 1. Urban transportation. 2. Urbanization. 3. Human geography. I. Cidell, Julie. II. Prytherch, David.
    HE305.T735 2015
    303.48'32—dc23   2015000232

ISBN: 978-1-138-89134-0 (hbk)
ISBN: 978-1-138-54642-4 (pbk)

Typeset in Sabon
by Apex CoVantage, LLC

From Julie to Ted, Jean, Joseph, and Josephine
From David to Kathleen, Eleanor, Vivian, and Mary Ann

Contents

List of Figures   xiii
List of Tables   xv

Approaching the City as Place of Flows

Foreword 1: Transportation Geographies and Mobilities
Studies: Toward Collaboration   3
SUSAN HANSON

Foreword 2: Mobilizing Transportation, Transporting
Mobilities   12
MIMI SHELLER

1   Introduction: Transportation, Mobilities, and Rethinking
Urban Geographies of Flow   19
DAVID PRYTHERCH AND JULIE CIDELL

PART I
Intersections: Everyday Places as Nodes

2   Rules of the Road: Choreographing Mobility in the Everyday
Intersection   45
DAVID PRYTHERCH

3   Concrete Politics and Subversive Drivers on the Roads of
Hyderabad, India   64
BASCOM GUFFIN

4   A Bridge Too Far: Traffic Engineering Science and the
Politics of Rebuilding Milwaukee's Hoan Bridge   81
GREGG CULVER

PART II
Corridors: Links in the Network

5 From Climate Fight to Street Fight: The Politics of Mobility
   and the Right to the City 101
   JASON HENDERSON

6 The Social Life of Truck Routes 117
   PETER V. HALL

7 Uncanny Trains: Cities, Suburbs, and the Appropriate Place
   and Use of Transportation Infrastructure 134
   JULIE CIDELL

PART III
Networks: Cities and Regions in Wider Context

8 Place-Making, Mobility, and Identity: The Politics and
   Poetics of Urban Mass Transit Systems in Taiwan 153
   ANRU LEE

9 Contesting the Networked Metropolis: The Grand Paris
   Regime of Metromobility 172
   THERESA ENRIGHT

10 Towards a City-Regional Politics of Mobility: In Between
   Critical Mobilities and the Political Economy of Urban
   Transportation 187
   JEAN-PAUL D. ADDIE

PART IV
Circulation: Assemblages and Experiences of Mobility

11 Selling the Region as Hub: The Promises, Beliefs, and
   Contradictions of Economic Development Strategies
   Attracting Logistics and Flows 207
   MARKUS HESSE

12 The Politics of Public Transit in Postsuburban Toronto 228
   CHRISTIAN METTKE

13 Place-Framing and Regulation of Mobility Flows
   in Metropolitan 'In-Betweens'   245
   SOPHIE L. VAN NESTE

14 'Peace, Love, and Fun': An Aerial Cable Car and
   the Traveling Favela   263
   BIANCA FREIRE-MEDEIROS AND LEONARDO NAME

Moving Forward

15 Rethinking Mobility at the Urban-Transportation-Geography
   Nexus   281
   ANDREW E. G. JONAS

   Contributors   295
   Index   299

Figures

2.1	Intersections like this in Hamilton, OH are ubiquitous and uniform settings for everyday mobility.	47
2.2	Intersections like Main/MLK are 'controlled' through signs, signals, and standard markings (shown here) (Source: FHA, 2012).	55
3.1	An attempt to calm traffic near Space Station's intersection via signage is unsuccessful.	72
4.1	Case study location, Milwaukee's Hoan Bridge.	87
4.2	Excerpts from WisDOT document depicting LOS impact of Alternative 1A as a "red flag" and "major concern."	90
6.1	Major Road Network (2014) mapped by the South Coast British Columbia Transportation Authority.	122
7.1	Map of the Chicago region, including CN and EJ & E lines.	136
7.2	At-grade crossing of EJ & E tracks at Old McHenry Road, Hawthorn Woods, IL.	141
8.1	"The Dome of Light," Kaohsiung MRT Formosa Boulevard Station.	165
11.1	The spatial representation of *Venlo central*.	215
11.2	The spatial representation of *Wallonia central*.	219
12.1	Conceptualizing the techno-urban development path.	232
12.2	Transit City and residents without a Canadian passport.	237
12.3	Techno-Urban Development Path of Public Transit in the GTA.	240
13.1	The Randstad, with its four main cities.	250
14.1	The favela from above.	271
14.2	Palmeiras, the last cable car station at Complexo do Alemão.	272

Tables

8.1 List of acronyms   155
13.1 Definitions of the key notions used to study place-framing.   248

Approaching the City as Place of Flows

Foreword 1
Transportation Geographies and Mobilities Studies
Toward Collaboration

Susan Hanson

I recently gave a talk about transportation to a group of highly accomplished social scientists from diverse fields (none from geography or transportation) and began with a simple observation: The essence of transportation is not planes, trains, and automobiles, but rather mobility and access. I then pointed out that transportation is implicated in just about everything else, especially those areas that people with a policy bent—like those in that audience—are interested in: employment, environment, climate change, sustainability, national security, inequality, health, education, energy, family life, information technology, cybersecurity, access to food, the political process itself. Is any major policy area untouched by transportation? After the talk, several in the audience told me these introductory remarks—which are simple, accepted basics in human geography—came as a 'eureka' moment for them. They'd never thought of transportation in this way before and confessed with some embarrassment that they found this 'new' view of transportation far more engaging than what they had thought the field was all about.

I don't really know where people get the idea that transportation *is* planes, trains, and automobiles (no doubt there's an interesting study lurking in that impression), but I wonder if the prevailing goals and strategies of traditional transport planning—along with the familiar materiality of transportation infrastructure—might understandably underwrite such a view. From its inception, transport planning has focused on 'saving' time, through its almost exclusive emphasis on motorized travel and on increasing speed, which has been accomplished via designing ever faster modes of travel (planes, trains, and automobiles) and building ever larger and straighter infrastructure (airports, bridges, tunnels, superhighways) to accommodate high-speed travel. That simply walking to a neighborhood store might also be considered transportation, and might even be considered more pleasurable than driving to a supermarket, can get lost in a fixation on speed and the technologies that support and promote it.

Transportation—in all its forms—is woven into the fabrics of our lives and the places we live, from the dwelling-unit scale to the scales of the neighborhood, region, and globe. Because transportation permeates places and lives, it is far more than just a means of reaching a destination. It is deeply implicated in the economic, social, political, and cultural well-being of places. It can also be a source of pleasure and annoyance, a wellspring for feelings of pride or insecurity, the basis of a sense of entitlement or exclusion; it can be enabling and simultaneously constraining. In fact, as editors Julie Cidell and David Prytherch point out in their introduction to this book, transportation is rife with dialectics and contradictions: movement/stability, nodes/links, dwelling/mobility, winners/ losers, self/other, flows/spaces, among others.

Transportation geographies and mobilities studies have both explored these contradictions, sometimes pursuing different goals, often using different methodologies, and frequently competing for intellectual territory. For more than 50 years transportation geographers and other social scientists have analyzed transportation systems and processes, primarily using the tools of quantitative social science and engineering. More recently, the field of critical mobilities studies has emerged to focus on the experiential and cultural dimensions of movement. My aim in this brief essay is to illuminate some of the common threads in transportation geographies and mobilities studies, to point to some of the ways in which mobilities studies grow out of transportation geographies, and to contend that close collaboration between the two is sorely needed to tackle a suite of pressing problems, such as sustainability. I use the plural 'geographies' and 'studies' intentionally to signal that neither endeavor is monolithic or unitary; each category masks considerable diversity, some associated with change over time, some with differences in goals and foci. To lay the groundwork for understanding the evolution of these two intersecting fields, the emphasis here will be on transportation geographies, particularly the early decades of that field. Because the editors delve into the history of transportation geographies in some detail in their introduction, this preface will sketch only broad outlines. The changing contexts within which each of these fields has developed are a key part of the story.

One need hardly note that since WWII much has changed within the social sciences and the discipline of geography, as well as in society writ large in the United States and around the world. These interwoven societal and disciplinary changes are essential to understanding transportation geographies and their relation to emerging mobilities studies.

What was the context within which transportation geography emerged? Transportation principles were at the core of the economic geography of the 1950s and 1960s, which itself was grounded in nomothetic (generalized) theories as opposed to ideographic (particularized) case studies.[1] The so-called quantitative revolution of the late 1950s and 1960s was only in part about adopting quantitative methods; more important was the call to make geography a 'systematic' science, one focused on illuminating general principles of spatial organization (e.g., Berry and Marble 1968; Abler, Adams, and Gould 1971). Transportation in the form of connectivity between places played a starring role.

Also important to the early days of transportation geography was the larger post–World War II context of escalating car ownership, mobility, and urbanization in North America and elsewhere. These changes, prompted by growing affluence and increasing suburbanization, posed very real challenges for planners: how to accommodate all this urban growth and how to keep automobiles moving so as to prevent strangulation of the cities. The field of modern urban transportation planning emerged in the U.S. to take up this challenge, with the primary responsibility falling on civil engineers, the group traditionally involved in designing and building infrastructure, including roads (e.g., the Chicago Area Transportation Study, completed in the late 1950s and described in Black 1990). In the calculus of traditional urban transportation planning, congestion is the enemy and more infrastructure, known in the planning realm as more 'capacity,' is the solution (Mitchell and Rapkin 1954). Planners and engineers now know that they cannot build their way out of congestion, because more capacity tends to generate more travel.

Civil engineers are not social or behavioral scientists, yet designing transportation systems to accommodate growth in mobility while keeping congestion to a minimum (the traditional goal) requires at least some understanding of social and behavioral processes. Transportation geographies are not the same thing as urban transportation planning, but the two have become intertwined, in large part because transportation geographers *are* trained in the social and behavioral sciences. Concepts embedded in behavioral geography (e.g., Cox and Golledge 1969), which emerged in the late 1960s, are one example of how transportation geographers have influenced transportation planning.

From its inception transportation planning has used zones (areas) as the units of analysis (the origins and destinations of trips) to measure and model flows of people and goods so as to estimate the location and sizing of infrastructure. Data for these zones represent aggregations of individuals and households for that area—individual differences within zones are lost. Behavioral geography, however, focuses on the individual or the household as units of analysis, and (more important) shifted attention away from the unitary understanding of humanity embodied in the concept of 'economic man.' In other words, a major contribution of behavioral geography was to recognize the importance of differences among people and in how people make mobility-related decisions, an insight that enables consideration of perceptions, values, preferences, and the like, and queries the sources of those differences (e.g., Pipkin 1986).

In the early 1970s behavioral geographers—along with some economists and civil engineers—with interests in transportation began to point to the value of adopting disaggregate approaches in understanding human mobility and in transportation planning. Such approaches are now an accepted part of transportation planning practice, not so much in the macromodels for urban regions (although they are increasingly being used there), but, for example, to identify people who lack adequate mobility to meet their daily needs or to design intra-neighborhood-scale infrastructure to foster walking and biking. One reason that behavioral geographers were able to communicate effectively with economists and civil engineers in the transportation arena is that they were/are grounded in a quantitative-analytic, data-rich research tradition.

For these and other reasons, the early 1970s saw the dawning of the explicit recognition that transportation was not just about infrastructure *but also about people*. Torsten Hagerstrand's influential 1970 article, "What About People in Regional Science?," not only introduced the concepts of time-geography but could also have been called "What About People in Transportation?" (Hagerstrand 1970). The precepts of time-geography are powerful, but proved difficult to operationalize empirically (i.e., to examine the space-time paths of large samples of individuals or of groups) until the advent of well-developed GISs and high-speed computers. With these technologies now well in place, time-geography is increasingly used as a framework for understanding urban mobility, especially new forms of mobile device-based mobility and other emerging mobility trends.

Mention of computing power raises another key dimension of the societal context within which transportation geography first emerged: Digital computing was in its infancy, or perhaps still in gestation, so from a research standpoint analyzing large data sets, especially those with spatially explicit variables, was exceedingly cumbersome. And of course the Internet, social media, and mobile devices—all of which play increasingly important roles in shaping human mobility—were barely imagined. Indeed, the first scholars to ask how communications technology might affect travel were thinking of the impact of the telephone (e.g., Abler 1975).

The expert-driven way in which transportation planning was carried out in the early years (1950s and 1960s), and the changes that were set in motion as a result of this top-down approach, set the context for shifts within transportation geography in the 1970s. It is significant in the context of this book, which is conceptually grounded in urban geography, that—within geography—it was scholars with strong backgrounds in intraurban, and especially urban-social, geography that initiated these fundamental shifts in mobility-oriented studies. With its long held focus on multifaceted dimensions of cities in terms of those dimensions' interrelatedness in place and across space (e.g., residential, industrial, commercial land uses; the social, economic, and political aspects of urban life; the influences of connectivity between and among settlements), the theoretical framework of urban geography was perhaps not an unlikely midwife for these shifts in transportation geography that occurred in reaction to early urban transportation planning efforts. In particular, the social and environmental externalities of major transportation infrastructure projects—especially urban freeways—were essentially ignored in the early days of highway building. Highways were built through urban neighborhoods, usually in places where the residents had little political power or voice; freeways significantly increased accessibility for some groups (e.g., those living near interchanges) while significantly disrupting urban life in many places (e.g., by increasing noise, decreasing air quality, reducing access, destroying neighborhoods, and highlighting disenfranchisement).

However, by the end of the 1960s freeway revolts were in full swing in the U.S., with activists focusing their agenda on ensuring public participation and voice in the planning process and minimizing adverse environmental impacts, such as the destruction of neighborhoods or animal habitat. The passage of the National Environmental Policy Act (NEPA) in 1970 made public participation and formal review of potential environmental impacts integral to the planning process for all projects involving federal funds in the U.S. freeway revolts. Along with citizen backlash against large-scale 'urban renewal' projects of the 1950s and 1960s in the U.S., this led, in the early 1970s, to advocacy planning, in which planners work pro bono on behalf of disenfranchised neighborhoods and groups, and to participatory planning, which aims to involve all voices in a community in the planning process. Clearly a reaction to earlier, experts-only approaches, such democratization highlighted key questions about inequalities, inequities, and power imbalances in the deeply political process of planning the transportation system. To whom do NEPA and other laws give voice? Who will participate? Whose values will guide the process? Whom will a project benefit? Whom will it affect adversely?

The research of social scientists, and particularly human geographers, interested in this nexus of issues put the social, political, ethical, and, to some extent, institutional dimensions of transportation on the research agenda in the 1970s and 1980s. Examples abound. Economist John Kain (1968) pointed out that members of minority groups trapped in urban ghettoes by residential segregation and inequities in housing markets lacked access to employment, which had been rapidly suburbanizing since WWII. Sociologist David Caplovitz (1967) documented that residents of low-income neighborhoods often pay more than others for the same or inferior goods—especially food—because they lack the needed mobility to travel outside their neighborhoods. In addition, geographers made diverse and important contributions. David Ley (1974) demonstrated the devastating impacts (especially on children and pedestrians) of a major highway through an impoverished African-American Philadelphia neighborhood. Roger Kasperson and Myrna Breitbart (1974) analyzed the complexities of giving voice to the powerless via advocacy and participatory planning. David Hodge (1981) showed that subsidies to the Seattle transit system were disproportionately borne by low-income urban residents, who also received inferior service compared with their suburban counterparts. David Banister (1980) examined variations in individuals' access and mobility in rural parts of Oxfordshire, UK. Jacky Tivers (1985) delved into the access and mobility issues facing women with young children, and others began to document gender inequalities in access: Sophie Bowlby (1979) in the context of access to grocery stores in Oxford, UK, and Susan Hanson and Perry Hanson (1980) for travel activity patterns in Uppsala, Sweden.

Beyond work in the economic geography tradition—from which transportation geography had emerged—by the early 1970s the social and political elements of transportation had become central to transportation studies itself. And they have remained so ever since. But such social and political dimensions did not then include the cultural. Cultural dimensions of transportation, mobility, and access received little attention until geography and other social sciences began to absorb the messages emanating from feminist and cultural studies. Likewise, quantitative approaches remained predominant in studies of mobility and access until the influences of the qualitative methods being used in sociology, anthropology, and feminist studies were felt. Against this historical backdrop, mobilities studies, spearheaded mainly by sociologists, emerged in the early 2000s to highlight the long absent—and hugely important—dimension of culture (and related elements) and the potential richness of qualitative approaches to questions of mobility, access, and transportation.

Consider for a moment how distinct the disciplinary and societal contexts of the middle of the 20th century were compared to the dawn of the 21st (when mobilities studies first appeared). Culture had been drummed out of the spatial science side of the discipline in the late 1950s as too 'soft' and therefore too hard to measure and model; but by 2000 its importance was once again widely appreciated within geography. Whereas in 1950 many parts of the U.S. and the world lacked electricity and telephones, mobilities studies came into a world in which cell phones were widespread, the Internet was diffusing, and social media were on the horizon. Once hailed as a marker of progress, the growth in vehicle miles traveled in the U.S. and around the world had become a source of multiple concerns. Additionally, the hegemony of the automobile—itself the product of intentional planning and policy—had come to prompt concerns, not only about the social and political inequities still permeating the transportation/mobilities arena and the adverse impacts of high-speed motorized mobility on quality of urban life, but also about relatively newer concerns surrounding resource (especially fossil fuel) consumption and associated assaults on the environment, health, safety, and the global climate; rising congestion; the geopolitical complexities of petroleum reliance; and the very sustainability of planet earth.

To these and other problems, mobilities studies have brought fresh approaches and different perspectives. In addition to welcoming qualitative methods and fully recognizing the significance of culture and values, mobilities studies explore people's experiences and associated meanings of travel and of immobility. Such studies appreciate the generative roles of symbolism and discourse in shaping the mobilities patterns and the transport infrastructure that have long interested transportation geographers.

These perspectives have enabled mobilities scholars to shine new light on the political processes and conflicts surrounding mobility-related projects throughout the world.

Clearly, mobilities studies have brought new dimensions of understanding to questions of access and mobility. At the same time, mobilities studies share a long history with transportation geographies, perhaps best illustrated by the long list of enduring themes that span and permeate both traditions: mobility, access, networks (nodes and links, places and flows), connections, connectivity, scale, social justice, externalities, politics, citizen involvement, activism, and governance, among others. Mobilities scholars have sometimes tried to distance themselves from this shared history, emphasizing differences over commonalities and shared interests. But the seriousness of the problems now confronting global society will require collaboration and all the intellect that can be brought to bear from every angle. The perspectives of mobilities studies *and* those of transportation geographies together have a much greater chance of charting ways forward than would either one flying solo. The conundrums demanding such collaboration are many, among which are: How will mobile technologies and new forms of mobility (e.g., car sharing) change urban life? What strategies will move the world most quickly, effectively, and justly to a low-carbon future? How best to unravel the complex relationships between transportation and prosperity? How best to accommodate a diversity of values about mobility and dwelling places in envisioning a viable future for humanity?

These and other pressing questions demand an active and robust collaboration between transportation geographies—with their still strong theoretical, quantitative tradition—and mobilities studies, with their orientation toward questions of culture, experience, meanings, and qualitative approaches. As I have argued elsewhere (Hanson 2010), transportation policies aimed at promoting sustainability (as just one example) cannot be based simply on the large sample, quantitative studies of mobility patterns that show, inter alia, that some population groups engage in less vehicular travel (and thereby demonstrate 'more sustainable' travel patterns). These studies alone are an insufficient policy guide because they leave aside the question of what reduced travel means to, and for, those travelers: Is it the outcome of choice, a reflection of, say, a conscious effort to reduce fossil fuel consumption? Is it an indicator of deprivation, the result, say, of an inability to drive? Or does the reason lie elsewhere? Answering these questions of meaning, usually via in-depth, qualitative studies with relatively small samples of people (by necessity), is imperative to crafting policies that will be equitable while moving the transportation system toward a more sustainable future. As this example illustrates, both wellsprings of expertise—transportation geographies and mobilities studies—and no doubt others as well will be essential for moving forward. Other policy fronts that require a joint effort are legion—as just three examples illustrate: education (how most effectively to match school size with the distances students must travel to maximize the benefits of education), energy (how to reduce fossil fuel consumption in transportation without compromising mobility needs), and responding to climate change (how to build resilient communities, in which access to essential goods and services is maintained when transportation infrastructure fails).

With its sweeping geographic scope, representing many corners of the globe, and its diverse, wide-ranging topics, this book is an excellent starting place for the collaborative work such problems demand. The emplacement of infrastructure is at the core of several chapters here, while people's perceptions, experiences, values, and meanings also figure largely. The editors rightly ground the integrative goals and function of the book in urban geography, a field that by its very nature practices integration in place: between, for example, the built environment and social/cultural/political environments, between flows and dwelling, between the policy world and the lived experience of everyday life, between material infrastructures and cultural meanings, between a local place and distant places. I hope that this book will be the first of many such efforts that will help the world to reap the benefits of combining transportation geographies and mobilities studies.

ACKNOWLEDGMENTS

My thanks to editors Julie Cidell and David Prytherch, and to Sophie Bowlby for helpful suggestions on an earlier draft.

NOTE

1 Transportation had also been a part of regional geography, the dominant paradigm into the 1950s.

WORKS CITED

Abler, R. (1975) "Effects of space-adjusting technologies on the human geography of the future," in R. Abler, D. Janelle, A. Philbrick, and J. Sommer (eds) *Human Geography in a Shrinking World*, North Scituate, MA: Duxbury Press, pp. 35–56.
Abler, R., Adams, J. S., and Gould, P. (1971) *Spatial Organization: The Geographer's View of the World*, Englewood Cliffs, NJ: Prentice Hall.
Banister, D. (1980) "Transport mobility in interurban areas: A case study approach in South Oxfordshire," *Regional Studies*, 14(4):285–296.
Berry, B.J.L., and Marble, D. F. (1968) *Spatial Analysis: A Reader in Statistical Geography*, Englewood Cliffs, NJ: Prentice Hall.
Black, A. (1990) "The Chicago area transportation study: A case study of rational planning," *Journal of Planning Education and Research*, 10(1):27–37.
Bowlby, S. (1979) "Accessibility, mobility and shopping provision," in B. Goodall and A. Kirby (eds) *Resources and Planning*, Oxford: Pergamon Press, pp. 293–324.

Caplovitz, D. (1967) *The Poor Pay More*, New York: Free Press.
Cox, K. R., and Golledge, R. G. (eds) (1969) *Behavioral Problems in Geography*, Evanston, IL: Northwestern University Press.
Hagerstrand, T. (1970) "What about people in regional science," *Papers of the Regional Science Association*, 24:7–21.
Hanson, S. (2010) "Gender and mobility: New approaches for informing sustainability," *Gender, Place, and Culture*, 17(1):5–23.
Hanson, S., and Hanson, P. (1980) "Gender and urban activity patterns in Uppsala, Sweden," *Geographical Review*, 70:291–299.
Hodge, D. (1981) "Modelling the geographic component of mass transit subsidies," *Environment and Planning A*, 13:581–599.
Kain, J. (1968) "Housing segregation, negro employment, and metropolitan decentralization," *Quarterly Journal of Economics*, 82:175–197.
Kasperson, R. E., and Breitbart, M. (1974) *Participation, Decentralization and Advocacy Planning*, Washington, DC: Association of American Geographers Commission on College Geography.
Ley, D. (1974) *Black Inner City as Frontier Outpost*, Washington, DC: Association of American Geographers Commission on College Geography.
Mitchell, R. B., and Rapkin, C. (1954) *Urban Traffic—A Function of Land Use*, New York: Columbia University Press.
Pipkin, J. S. (1986) "Disggregate travel models," in S. Hanson (ed) *The Geography of Urban Transportation*, New York: Guilford Press, pp. 179–206.
Tivers, J. (1985) *Women Attached: The Daily Lives of Women with Young Children*, London: Croom Helm.

Foreword 2
Mobilizing Transportation, Transporting Mobilities
Mimi Sheller

The fields of transport history, transportation research, and transportation geography have all been influenced by the burgeoning and diverse perspectives of critical mobilities research, while mobilities research itself has widened and deepened its theoretical and methodological approaches to transportation systems, experiences, meanings, politics, and social practices. Mobilities research is increasingly recognized as an important addition to the fields of transportation research (Knowles et al. 2008; Shaw and Docherty 2014), transport geography, and transportation planning, in part because it can help to "bridge the quantitative–qualitative divide" (Goetz et al. 2009; and cf. Jensen et al. 2014). Even more importantly, the recent productive commingling of research in these adjacent fields has refreshed the ways in which we approach a whole range of classic topics in urban studies: e.g., the design, building, and appropriation of urban infrastructure systems; the processes of large-scale technological change, especially sustainability transitions; scalar politics and the production of spatial relations through transportation investment or disinvestment; the relation between systems complexity, risk, failure, and resiliency; the embodied experience of streets, stations, and various kinds of vehicles and ways of moving; and the concerns with accessibility, social exclusion, and the dynamic contestation of the right to the city.

There has been a mobilizing of transport studies, so to speak, and a transporting of mobilities studies. At the heart of this transformation is a theoretical shift that seeks to understand spatiality in more relational ways, and to understand the relations enabled by transport in more mobile ways. The methodological diversity of these new hybrid perspectives and the move towards relational ontologies of the "new mobilities paradigm" (Hannam, Sheller, and Urry 2006; Sheller 2014) have far reaching implications for how we understand transportation and urban geographies. Mobilities research builds on a range of philosophical perspectives to radically rethink the relation between bodies, movement, and space; thus, it can inform research on the production of space, the politics of transport, and the subtle meanings and diverse experiences of (im)mobilities in the city. It includes new ways of thinking about walking, driving, passengering, flying, and other modes of movement; the social and political dimensions of the production and consumption of the built environments that afford such mobilities; and the enacted spaces, cultural meanings, and diverse lived experiences of moving through such mobile (and immobile) places.

This volume, *Transportation, Mobility, and the Production of Urban Space*, brings a welcome addition to these conversations through its careful case studies of transportation politics at the interface of urban geography and mobilities studies in specific cities located in Asia, Europe, and North and South America, including Chicago, Kaohsiung, Milwaukee, Hyderabad, San Francisco, Grand Paris, Rio, Rotterdam and The Hague, Toronto, and Vancouver. These diverse cases illustrate how every transportation planning decision is grounded in, and has longlasting impacts on, urban (and suburban) spatiality, urban inequality, and urban culture. What we learn here is that every transportation decision entails both connecting places together and excluding or bypassing other places; and every transportation investment adds value to some places, while destroying value in others, or sometimes destroying place itself. Urban form is shaped by transport design, and transport choices, plans, and investments are shaped by urban politics.

Cities are formed by mobilities: Often located at the confluence of rivers, roadways, ports, rail termini, and airports, they orchestrate flows of people, goods, information, and ideas (Sheller and Urry 2000). At the same time, everyday mobility practices and associated mobility regimes are in turn formed by urban dynamics and political contestation, such as the conflict between automobility and bicycling (e.g., Furness 2010); by public policies concerning urban migration, urbanization, and right to the city (Mitchell 2003); by forms of urban governance and policy that shape transport and communication infrastructures, and access to them; and by urban technological innovation and regional agglomerations that shape the spatiality and scale of mobility systems and infrastructures such as highway systems (Merriman 2007, 2009). Mobility systems persist and combine into local, national, and even transnational cultural assemblages of mobility that remain very durable over long periods of time (Mom 2014). Living in the midst of a deeply "dominant system of automobility" (Urry 2004), it is difficult to see how we will move beyond it. Nevertheless, the past teaches us that even mobility systems that have been around a long time will eventually be replaced.

What will come next, "after the car" (Dennis and Urry 2008)? Where are the openings for new transportation systems or mobility regimes to emerge (Dudley et al. 2011)? Can such transitions in mobilities be accelerated, directed, guided, or fostered? And what can we learn from the current transitions that are taking place in some cities? Mobilities research suggests that we cannot look at transportation in isolation, but must also consider how systems such as mobility and communication interact. Certainly the emergence of autonomous and connected vehicles will be highly disruptive to existing transport planning paradigms, such as the prediction of traffic volumes; and mobile locative media are already challenging existing models for multimodality (with vehicle-sharing and ride-sharing technologies already pushing public transit and taxi services to modernize their operating systems). Complexity and system dynamics are also crucial; for example, the impact that higher-speed rail in the United States will have on air transport networks for intercity travel in sprawling urban areas such as the Northeast, not to mention how drone delivery systems, 3D printing, and additive manufacturing systems might reconfigure freight logistics (Birtchnell and Urry 2014).

At this lively intersection of transportation studies and mobilities research, many researchers are concerned not only with delineating and describing the emergence of historical and contemporary mobility regimes, technologies, and practices, but also with critically addressing normative issues of mobility justice (such as movements for sustainable mobility and mobility rights) and mobility capabilities (such as the demands of social movements for rights of access to the city and transportation justice). The chapters in this volume help us to really get down into the nitty-gritty of urban politics and transport geographies, to see some of the processes driving these slow transitions.

There is, in a sense, a mobilization of space itself, as places are moored or unmoored from different transport infrastructures. And this should not be a passive verb: Specific actors do the conceptual, technical, and physical work of (un)mooring urban space through transportation system plans, designs, standards, measures, rhetorics, marketing, and decisions. Sometimes those actors are specific individuals or groups, but other times they are agreed upon forms of 'objectivity' and rationality such as cost-benefit analyses, or 'level of service' measurements. Thus, the contributors to this volume remind us that it is important to attend to the logics of justification and to pick apart how different social actors use narratives to frame transportation issues and shape decision-making contexts, even before any decisions are made. This dovetails with critical geographies of mobility that focus on the history of mobility, its modes of regulation, and the power relations associated with it—in short, the politics of mobility (Cresswell 2006, 2010; Adey 2009).

Whether to promote existing patterns of automobility or enable healthier modes of active transport, whether to build light rail systems or high-speed rail, to extend metro lines or change local zoning codes, to disinvest from public transit or privatize bus systems—these are all political decisions framed by competing constituencies exposed to differing costs and benefits, and more or less included or excluded from decision-making. Such decisions also have huge impacts on quality of life, levels of pollution, the health of entire communities, and the lock-in of massive infrastructures that are difficult to remove once built. As cities across the world face the effects of climate change, better understanding of these decision-making processes will be crucial to both mitigation and adaptation strategies. Insofar as our old ways of transporting goods and moving people are broken, or at the very least are reaching limits of capacity and sustainability, then our old ways of planning transport, and indeed of doing transport studies, are also in need of updating (Grieco and Urry 2012). We need to be more innovative, more creative, more multidisciplinary, more humanistic, more empathic, more exacting, and more critical in the standards we hold ourselves to, the methods we employ, and the theories we advance.

The field of mobilities research consists of efforts to do just this, in order to push the conversation about transportation and urban geography in new directions. It is with these kinds of question in mind that mobilities research puts emphasis on the relations between mobilities and immobilities, scapes and moorings, movement and stillness (Hannam et al. 2006, p. 3). These co-constitutive frictions of differential mobilities and relative velocities are at the heart of recent mobilities research (Adey et al. 2014; Cresswell 2014; Vannini 2014), and have much to contribute to our understanding of transport geographies and urban geography. Mobilities research furthermore reminds us that culture, lived experience, and meanings are also crucial elements of technological systems (Cresswell 2006; Freudendal-Pedersen 2009). Any city is made up of technologies, practices, infrastructures, networks, and assemblages of all of these—as well as narratives, images, and stories about them—which together inform its mobility culture. Transitions in mobility systems rest not just on individual choices, technological transformations, or economic forces, but on transitions in mobility cultures and the ways in which practices and networks are culturally assembled in producing and performing the mobility space of the city.

Critical mobility thinking in the field of urban studies also calls for "re-conceptualising mobility and infrastructures as sites of (potential) meaningful interaction, pleasure, and cultural production" (Jensen 2009), where people engage in "negotiation in motion" and "mobile sense making" (Jensen 2010). Histories of mobility and place-making emphasize the rhythms, forces, atmospheres, affects, and materialities of various modes of transport (Edensor 2014; Merriman 2012; Adey 2010). Building on Georg Simmel's ideas of 'urban metabolism' and Henri Lefebvre's 'rhythmanalysis' (2004), mobility theorists argue that bodies and objects shape cities, and in turn are shaped through their rhythms of movement, their pace and synchrony (Edensor 2011, 2014). The recent *Routledge Handbook of Mobilities* (Adey et al. 2014), for example, is organized around the categories of qualities, spaces and systems, materialities, subjects, and events, rather than more traditional 'transportation-related' topics. Along with spatiality and materiality there is also a growing interest in temporalities. Temporalities of slowness, stillness, waiting, and pauses are all part of a wider, sensuous geography of movement and dwelling in which human navigation of embodied, kinaesthetic, and sensory environments are crucial (Merleau–Ponty 1962; Jensen 2010).

Where, then, are emergent new cultures of mobility reshaping transportation and potentially urban space? To take just one example from my own context, across a range of different measures it is now empirically evident that in the U.S., as a whole, there has been a 10-year decline in driving, indicated by fewer car trips per driver and per household; fewer miles driven per driver, per vehicle, and per household; and significantly fewer young people getting drivers' licenses (Pickrell and Pace 2013; Short 2013; Sivak 2011, 2013; Sivak and Schoettle 2011). In terms of individual drivers, the average number of miles driven peaked in July 2004 (at more than 900 miles per month) and has continued to decline since then, reaching around 820 miles per month in July 2012, a 9% drop in eight years, with an especially sharp 23% decline in vehicle miles traveled (VMT) for 16–25-year-olds. These declines since 2004 can be noted in data measuring distances driven per person, per licensed driver, per household, and per registered vehicle (Sivak 2013). At the same time, there are more reported trips being taken by bike, by walking, and by public transit, more people reporting working from home on one or more days per week, and more use of information technology for 'virtual mobility' as trip replacement. This has led a number of analysts to ask whether motorization in the U.S. (and other industrialized countries) has peaked, and to predict whether it may possibly continue to decline further.

However, the traditional transport geography approach can only take us so far in understanding these trends. It helps us to parse the national travel survey data on different modes of transport, compare commuting patterns in different cities, and perhaps generate Geographic Information Systems (GIS) visualizations of different time-space prisms (Schwanen and Kwan 2012). But we also need to ask why it is happening, how and for whom it is happening, and whether it will last. In which cities and regions is the decline in driving strongest and why? For whom are these trends becoming permanent, and how are they connected to lifestyles, choice of residence, patterns of family formation, use of new communication technologies, and changing patterns of employment? How do the patterns diverge according to variables such as age, race, gender, and class in ways that are not simply statistical artifacts, but reflect, for example, something of the changing racing and gendering of urban space and connect to wider mobilities generated by migration and urbanization patterns? And, if the trends continue, how will the decline in driving and the rise of 'connected mobility' and the digital street reshape urban space and transport infrastructure, as decisions are taken now about where to invest transport funding, where to live, and where to locate companies? What new exclusions or bypassing effects, frictions, and deceleration might occur if better connected city centers become too expensive for most people to live in, with high-quality transport infrastructure increasingly accessible to elites only? How, indeed, can the quantitative description of the trends be mobilized into qualitative narratives that contest business as usual in the field of transportation planning by framing issues in new ways informed by a mobilities perspective?

The chapters in this volume offer us some excellent examples of how to proceed in answering these kinds of questions, unpacking the urban politics of contested mobility in a range of specific cities while maintaining an awareness of the qualitative and experiential dimensions of diverse mobility practices and their production of space and bodies. In doing so, they offer a productive way forward for transportation geography as a crucial tool for understanding 21st-century urban spatial production.

WORKS CITED

Adey, P. (2009) *Mobility*, London: Routledge.
Adey, P. (2010) *Aerial Life: Spaces, Mobilities, Affects*, Oxford: Wiley-Blackwell.
Adey, P., Bissell, D., Hannam, K., Merriman, P., and Sheller, M. (eds) (2014) *The Routledge Handbook of Mobilities*, London: Routledge.
Birtchnell, T., and Urry, J. (2014) "The mobilities and post-mobilities of cargo," *Consumption, Markets and Culture*. doi:10.1080/10253866.2014.899214
Cresswell, T. (2006) *On the Move: Mobility in the Modern Western World*, New York: Routledge.
Cresswell, T. (2010) "Towards a politics of mobility," *Environment and Planning D: Society and Space*, 28(1):17–31.
Cresswell, T. (2014) "Friction," in P. Adey et al. (eds) *The Routledge Handbook of Mobilities*, London: Routledge, pp. 107–115.
Dennis, K., and Urry, J. (2008) *After the Car*, Cambridge: Polity.
Dudley, G., Geels, F., and Kemp, R. (eds) (2011) *Automobility in Transition? A Socio-Technical Analysis of Sustainable Transport*, London: Routledge.
Edensor, T. (2011) "Commuter: Mobility, rhythm, commuting," in T. Cresswell and P. Merriman (eds) *Geographies of Mobilities: Practices, Spaces, Subjects*, Farnham: Ashgate, pp. 189–204.
Edensor, T. (2014) "Rhythm and arrhythmia," in P. Adey et al. (eds) *The Routledge Handbook of Mobilities*, London: Routledge, pp. 163–171.
Freudendal-Pedersen, M. (2009) *Mobility in Daily Life: Between Freedom and Unfreedom*, Farnham: Ashgate.
Furness, Z. (2010) *One Less Car: Bicycling and the Politics of Automobility*, Philadelphia: Temple University Press.
Goetz, A., Vowles, T., and Tierney, S. (2009) "Bridging the qualitative-quantitative divide in transport geography," *The Professional Geographer*, 61(3):323–335.
Grieco, M., and Urry, J. (eds) (2012) *Mobilities: New Perspectives on Transport and Society*, Aldershot: Ashgate.
Hannam, K., Sheller, M., and Urry, J. (2006) "Mobilities, immobilities, and moorings," *Mobilities*, 1(1):1–22.
Jensen, O.B. (2009) "Flows of meaning, cultures of movements—Urban mobility as meaningful everyday life practice," *Mobilities*, 4(1):139–158.
Jensen, O.B. (2010) "Negotiation in motion: Unpacking a geography of mobility," *Space and Culture*, 13(4):389–402.
Jensen, O.B., Sheller, M., and Wind, S. (2014) "Together and apart: Affective ambiences and negotiation in families' everyday life and mobility," *Mobilities*. doi:10.1080/17450101.2013.868158
Knowles, R., Shaw, J., and Docherty, I. (eds) (2008) *Transport Geographies: Mobilities, Flows, And Spaces*, Oxford: Blackwell.

Lefebvre, H. (2004) *Rhythmanalysis: Space, Time and Everyday Life*, New York: Continuum.
Merleau–Ponty, M. (1962) *Phenomenology of Perception*, New York: Routledge.
Merriman, P. (2007) *Driving Spaces: A Cultural–Historical Geography of England's M1Motorway*, Malden: Wiley-Blackwell.
Merriman, P. (2009) "Automobility and the geographies of the car," *Geography Compass*, 3(2):586–589.
Merriman, P. (2012) *Mobility, Space and Culture*, London: Routledge.
Mitchell, D. (2003) *The Right to the City: Social Justice and the Fight for Public Space*, New York: Guilford Press.
Mom, G. (2014) *Atlantic Automobilism: The Emergence and Persistence of the Car, 1895 to 1940*, New York: Berghahn Books.
Pickrell, D., and Pace, D. (2013, May) "Driven to extremes: Has growth in automobile use ended?." Washington, DC: John A. Volpe National Transportation Systems Center, Research and Innovative Technology Administration, U.S. Department of Transportation. Retrieved January 4, 2014 from http://www.volpe.dot.gov/noteworthy/2013/has-growth-in-automobile-use-ended.html.
Schwanen, T., and Kwan, M.-P. (eds) (2012) *Critical Space-Time Geographies*, special issue of *Environment and Planning A*, 44(9).
Shaw, J., and Docherty, I. (2014) "Geography and transport," in P. Adey et al. (eds) *The Routledge Handbook of Mobilities*, London: Routledge, pp. 25–35.
Sheller, M. (2014, May 23) "The new mobilities paradigm for a live sociology," *Current Sociology*. doi:10.1177/0011392114533211
Sheller, M., and Urry, J. (2000) "The city and the car," *International Journal of Urban and Regional Research*, 24:737–757.
Short, D. (2013, September 21) "Vehicle miles driven: Population-adjusted fractionally off its post-crisis low." Retrieved January 4, 2014 from http://www.advisorperspectives.com/dshort/updates/DOT-Miles-Driven.php.
Sivak, M. (2013) *Has Motorization in the U.S. Peaked? Part 2: Use of Light-Duty Vehicles*, Detroit, MI: University of Michigan Transportation Research Institute.
Sivak, M., and Schoettle, B. (2011, June) *Recent Changes in the Age Composition of U.S. Drivers: Implications for the Extent, Safety, and Environmental Consequences of Personal Transportation*. Detroit, MI: University of Michigan Transportation Research Institute.
Urry, J. (2004) "The 'system' of automobility," *Theory, Culture and Society*, 21(4/5): 25–39.
Vannini, P. (2014) "Slowness and deceleration," in P. Adey et al. (eds) *The Routledge Handbook of Mobilities*, London: Routledge, pp. 116–124.

1 Introduction
Transportation, Mobilities, and Rethinking Urban Geographies of Flow

David Prytherch and Julie Cidell

URBAN INQUIRY AT AN INTERSECTION

Asked to imagine the city, most of us cannot help but think of place. We conjure the landscape: buildings, streets, objects, people, and the land they occupy, whether a neighborhood or district, bustling street or central square. Geographers and others contribute much to our understanding of the city, including a cartographic imaginary that underlies urban policy and planning. Making sense of such assemblages—their processes, patterns, and experiences in space—is the defining challenge of urban inquiry across disciplines. For nearly a century, urban geographers have made important contributions to understanding how environmental, economic, political, and cultural processes combine to produce the built environment and dynamic social relations within it.

But our urban imaginary has often neglected to capture a most essential element of this same urban scene: motion. Urban places, despite their apparent solidity in space and our imaginations, pulse with flows of people, objects, energy, and ideas. Stand at the corner of a busy intersection, and one is reminded how the built environment is but conduit and channel for circulating matter and energy, flowing in networks from the molecular to the global scale. To the degree we are oriented to the material landscape and its social organization, and conditioned by the visual and cartographic, our theories and representations of urban life often can't help but fix space and time. Indeed, that's what our theories and maps are designed to do: abstract the complexity of urban process and objectify it in the concreteness of epistemology or cartography. But all the objects of urban and geographic inquiry, mapping, and theorization are but immanences temporarily embodying dynamic flows or relations. And the structures we trace are just the contours of networked material flows and social relations in space, including the intersections that unite them and the boundaries that divide. No matter how nuanced, theories and maps offer compelling snapshots of a dynamic world, but in doing so impart a stasis in place of the city's dynamism.

The 21st-century city thus presents a major epistemological challenge: How best to comprehend and theorize the city as both space and circulatory system? No topic presents this challenge more clearly than transportation, the most explicitly motive force in urban life. Transportation technologies and systems structure the movement of people, objects, and energy through urban space, and have long been an object of urban analysis. But interest in transportation has waxed and waned in urban geography and related fields. Questions of spatial interaction and movement once central to the subdiscipline's early intellectual history, and integral to once dominant positivist traditions that modeled urban flows, were progressively marginalized with late 20th-century critical and cultural turns, their theorization largely ceded to quantitative and policy-oriented debates among regional scientists, economists, and engineers.

Faced with surging scholarly and popular interest in urban transportation and mobility, however, urban scholars are considering how to re-approach flow. This means engaging those literatures for which circulation has always been central. In transportation geography, for example, one encounters well-developed theories and techniques for understanding accessibility and spatial interaction, defined by traditional strengths in quantitative and spatially explicit modeling (Hanson and Kwan 2008) and links to civil engineering. Such largely positivistic approaches have made transportation geography policy relevant (Goetz 2006), but at times insular (Hanson 2003). Meanwhile, critical scholarship in 'new mobilities' has emerged to assess the political, social, and cultural forces shaping the experience of movement (Cresswell 2006a; Hannam et al. 2006; Sheller and Urry 2006; Blomley 2011, 2012) and the "politics of mobility" (Cresswell 2010). These new critical and qualitative approaches, however, can neglect the larger spatial contexts, indeed infrastructures, of transportation and its established theories. Fortunately, exciting work is emerging at the overlapping boundaries of transportation geography and mobilities scholarship, engaging their relative strengths to address what they individually neglect. Focus on the urban landscape reveals how much potential remains for such engagement.

In the journey towards a more integrative perspective on the contemporary city, urban studies and geography thus stand at the intersection of key questions and debates:

- How do urban places intersect and function within wider transport networks and flows?
- How is transportation experienced, represented, and contested within networked urban places?
- How are mobility/immobility spatially and socially distributed within (urban) transportation systems and their governance?
- How are transportation places and networks co-constructed through mobility and circulation?

To answer these questions, this book critically examines how urban space is structured and articulated through transportation infrastructure and everyday flows, exploring connections in spatial theory and practice between transport geographies and new mobilities in the production of networked urban places. It joins other recent work that attempts to rethink transportation geographies, whether by engaging with mobilities perspectives (Keeling 2007; Bergman and Sager 2008; Canzler et al. 2008; Knowles et al. 2008; Adey 2010; Cresswell and Merriman 2011; Shaw and Hesse 2010; Shaw and Sidaway 2010; Henderson 2013; Jensen 2013, 2014), the materialities of flow (Hall and Hesse 2013; Graham and McFarlane 2015), or the larger political and economic systems that structure the movement of people, goods, and fuel (Aaltola 2005; Cowen 2010; Huber 2009; Siemiatycki 2013; Moraglio and Kopper 2015). It seeks to contribute a fresh, diverse, and international perspective on the spatialities and power-geometries (Massey 1994) of urban *places of flows* (akin to Castells 1999, 2000; see also Zavestoski and Agyeman 2015). Transportation geography and mobilities studies offer distinct and complementary contributions to understanding movement and urban space, but each needs to be better linked to the other and to the urban places where movement is practiced, experienced, and governed. Urban geography, with its long-standing emphasis on urban places and the networked spatial relations, and a history of integrating positivist and critical approaches, offers one venue for linking sometimes disparate debates to the everyday experience and governance of urban flow.

Standing at major intersections, like where Michigan Avenue and Wacker Drive join in Chicago's Loop, one is struck less by surrounding skyscrapers than the incessant, negotiated circulation of cars, bikes, and buses on the streets, pedestrians along sidewalks and up staircases and elevators, elevated trains rumbling above and the subway below, passenger and cargo planes crisscrossing the sky, and boats traveling along the Chicago River, whose portage between the Great Lakes and Mississippi watersheds made the city possible (as portrayed so effectively by Cooke and Lewis 2009). How such complex flows are negotiated, planned, and structured in and through urban places is central to urban life. Only by engaging such spaces with empirical sensitivity and theoretical rigor can scholars and practitioners fully rethink how urban places function as nodes/intersections and corridors/links within networks of potentially global scope, and the experiences of people and materials and energy within them. To stand at such an intersection is, therefore, to stand at a key crossroads in 21st-century geographical debates.

URBAN GEOGRAPHIES OF TRANSPORTATION AND MOBILITY: FOUND AND LOST AND FOUND AGAIN

It is worth recalling that urban studies and geography have always been concerned with mobility and networked spatial relations of movement and flow, if not consistently so. Indeed, urban geography's origins can be traced to the Chicago School's dynamic systems metaphor of urban ecology

(Fyfe and Kenny 2005), which modeled neighborhood and change as processes of 'invasion' and 'succession' among people and land uses (Burgess 1925). Descriptive regional geographers emphasized the role of cities within regions as "element complexes," noting "the face of the earth includes moving objects that are constantly connecting its various parts" (Hartshorne 1939). Relations among urban places within a geometrical matrix of spatial relations, such as transportation and its costs, defined the locational theories of Von Thünen, Weber, Lösch, and Christaller.

A network perspective was also at the heart of urban geography's quantitative revolution, as the locational analysis of Isard (1956) and Haggett (1965) attempted "logically and empirically rigorous investigation of the spatial relations of phenomena and related flow patterns" (Barnes 2009). Statistical and modeling approaches understood spatial structure in terms of movement (interaction between points), networks (lines of linkage between points), nodes (convergence of links), hierarchies (differential role played by nodes), and surfaces (spaces between nodes) (ibid.). Urban geographers led the application of locational analysis (Berry 1973) as urban geography and regional science became deeply entwined (Wheeler 2001a). At the same time, the time geographies of Hägerstrand also theorized the trajectories of people and other objects in space and time (Hägerstrand and Lenntorp 1974), conceived as a web of paths or "continuity in the succession of situations" (Hägerstrand 1982, p. 323). Within these approaches, movement and circulation, and very often transportation itself, were essential to conceptualizing urban process.

Later critical, humanist, and structuration alternatives defined a more "plural" urban geography (Gauthier and Taaffe 2002), but these too considered the city a place of flows. Marxian political economy (Harvey 1973, 1978) approached the structural logics of capitalist urban processes through circuits of capital in the city as built environment for production, circulation, exchange, and consumption (Harvey 1978, 1999). Others emphasized social practice in place *and* time as a way of reconciling structure and agency/behavior towards a theory of structuration (Giddens 1984; Pred 1984). Even humanist geographers considered streets as key elements in the cultural landscape (Jakle 1977, 1994; Lewis 1979), part of the city as an "ensemble which is under continuous creation and alteration" (Meinig 1979, p. 6). But the growing emphasis on the social/cultural over the urban/economic in urban geography (Wheeler 2002) frankly came to marginalize transportation as a concern, rejecting "neat models of sociospatial structure" (Knox 2003) to focus on "place, consumption, and cultural politics" (ibid., p. 276). In a postmodern stance that "unhook(ed) the city from the bedrock of analytic categories" (Lake 2003, paraphrasing Beauregard 1991), however, network-based epistemologies fell out of favor. As a result, late 20th-century urban geography came largely to neglect the spatialities and experiences of mobility and transportation, albeit with notable exceptions.[1]

Fortunately, recent years have witnessed awakening interest in transportation among urban geographers, who have joined a new mobilities literature to talk explicitly of "politics of mobility" (Henderson 2006). New work explores questions about "how the city should be configured and by whom" (ibid., p. 193), in relation to 'automobility' (Urry 2004; Argenbright 2008), neoliberalism and transit planning (Addie 2013), or social inequality (Golub et al. 2013; Zavestoski and Agyeman 2015). Others increasingly examine the legal and engineering geographies of urban circulation, on and beyond the sidewalk (Henderson 2006, 2013; Mitchell 2005; Carr 2010; Blomley 2011, 2012). Work on global cities has gained still more relevance in approaching entrepôt cities as "relational cities" (Sigler 2013) and the complex, local/global dialectics shaping freight networks and spaces (Cidell 2011, 2012; Antipova and Ozdenerol 2013). And seeking to understand the increasingly uneven distribution of networks of various kinds, including communication and transportation, Graham and Marvin (2001) have provided influential insight into processes of 'splintering urbanism' (see also Graham and McFarlane 2015), through a focus on cities and regions as "staging posts in the perpetual flux of infrastructurally mediated flow, movement and exchange" (Graham and Marvin 2001, p. 8).

Despite such advances, urban geographic inquiry still lacks the sustained focus and coherent framework necessary for approaching transportation, having in many ways ceded the spatiality and functionality of mobility to transportation geographers and leaving mobilities scholars to focus upon transportation as the "interplay of culture, power and the urban"(Lake 2003, p. 462). Urban geography maintains historic strengths that can contribute much to understanding the city's functionality as a networked system of flows, imagining "the massive technical systems that interlace, infuse and underpin cities and urban life" (Graham and Marvin 2001, p. 8). But to fully reengage questions of urban transport and mobility, urban geographers may gain much by engaging those distinct literatures most fully dedicated to them.

TRANSPORTATION GEOGRAPHY: A TRADITION OF RELEVANCE, EMERGING BEYOND THE QUANTITATIVE

Interest in movement may have waxed and waned elsewhere, but it has remained central to the coherent and vigorous subfield of transportation geography (Black 2003; Hanson 1986; Hoyle and Knowles 1999; Rodrigue 2006; Rodrigue et al. 2013; Taaffe and Gauthier 1973; see also Hanson, this volume). While inquiry into urban, cultural, economic, and environmental processes is distributed widely across the social sciences and humanities, transportation has consistently remained a domain of quantitative geographers, economists, and engineers. Strengths in quantitative modeling and interaction with engineering have earned transportation geographers an uniquely high policy relevance, contributing to a subdisciplinary body of knowledge most explicitly concerned with transport networks, infrastructures, and processes. But as critical interest in mobilities has exploded in the past decade, transportation geography's positivist emphasis has to some degree distanced it from vibrant debates about the social and political dimensions of mobility. Transport geography has much to contribute, though, perhaps through still greater engagement with the social complexity of urban places.

Transportation geography shares with urban geography strong roots in the quantitative revolution, when transportation studies enjoyed a "heyday" (Hanson and Kwan 2008, p. xii) amidst geography's shift away from a chorographic to a systematic understanding of spatial processes (Sheppard 2001). Economic and transport geographies evolved in tandem to analyze interactions between places and transport systems (Garrison et al. 1959), developing influential spatial analysis techniques for studying intra- and interurban transportation (Berry 1964; Morrill 1967; Taaffe et al. 1963; Ullman 1954). These approaches considered transportation in terms of spatial separation, networks, spatial interaction, friction of distance, and "the relationship between spatial interaction and place" (Hanson and Kwan 2008, p. xiii). The application of quantitative and model-based perspectives, in partnership with civil engineers and economists (ibid., p. xiv), has enabled transport geographers to play an important role in policy and investment decisions.

Transportation geography, defined early by positivist and modernist assumptions that endure at the heart of the subdiscipline, has nonetheless recognized the ways geographic contexts for transportation involve "meanings, institutions, and social relations" (Hanson and Kwan 2008, pp. xiii–xiv). For example, as early as the 1970s, growing awareness of social and environmental issues prompted some to apply familiar quantitative methods to new subjects like alternative transportation (Hanson and Hanson 1976) and the role of difference in structuring socio-spatial relations and planning policies (Hanson 1980, 1982; Kwan 1999). And key scholars have advocated the progressive potential of transportation geography—including positivist and quantitative approaches (Sheppard 2001)—to address social differences like gender, ethnicity, class, and life course (McLafferty and Preston 2010; Monk 2010; Monk and Hanson 1982).

Although increasingly diverse, transportation geography remains defined by a particular set of key themes and related concepts (Hanson and Kwan 2008), including an emphasis on the materiality of transportation practices and infrastructure, their role in the urban process, and how policies shape them. Within North America, much research has focused upon accessibility and mobility (Buliung and Kanaroglou 2006; Farber and Páez 2011; Farrington 2007; Kenyon et al. 2002; Kwan 1999; Weber and Kwan 2002), whether defining accessibility as the number of opportunities (or activity sites) within a given distance, or travel time or mobility as something with "its own positive utility" (Hanson and Kwan 2008, p. xvi) that is strongly "related to the fate of places" (ibid., p. xvii). Much work explores how the networked structure of transportation—and its influence on relative transportation costs—influences urban development (Bowen 2008; Fowler 2006; Keeling 2009; Knowles 2006; Moon 1990; O'Connor 2010), and the global structure of passenger and freight flows (Pederson 2001; Taylor, Catalano, and Gane 2003; Taylor, Derudder, and Witlox 2007; Urry 2004). Together, these concepts—and their positivist and econometric heredity—have guided North American geographic analysis of urban transportation, interregional transport, and ensuing policy. Such research has been influential in planning, focusing attention on increasing accessibility to opportunities such as jobs, shopping, and recreation, rather than enabling mobility for its own sake (Bertolini 2006; Farrington 2007; Curtis and Scheurer 2010). And emphasis on uneven access to transportation networks (as a public good) has prompted vigorous debates on issues of equity (Hanson 1998; Keeling 2007; Lucas 2012; Tomaney and Marques 2013).

As critical interest in mobilities has emerged across disciplines, transportation geographers are also looking beyond traditional emphases to engage approaches from sociology, critical geography, cultural studies, etc. Indeed, Curl and Davidson argue, "it is the ability of geographers to be critical, in drawing attention to inequalities, ethical and social issues, which distinguishes Transport Geography from transport studies more broadly" (2014, p. 101; see also Shaw and Hesse 2010). Emerging critical scholarship in transport geography—particularly in the United Kingdom, continental Europe, and Australia/New Zealand—increasingly focuses on identity (Aldred 2013) and social exclusion (Lucas 2011, 2012; Mullen et al. 2014), offering new perspectives on infrastructure and usage (Vreugdenhil and Williams 2013) and practice and habit (Kent and Dowling 2013; Schwanen et al. 2012). Such emerging critical work represents a significant shift within transportation geographies, and creates new opportunities for engaging the subdiscipline with others that share its interest in processes of circulation.

Despite its sustained and valuable inquiry into the movement of people and goods, however, transportation geography's impact on growing debates has remained limited (Shaw and Sidaway 2010). Lamentably, social, urban, and political geographies of globalization too often taken transportation for granted, theoretically and empirically (Aoyama et al. 2006; Cidell 2006; Hesse 2010), and the study of global value chains, transnational families, and creative classes can neglect the political, social, and material aspects of transportation and mobility. At the same time, some see transportation geography's declining "disciplinary centrality" (Hanson 2003, p. 481) in missed opportunities to contribute to major debates about globalization, politics of scale, or multiple spatialities (Leitner et al. 2008). And despite considerable research on the deregulation of airlines, railroads, and trucking, the links between neoliberalism and transportation have been neglected by all but a few (Addie 2013; Docherty and Shaw 2011; Enright 2013;

Farmer 2011). An 'anthropology of mobility' is beginning to develop, but its impact is yet hard to assess (Dalakoglou and Harvey 2012; Lugo 2013).

As transportation practices and systems become ever more pressing issues, however, there is only greater need for scholarship on the place of transport networks and infrastructures in urban life. Transport geographies can offer fundamental insight into accessibility, mobility, networks, and spatial interactions in the city, particularly through expanded research into transportation practices, planning, and infrastructure (Aldred and Jungnickel 2013), the connection of global production and urban networks (Hesse 2010), and their significance to geopolitics (Derudder et al. 2013). Yet exploring their relevance to the city—and urban studies and geography—still requires what Hanson calls a yet greater ability to "imagine questions, methodologies, and epistemologies beyond those bequeathed . . . by economists and civil engineers" (Hanson 2006, p. 232). Understanding mobility in a 'splintered city' (Graham and Marvin 2001; Miciukiewicz and Vigar 2012) requires a still more plural transportation geography. Continued engagement with the emerging subfield of mobilities, and their relation to the urban, offers one such route.

CRITICAL MOBILITIES: NEW PERSPECTIVES ON THE POLITICS OF MOVEMENT

In the past decade, it can be argued that the most vibrant scholarship on the politics and practices of being mobile has emerged from outside urban and transport geography, led by critical scholars in sociology, geography, and anthropology. Building on early interest in mobility among critical scholars (Augé 1995; Graves-Brown 1997; Miller 2001; Urry 2000; Whitelegg 1997), a defined subfield burst onto the academic scene with the founding of Lancaster University's Centre for Mobilities Research in 2003 and the new journal *Mobilities* in 2006. Calling for a new kind of social science that does not presuppose being still as the default and mobility as the exception, these scholars argued for critical attention to mobility as a defining feature of the 21st-century experience of rapid movement of information, people, and goods (Hannam et al. 2006; Sheller and Urry 2006; see also Sheller, this volume). Foundational work explored how vehicles like cars are more than technology, but forces structuring cultural, political, and social life (Urry 2004). Academics across disciplines have rapidly embraced this new paradigm, expressing a latent desire to understand the social and cultural aspects of transportation, dissatisfaction with existing and largely quantitative approaches, and opportunity to contribute to theory building and publication in an entirely new subfield (e.g., Adey 2009; Adey et al. 2014).

Mobilities studies share the same object of study as transportation geography, but emphasize meaning in transportation; interconnections among mobile people, things, *and* information across scales; uneven relations between immobility and mobility; and the use of new, mobile methodologies (Cresswell 2011, 2012). Perhaps the key distinguishing characteristic is the emphasis on the 'politics of mobility,' and its attention to "the ways in which mobilities are both productive of such social relations and produced by them" (Cresswell 2010, p. 21). This particularly includes the "relation between human mobilities and immobilities, and the unequal power relations which unevenly distribute motility, the potential for mobility" (Hannam et al. 2006, p. 15), such as heretofore understudied, coerced mobilities (themes that have just recently appeared in transportation geography; e.g., Kelly et al. 2011; Olvera et al. 2013).

Mobilities scholarship contributes fresh attention to the "political, cultural and aesthetic implications and resonances of movements . . . the meanings ascribed to these movements, and the embodied experiences of mobility" (Cresswell and Merriman 2011, p. 11). Whereas much transportation scholarship focused on material processes and infrastructure, mobilities scholarship highlights the practices and representational schemes surrounding different subject positions of the mobile (Cresswell and Merriman 2011). Applying theoretical insights from science and technology studies, psychoanalytic theory, and nonrepresentational theory, among others (Bissell 2009; Van Duppen and Spierings 2013; Vannini 2011; Wilson 2011), work in mobilities addresses the ways in which race, physical ability, and gender mark mobile bodies and shape the experience of users of transportation systems (Akyelken 2013; Golub et al. 2013; Henderson 2006; Imrie 2012; Rodriguez 1999). At the same time, mobilities is also concerned with more than representational "analysis of the relation between mobility systems and infrastructural moorings" (Hannam et al. 2006). These moorings can be policy and techniques applied to transport infrastructure, such as the political contests between the "transportation engineer's focus on 'traffic flow' in allowing motor vehicles to move as efficiently as possible versus the pedestrian advocate's desire for 'place' as the intimate context of urban life" (Patton 2007, p. 928; see also Henderson 2013). Or they can be applied to the wider political institutions that govern movement: Enright suggests that the stakes in major transportation projects are "huge, and . . . reach to the very essence of urban politics today" (2013, pp. 798–799; see also Enright, this volume). These bring diverse theoretical approaches to bear on 'traditional' objects of transportation geography, including freight (Cidell 2011; Hesse 2013), air travel (Adey et al. 2007; Adey 2010, Frétigny 2013), and motorways (Merriman 2007), as well as social issues like migration (Benson 2011), legal rights and race (Hague 2010), dance (Cresswell 2006b), militarism (Kaplan 2006), and surveillance (Molz 2006). And mobility scholarship, drawing from critical traditions, is well positioned to contribute new insights into enduring and value-laden questions of equity (Jocoy and Del Casino Jr. 2010; Rodriguez 1999; Zavestoski and Agyeman 2015).

As many have rushed to this emerging field, others caution against neglecting decades of existing work on transportation and travel

(Cresswell 2011). But mobilities have nonetheless emerged as a distinct subdiscipline, with distinct emphases and approaches. Mobilities articles rarely cite existing work from transportation studies, despite calls to more closely connect the two fields (Shaw and Hesse 2010). And as mobilities scholarship emphasizes social and cultural meanings and individual experience, such work often neglects the infrastructural systems (both political and material) that enable travel. Mobile methodologies (Büscher and Urry 2009; Hein et al. 2008) add new insights into the hypermobile world familiar to researchers (Lassen 2006), but these methods have limits when applied to broader transportation systems, especially outside the developed world (Cwerner 2006; Kenny 2007; Collins 2009; and Aouragh 2011 notwithstanding).

Continued effort to engage the methods and subjects of study from transportation geography and mobilities can overcome the remaining limitations of each while building on their respective strengths (Curl and Davison 2014). Such synthesis can help us better understand how the everyday practices and experiences of individual travelers, within their specific social and political contexts, both shape and are shaped by cities. But that requires close attention to the everyday transportation places, landscapes, and networks where the interests of urban geography, transport geography, and mobilities intersect.

TOWARDS AN URBAN GEOGRAPHY OF NETWORKED FLOW?

Integrating the largely parallel literatures in transport geographies and mobilities requires an approach that focuses equally on the materialities and experiences of transportation in the urban landscape, towards "a fully networked urbanism" (Graham and Marvin 2001, p. 414). Urban geographers have long experience linking everyday places to networked processes, and a subdisciplinary history that accommodates (and sometimes integrates) both spatial modeling and critical approaches. Urban geographies of transport have the potential to integrate flow and concreteness, the material and experiential, and the functional and political.

One potential point of intersection may perhaps be found in Manuel Castells's landmark *The Rise of Network Society*, which approached "the emergence of a new social structure" (2000, p. 26) of cities in the information age through a social theory of space *and* time. If social life may often be defined by a *space of places* or "locale[s] whose form, function, and meaning are self-contained within the boundaries of physical contiguity" (ibid., p. 453), a "new spatial logic" (ibid., p. 408) or "material organization of time-sharing practices that work through flows" (ibid., p. 442) defines what he called a *space of flows*. Flows are not just one element of social organization, he argued, but "the expression of processes *dominating* our economic, political, and symbolic life" (ibid.). The "purposeful, repetitive, programmable sequences of exchange and interaction" (ibid.) are both embedded in and constitutive of social consciousness *and* dominant social structures, including the circuits, nodes and hubs, and socio-spatial organizations constituting the "material supports for dominant societal processes and functions" (ibid.). Acknowledging a simultaneous relationship between the space of flows and the space of places, Castells argued the "geography of the new history will not be made, after all, of the separation between places and flows," but "out of the interface between places and flows and between cultures and social interests" (1999, p. 302).

An explicit focus on transportation as that most material form of communication invites us to take up Castells's vocabulary while unbinding it from purported distinctions between the spatiality of flow and place, towards a more fully networked urbanism in theory and practice. Urban geographers and mobilities scholars understand transportation spaces to be places themselves, as locales whose form, function, and meaning are not self-contained but woven with the networked social organization of flow. And while we may recognize a tension between the politics of flow and place, too stark a dichotomy "ignores the role of place-based agents in making flows" (Hall and Hesse 2013, p. 6). A focus on urban transportation highlights the inextricability of flow *and* place—and cities as *places of flow*—and may help us better appreciate the ways the logic and meaning of places are indeed "absorbed in the network" and constitutive of it (Castells 1999, p. 443). Rather than seeing places transcended by "the space of flows and timeless time" (ibid., p. 507), or the space of flows as "non-places" (Augé 1995), the logics of place and flow are always inextricable in the urban landscapes, processes, and experiences. An urban geographical perspective, which has long sought to reconcile the spatial/quantitative and place-based/qualitative, may contribute to an integrated understanding of transportation spaces as place and flow, fixity and motion, practiced and represented, and often subject to fierce political contestation. Urban geographers are well poised to understand how urban transportation spaces are *intersecting places* that both structure and are structured by mobility as both *networked flows* and human experience (Cidell 2012).

What would such an integrative perspective on urban places of flow look like? It might focus on four related questions shared among these linked subdisciplines.

How do urban places intersect and function within wider transport networks and flows? If a fundamental basis for understanding mobility is the "fact of physical movement" (Cresswell 2010, p. 15), an urban geographical perspective—like that of transport geography—can contribute particular insights into how urban places function materially as nodes or corridors/pathways within networked flows, and how the assemblage of such places constitutes wider urban and transportation networks. With its long-standing emphasis on the built environment, informed with the similarly material perspectives of transportation geography, an urban geographical perspective offers a valuable contribution to understanding the material and spatial referents to a wider politics of mobility.

How is transportation experienced, represented, and contested within networked urban places? Despite sharing with transport geography the long-standing traditions of empirical positivism and quantitative methodology, urban geographers have equally strong traditions of exploring the behavioral and representational aspects of urban life and cultural landscapes. A fully nuanced understanding of urban places of flow will require, however, going beyond extending the 'cultural turn' in urban geography to transportation (Weber 2004) by incorporating the new insights of mobilities scholarship. Doing so more fully, but in ways fully embedded in the materiality of the urban landscape and transportation systems, is a key contribution that urban geographers of transport are well positioned to make.

How are mobility and immobility spatially and socially distributed within urban transportation systems and their governance? Urban geographers, with their expertise in questions of urban governance, can contribute much to how planning and policy manage the uneven benefits of mobility as "network capital"(Urry 2012). Urban geographers are expert in assessing the "infrastructural moorings" that shape urban mobility (Hannam et al. 2006), as transportation geographers understand how flows are "embedded in structure and agency" and not merely derived from locational factors. Transportation development is a process of "working out of structure and agency" (Ng et al. 2013, p. 3; see also Siemiatycki 2006). Urban and transport geographies can together contribute valuable insight into the ways policy and planning shape urban places and networks, and thus the practices and experiences of mobility.

How are urban places and transportation networks co-constructed through the practices of mobility and circulation? A more fully relational and dialectical understanding of how places structure everyday life and mobility—and are structured by them—will require greater emphasis not only on the infrastructural spaces that constitute transportation systems, but also on how the entire urban environment is inextricable from wider circulation networks (Cooke and Lewis 2009). Transport shapes space (Knowles 2006), but urban space also shapes transport. A focus on urban processes and patterns can provide insight into the simultaneous co-construction of the city's spaces of places and flows. Working out the appropriate relationship between flow and place is therefore not only an academic matter, but has significant implications for how we build and inhabit our cities.

THE BOOK'S STRUCTURE AND CONTENT

The following chapters explore intersecting urban places of flow by seeking to join transportation geography's insights on the materiality of networked transport systems and critical understanding of the politics and experiential qualities of movement. Together in one volume they join a growing rank of exciting works exploring mobilities, transport, and/or the urban (Bergmann and Sager 2008; Blomley 2011; Cresswell and Merriman 2011; Canzler, Kaufmann, and Kesselring 2012; Hall and Hesse 2013; Moraglio and Kopper 2015; Zavestoski and Agyeman 2015). They engage the urban landscape in fresh ways as an ideal object through which to understand both the infrastructural materiality and social relations of transportation/mobility in local places and wider networks, and cities as places of flow.

The first section, "Intersections: Everyday Places as Nodes," explores everyday urban places as nodes in networks of intersecting flows, from the scale of urban intersections to regional ports and airports, as well as the dialectical relationship between the social and spatial structure of urban places and the different modes and flows that pass through them. David Prytherch examines street intersections as both key infrastructure nodes and central arenas for the high-speed drama of urban life, where particular 'rules of the road'—like statutory law, roadway geometry and engineering, traffic control, police enforcement, and judicial adjudication—choreograph everyday mobility, helping order bodies in movement in space (Merriman 2012). Bascom Guffin's study of Hyderabad considers how symbolic or 'mediate' elements of traffic control are contested (and sometimes fail) in the context of a city of new drivers, leading authorities to regulate spaces of traffic through a more immediate 'politics of the concrete,' which 'subversive drivers' navigate. Gregg Culver explores how engineering tools like level of service (LOS) reflect and reinforce normative spatial visions of mobility, obscuring and naturalizing automobility through a discourse of scientific objectivity, and stifling more plural politics and practices of mobility in the redecking of Milwaukee's iconic Hoan Bridge.

Section two, "Corridors: Links in the Network," traces the spatiality and politics of transportation corridors and pathways as urban places of flows, exploring how links in urban and transport networks connect or divide places and mobility within/between them. Jason Henderson argues that systemic problems like mobility and climate—and the 'climate fight'—must be brought down to streets and bus stops, in which roadway engineering and transit planning become 'street fights' over the reallocation of street space, whether for alternative modes or neoliberal transit like Google buses. Peter V. Hall asks why and how certain city roads become accepted as routes for truck movement in Vancouver, in a contested politics of planning and regulation through which the mobility of goods in urban spaces "is 'channelled' into acceptable conduits" (Cresswell 2010, p. 24). Julie Cidell explores conflicts surrounding the restructuring of freight rail networks in suburban Chicago, where a politics of 'the uncanny'—in which predictable daily transportation practices become unpredictable—imbues infrastructure with wider fears about exclusion, disruption, and the nature of the city itself.

Section three, "Networks: Cities and Regions in Wider Context," expands the book's scope to the larger networks that structure transport and mobility within cities/regions and the spatial politics and planning surrounding them. Anru Lee's study of the symbolic and cultural meanings of Taiwan's mass transit system reminds us of the need to consider what transportation systems *mean* in addition to the provision of accessibility and mobility, how mass transit can become a vehicle of not only physical movement but also of change, breaking away, and becoming. Theresa Enright examines how conflicts over transit projects, like the signature Grand Paris Express (GPE) rapid transit network, are also conflicts over the essence of the metropolis, in which transit systems function as a horizon of movement and connection, a set of multiscalar social and spatial relationships constituting the dynamic flux of metropolitan life. Jean-Paul D. Addie engages dialectical-materialist urban analysis and critical mobilities to explore how transport networks function within a "relationality/territoriality dialectic" (McCann and Ward 2010) to transform the postsuburban, global city-region of Toronto, Canada.

Section four, "Circulation: Assemblages and Experiences of Mobility," explores the broader discourses and imaginaries that frame travel across urban places and networks. Markus Hesse examines the regional politics of freight distribution as torn between selling and restricting logistical hubs/networks, through a policy discourse that increasingly frames logistics in terms of modernity, growth, and prosperity. Christian Mettke analyzes the challenges of the politics of mobility within postsub/urban settings, considering how public transit provision within the Greater Toronto Area (GTA) highlights this spatial shift and dialectic of postsuburbanization and urban mobility to both fragment access to public services and fulfill new city-regional functions. Sophie L. Van Neste considers how new infrastructure projects in the 'in-between' places of the Randstad (Netherlands) are strategically framed and counterframed by authorities and local residents, particularly around a controversial highway project and its alternatives. Bianca Freire-Medeiros and Leonardo Name explore linkages between new transportation technologies and city branding in the Global South, examining how a cable car in Rio de Janeiro not only provides connections to a local *favela* but also transnational arrangements that invent, market, and sell territories associated with poverty and violence as tourist attractions.

In the book's conclusion, Andy E. G. Jonas reflects upon the challenges and opportunities of rethinking our approach to mobility through a more relational approach to the coproduction of transportation networks and urban spaces, suggesting the importance of still more research at what he calls "the urban-transportation-geography nexus" (Jonas, this volume). He ponders future directions for geographic thinking on the contemporary city of flows, particularly around such themes as ordinary urbanism; transportation, mobility, and the environment; the relationship between connected cities/competitive states amidst the 'geopolitics of capitalism;' and the 're-worlding' of the city and its meanings.

MOVING FORWARD

Urban transportation and places of flow are the very embodiments of the fundamental networked material and social processes that constitute city life (Siemiatycki 2006). They provide a literal and metaphorical intersection where we may witness and theorize the dialectical relations between place, transport infrastructure, and mobile bodies. Taking an urban geographical perspective on movement-in-place thus provides an ideal place to connect largely parallel debates in transportation geography and mobilities, towards an integrative and grounded perspective that helps scholars and practitioners fully rethink urban places from the neighborhood street to global city networks, and the key conceptual and policy questions surrounding them. By standing at the intersection of key debates in urban, transportation, and mobilities studies, we hope this book and its contributions may provide an integrative way to approach some of the major conceptual and practical challenges of 21st-century urban life.

ACKNOWLEDGMENTS

We would like to thank Max Novick at Routledge for his support of this project, and four anonymous reviewers for very thoughtful and constructive critiques. And we could not have done this project without the work of our wonderful contributors.

NOTE

1 A review of the past 15 years in the journal *Urban Geography* shows continued regional science research into transportation access and community (Rain 1999; Ong and Houston 2002; Sultana 2002; Weber and Kwan 2003; Horner 2004; Clark 2006; Shearmur 2006; Clark and Wang 2010), as well as network-based analysis of economic services (Taylor et al. 2003), airline connectivity (Derudder et al. 2007; Taylor et al. 2007), and telecommunications (Shen 1999, 2000; Walcott and Wheeler 2001; Palm 2002). Others take a place-based approach to transport infrastructure, like streets and alleys (Ford 2001). But the most intense debates in urban geography had little connection to mobility, focusing on the social construction and politics of scale (e.g., Marston 2000; Brenner 2000, 2001) and its spatial imaginary of boundedness and hierarchy (Marston, Jones III, and Woodward 2005).

WORKS CITED

Aaltola, M. (2005) "The international airport: The hub-and-spoke pedagogy of the American empire," *Global Networks*, 5(3):261–278.

Addie, J. P. (2013) "Metropolitics in motion: The dynamics of transportation and state reterritorialization in the Chicago and Toronto city-regions," *Urban Geography*, 34(2):188–217.
Adey, P. (2009) *Mobility*, London: Routledge.
Adey, P. (2010) *Aerial Life: Mobilities, Spaces, Affects*, London: Wiley-Blackwell.
Adey, P., Bissell, D., Hannam, K., Merriman, P., and Sheller, M. (eds) (2014) *The Handbook of Mobilities*, London: Routledge.
Adey, P., Budd, L., and Hubbard, P. (2007) "Flying lessons: Exploring the social and cultural geographies of global air travel," *Progress in Human Geography*, 31(6):773–791.
Akyelken, N. (2013) "Development and gendered mobilities: Narratives from the women of Mardin, Turkey," *Mobilities*, 8(3):424–439.
Aldred, R. (2013) "Who are *Londoners on Bikes* and what do they want?," *Journal of Transport Geography*, 30:194–201.
Aldred, R., and Jungnickel, K. (2013) "Matter in or out of place? Bicycle parking strategies and their effects on people, practices and places," *Social and Cultural Geography*, 14(6):604–624.
Antipova, A., and Ozdenerol, E. (2013) "Using longitudinal employer dynamics (LED) data for the analysis of Memphis Aerotropolis, Tennessee," *Applied Geography*, 42(1):48–62.
Aouragh, M. (2011) "Confined offline, traversing online Palestinian mobility through the prism of the Internet," *Mobilities*, 6(3):375–397.
Aoyama, Y., Ratick, S., and Schwarz, G. (2006) "Organizational dynamics of the US logistics industry: An economic geography perspective," *Professional Geographer*, 58(3):327–340.
Argenbright, R. (2008) "*Avtomobilshchina*: Driven to the brink in Moscow," *Urban Geography*, 29(7):683–704.
Augé, M. (1995) *Non-Places: Introduction to an Anthropology of Supermodernity*, London: Verso Books.
Barnes, T. (2009) "Locational analysis," in D. Gregory, R. Johnston, G. Pratt, M. Watts, and S. Whatmore (eds) *The Dictionary of Human Geography*, Malden, MA: Wiley-Blackwell, pp. 428–429.
Beauregard, R. (1991) "Without a net: Modernist planning and the postmodern abyss," *Journal of Planning Education and Research*, 10:189–194.
Benson, M. (2011) "The movement beyond (lifestyle) migration: Mobile practices and the constitution of a better way of life," *Mobilities*, 6(2):221–235.
Bergmann, S., and Sager, T. (eds) (2008) *The Ethics of Mobilities: Rethinking Place, Exclusion, Freedom and Environment*, Aldershot: Ashgate.
Berry, B. (1964) "Cities as systems within systems of cities," *Papers in Regional Science*, 13:147–164.
Berry, B.J.L. (1973) *The Human Consequences of Urbanization*, New York: St. Martin's Press.
Bertolini, L. (2006) "Fostering urbanity in a mobile society: Linking concepts and practices," *Journal of Urban Design*, 11(3):319–334.
Bissell, D. (2009) "Conceptualising differently-mobile passengers: Geographies of everyday encumbrance in the railway station," *Social and Cultural Geography*, 10(2):173–195.
Black, W. (2003) *Transportation: A Geographical Analysis*, London: Guilford.
Blomley, N. (2011) *Rights of Passage: Sidewalks and the Regulation of Public Flow*, New York: Routledge.
Blomley, N. (2012) "2011 Urban Geography Plenary Lecture—Colored rabbits, dangerous trees, and public sitting: Sidewalks, police, and the city," *Urban Geography*, 33(7):917–935.

Bowen, J. T. (2008) "Moving places: the geography of warehousing in the US," *Journal of Transport Geography*, 16(6):379–387.
Brenner, N. (2000) "The urban question as a scale question: Reflections on Henri Lefebvre, urban theory, and the politics of scale," *International Journal of Urban and Regional Research*, 24:360–377.
Brenner, N. (2001) "The limits to scale? Methodological reflections on scalar structuration," *Progress in Human Geography*, 25(4):591–615.
Buliung, R., and Kanaroglou, P. (2006) "Urban form and household activity-travel behavior," *Growth and Change*, 37(2):172–199.
Burgess, E. W. (1925) "The growth of a city: An introduction to a research project," *Publications of the American Sociological Society*, 18:85–97.
Büscher, M., and Urry, J. (2009) "Mobile methods and the empirical," *European Journal of Social Theory*, 12(1):99–116.
Canzler, W., Kaufmann V., and Kesselring, S. (2012) *Tracing Mobilities: Towards a Cosmopolitan Perspective*, Aldershot: Ashgate.
Carr, J. (2010) "Legal geographies—Skating around the edges of the law: Urban skateboarding and the role of law in determining young peoples' place in the city," *Urban Geography*, 31(7):988–1003.
Castells, M. (1999) "Grassrooting the space of flows," *Urban Geography*, 20(4): 294–302.
Castells, M. (2000) *The Rise of Network Society* (2nd ed.), Cambridge, MA: Blackwell.
Cidell, J. (2006) "Air transportation, airports, and the discourses and practices of globalization," *Urban Geography*, 27(7):651–663.
Cidell, J. (2011) "Flows and pauses in the urban logistics landscape: The municipal regulation of shipping container mobilities," *Mobilities*, 7(2):233–246.
Cidell, J. (2012) "Fear of a foreign railroad: Transnationalism, trainspace, and (im) mobility in the Chicago suburbs," *Transactions of the Institute of British Geographers*, 37(4):593–608.
Clark, W. (2006) "Race, class, and space: Outcomes of suburban access for Asians and Hispanics," *Urban Geography*, 27(6):489–506.
Clark, W., and Wang, W. W. (2010) "The automobile, immigrants, and poverty: Implications for immigrant earnings and job access," *Urban Geography*, 31(4):523–540.
Collins, G. (2009) "Connected: Exploring the extraordinary demand for telecoms services in post-collapse Somalia," *Mobilities*, 4(2):203–223.
Cooke, J., and Lewis, R. (2009) "The nature of circulation: The urban political ecology of Chicago's Michigan Avenue Bridge, 1909–1930," *Urban Geography*, 31(3):348–368.
Cowen, D. (2010) "A geography of logistics: Market authority and the security of supply chains," *Annals of the Association of American Geographers*, 100(3):600–620.
Cresswell, T. (2006a) *On the Move: Mobility in the Western World*, London: Routledge.
Cresswell, T. (2006b) "'You cannot shake that shimmie here': Producing mobility on the dance floor," *Cultural Geographies*, 13(1):55–77.
Cresswell T. (2010) "Towards a politics of mobility," *Environment and Planning D*, 28(1):17–31.
Cresswell, T. (2011) "Mobilities I: Catching up," *Progress in Human Geography*, 35(4):550–558.
Cresswell, T. (2012) "Mobilities II: Still," *Progress in Human Geography*, 36(5): 645–653.
Cresswell, T., and Merriman, P. (eds) (2011) *Geographies of Mobilities: Practices, Spaces, Subjects*, Aldershot: Ashgate.

Curl, A., and Davison, L. (2014) "Transport geography: Perspectives upon entering an accomplished research sub-discipline," *Journal of Transport Geography*, 38:100–105.
Curtis, C., and Scheurer, J. (2010) "Planning for sustainable accessibility: Developing tools to aid discussion and decision-making," *Progress in Planning*, 74(2):53–106.
Cwerner, S. (2006) "Vertical flight and urban mobilities: The promise and reality of helicopter travel," *Mobilities*, 1(2):191–215.
Dalakoglou, D., and Harvey, P. (2012) "Roads and anthropology: Ethnographic perspectives on space, time, and (im)mobility," *Mobilities*, 7(4):459–465.
Derudder, B., Bassens, D., and Witlox, F. (2013) "Political-geographic interpretations of massive air transport developments in Gulf cities," *Political Geography*, 36:A4–A7.
Derudder, B., Witlox, F., and Taylor, P. (2007) "U.S. cities in the world city network: Comparing their positions using global origins and destinations of airline passengers," *Urban Geography*, 28(1):74–91.
Docherty, I., and Shaw, J. (2011) "Transport in a sustainable urban future," in J. Flint and M. Raco (eds) *The Future of Sustainable Cities: Critical Reflections*, Bristol: Policy Press, pp. 131–152.
Enright, T. (2013) "Mass transportation in the neoliberal city: The mobilizing myths of the Grand Paris Express," *Environment and Planning A*, 45:797–813.
Farber, S., and Páez, A. (2011) "Running to stay in place: The time-use implications of automobile oriented land-use and travel," *Journal of Transport Geography*, 19(4):782–793.
Farmer, S. (2011) "Uneven public transportation development in neoliberalizing Chicago, USA," *Environment and Planning A*, 43(5):1154–1172.
Farrington, J.H. (2007) "The new narrative of accessibility: Its potential contribution to discourses in (transport) geography," *Journal of Transport Geography*, 15(5):319–330.
Ford, L.R. (2001) "Alleys and urban form: Testing the tenets of new urbanism," *Urban Geography*, 22(3):268–286.
Fowler, C. (2006) "Reexploring transport geography and networks: A case study of container shipments to the West Coast of the United States," *Environment and Planning A*, 38(8):1429–1448.
Frétigny, J.-B. (2013) "Les mobilités à l'épreuve des aéroports: Des espaces publics aux territorialités en réseau." Ph.D. dissertation, Université Paris I Panthéon-Sorbonne, UMR 8504, Géographie-Cités.
Fyfe, N.R., and Kenny, J. (eds) (2005) *The Urban Geography Reader*, London: Routledge.
Garrison, W., Berry, B., Marble, D., Nysteun, J., and Morrill, R. (1959) *Studies of Highway Development and Geographic Change*, Seattle: University of Washington Press.
Gauthier, H.L., and Taaffe, E.J. (2002) "Three 20thcentury 'revolutions' in American geography," *Urban Geography*, 23(6):503–527.
Giddens, A. (1984) *The Constitution of Society: Outline of the Theory of Structuration*, Berkeley, CA: University of California Press.
Goetz, A. (2006) "Transport geography: Reflecting on a sub-discipline and identifying future research trajectories. The insularity issue in transport geography," *Journal of Transport Geography*, 14:230–231.
Golub, A., Marcantonio, R., and Sanchez, T. (2013) "Race, space, and struggles for mobility: Transportation impacts on African-Americans in Oakland and the East Bay," *Urban Geography*, 34(5):699–728.
Graham, S., and Marvin, S. (2001) *Splintering Urbanism: Networked Infrastructures, Technological Mobilities and the Urban Condition*, London: Routledge.

Graham, S., and McFarlane, C. (eds) (2015) *Infrastructural Lives: Urban Infrastructure in Context*, Oxford: Routledge.
Graves-Brown, P. (1997) "From highway to superhighway: The sustainability, symbolism and situated practices of car culture," *Social Analysis*, 41(1):64–75.
Hägerstrand, T. (1982) "Diorama, path, and project," *Tijdschrift voor Economische en Social Geografie*, 73:323–339.
Hägerstrand, T., and Lenntorp, B. (1974) "Samhälls-organisation i tidsgeografiskt perspektiv (Social organization in a time geographical perspective)," *SOU (Swedish Government Official Reports)*, 2:221–232.
Haggett, P. (1965) *Locational Analysis in Human Geography*, London: Edward Arnold.
Hague, E. (2010) "'The right to enter every other state'—The supreme court and African American mobility in the United States," *Mobilities*, 5:331–347.
Hall, P., and Hesse, M. (eds) (2013) *Cities, Regions and Flows*, Abington: Routledge.
Hannam, K., Sheller, M., and Urry, J. (2006) "Editorial: Mobilities, immobilities, and moorings," *Mobilities*, 1(1):1–22.
Hanson, S. (1980) "The importance of the multi-purpose journey to work in urban travel," *Transportation*, 9:229–248.
Hanson, S. (1982) "The determinants of daily travel-activity patterns: Relative location and socio-demographic factors," *Urban Geography*, 3:179–202.
Hanson, S. (ed) (1986) *The Geography of Urban Transportation*, New York: Guilford.
Hanson, S. (1998) "Off the road? Reflections on transportation geography in the information age," *Journal of Transport Geography*, 6(4):241–249.
Hanson, S. (2003) "Transportation: Hooked on speed, eyeing sustainability," in E. Sheppard and T. Barnes (eds) *A Companion to Economic Geography*, Malden, MA: Blackwell Publishing, pp. 469–483.
Hanson, S. (2006) "Viewpoint: Imagine," *Journal of Transport Geography*, 14: 232–233.
Hanson, S., and Hanson, P. (1976) "Problems in integrating bicycle travel into the urban transportation planning process," *Transportation Research Record: Journal of the Transportation Research Board: The Bicycle as a Transportation Mode*, 570:24–30.
Hanson, S., and Kwan, M.P. (2008) "Introduction," in *Transport: Critical Essays in Human Geography*, Aldershot: Ashgate, pp. xiii–xxv.
Hartshorne, R. (1939) *The Nature of Geography*, Lancaster, PA: Association of American Geographers.
Harvey, D. (1973) *Social Justice and the City*, London: Edward Arnold.
Harvey, D. (1978) "The urban process under capitalism: A framework for analysis," *International Journal of Urban and Regional Research*, 2L:101–131.
Harvey, D. (1999) *The Limits to Capital*, London: Verso.
Hein, J., Evans, J., and Jones, P. (2008) "Mobile methodologies: Theory, technology and practice," *Geography Compass*, 2(5):1266–1285.
Henderson, J. (2006) "Secessionist automobility: Racism, anti-urbanism, and the politics of automobility in Atlanta, Georgia," *International Journal of Urban and Regional Research*, 30(2):293–307.
Henderson, J. (2013) *Street Fight: The Politics of Mobility in San Francisco*, Amherst, MA: University of Massachusetts Press.
Hesse, M. (2010) "Cities, material flows and the geography of spatial interaction: Urban places in the system of chains," *Global Networks*, 10(1):75–91.
Hesse, M. (2013) "Cities and flows: Re-asserting a relationship as fundamental as it is delicate," *Journal of Transport Geography*, 29:33–42.
Horner, M.W. (2004) "Exploring metropolitan accessibility and urban structure," *Urban Geography*, 25(3):264–284.

Hoyle, B., and Knowles, R. (eds) (1999) *Modern Transport Geography* (2nd ed.), New York: Wiley.
Huber, M. (2009) "The use of gasoline: Value, oil and the 'American way of life,'" *Antipode*, 41(3):465–486.
Imrie, R. (2012) "Auto-disabilities: The case of shared space environments," *Environment and Planning A*, 44(9):2260–2277.
Isard, W. (1956) *Location and Space Economy*, Cambridge, MA: The MIT Press.
Jakle, J.A. (1977) *Images of the Ohio Valley: A Historical Geography of Travel, 1740 to 1860*, New York: Oxford University Press.
Jakle, J.A. (1994) *The Gas Station in America*, Baltimore, MD: The Johns Hopkins University Press.
Jensen, O.B. (2013) *Staging Mobilities*, Abingdon: Routledge.
Jensen, O.B. (2014) *Designing Mobilities*, Aalborg: Aalborg University Press.
Jocoy, C., and Del Casino Jr., V. (2010) "Homelessness, travel behavior, and the politics of transportation mobilities in Long Beach, California," *Environment and Planning A*, 42(8):1943–1963.
Kaplan, C. (2006) "Mobility and war: The cosmic view of US 'air power,'" *Environment and Planning A*, 38(2):395–407.
Keeling, D. (2007) "Transportation geography: New directions on well-worn trails," *Progress in Human Geography*, 31(2):217–225.
Keeling, D. (2009) "Transportation geography: Local challenges, global contexts," *Progress in Human Geography*, 33(4):516–526.
Kelly, C., Tight, M., Hodgson, F., and Page, M. (2011) "A comparison of three methods for assessing the walkability of the pedestrian environment," *Journal of Transport Geography*, 19(6):1500–1508.
Kenny. E. (2007) "Gifting Mecca: Importing spiritual capital to West Africa," *Mobilities*, 2(3):363–381.
Kent, J., and Dowling, R. (2013) "Puncturing automobility? Car sharing practices," *Journal of Transport Geography*, 32:86–92.
Kenyon, S., Lyons, G., and Rafferty, J. (2002) "Transport and social exclusion: Investigating the possibility of promoting inclusion through virtual mobility," *Journal of Transport Geography*, 10(3):207–219.
Knowles, R., Shaw, J., and Docherty, I. (eds) (2008) *Transport Geographies: Mobilities, Flows and Spaces*, Malden, MA: Wiley-Blackwell.
Knowles, R. D. (2006) "Transport shaping space: Differential collapse in time-space," *Journal of Transport Geography*, 14:407–425.
Knox, P. L. (2003) "The sea change of the 1980s: Urban geography as if people and places mattered," *Urban Geography*, 24(4):273–278.
Kwan, M.-P. (1999) "Gender and individual access to urban opportunities: A study using space-time measures," *Professional Geographer*, 51(2):210–227.
Lake, R. (2003) "Introduction: The power of culture and the culture of power in urban geography in the 1990s," *Urban Geography*, 24(6):461–464.
Lassen, C. (2006) "Aeromobility and work," *Environment and Planning A*, 38(2):301–312.
Leitner, H., Sheppard, E., and Sziarto, K. M. (2008) "The spatialities of contentious politics," *Transactions of the Institute of British Geographers*, 33(2):157–172.
Lewis, P. (1979) "Axioms for reading the landscape," in D. W. Meinig (ed) *The Interpretation of Ordinary Landscapes*, Oxford: Oxford University Press, pp. 11–32.
Lucas, K. (2011) "Making the connections between transport disadvantage and the social exclusion of low income populations in the Tshwane region of South Africa," *Journal of Transport Geography*, 19(6):1320–1344.
Lucas, K. (2012) "Transport and social exclusion: Where are we now?," *Transport Policy*, 20:205–213.

Lugo, A. (2013) "CicLAvia and human infrastructure in Los Angeles: Ethnographic experiments in equitable bike planning," *Journal of Transport Geography*, 30:202–207.
Marston, S. (2000) "The social construction of scale," *Progress in Human Geography*, 24:219–242.
Marston, S., Jones III, J. P., and Woodward, K. (2005) "Human geography without scale," *Transactions of the Institute of British Geographers*, 30:416–432.
Massey, D. B. (1994) *Space, Place, and Gender*, Minneapolis: University of Minnesota Press.
McCann, E., and Ward, K. G. (2010) "Relationality/territoriality: Toward a conceptualization of cities in the world," *Geoforum*, 41(2):175–184.
McLafferty, S., and Preston, V. (2010) "Quotidian geographies: Placing feminism," *Gender Place and Culture*, 17(1):55–60.
Meinig, D. W. (1979) "Introduction," in D. W. Meinig (ed) *The Interpretation of Ordinary Landscapes*, Oxford: Oxford University Press, pp. 1–7.
Merriman, P. (2007) *Driving Spaces: A Cultural-Historical Geography of England's M1 Motorway*, London: Blackwell Publishing.
Merriman, P. (2012) "Human geography without time-space," *Transactions of the Institute of British Geographers*, 37(1):13–27.
Miciukiewicz, K., and Vigar, G. (2012) "Mobility and social cohesion in the splintered city: Challenging technocentric transport research and policy-making practices," *Urban Studies*, 49(9):1941–1957.
Miller, D. (ed) (2001) *Car Cultures*, Oxford: Berg.
Mitchell, D. (2005) "Property rights, the first amendment, and judicial anti-urbanism: The strange case of *Virginia v. Hicks*," *Urban Geography*, 26(7):565–586.
Molz, J. G. (2006) " 'Watch us wander': Mobile surveillance and the surveillance of mobility," *Environment and Planning A*, 38(2):377–393.
Monk, J. (2010) "Time, place, and the lifeworlds of feminist geographers: The US in the 1970s," *Gender Place and Culture*, 17(1):35–42.
Monk, J., and Hanson, S. (1982) "On not excluding half of the human in human geography," *The Professional Geographer*, 34:11–23.
Moon, H. (1990) "Land use around suburban transit stations," *Transportation*, 17(1):67–88.
Moraglio, M., and Kopper, C. (2015) *The Organization of Transport: A History of Users, Industry, and Public Policy*, Oxford: Routledge.
Morrill, R. (1967) "The movement of persons and the transportation problem," in W. Garrison (ed) *Quantitative Geography*, Evanston, IL: Northwestern University Press, pp. 83–94.
Mullen, C., Tight, M., Whiteing, A., and Jopson, A. (2014) "Knowing their place on the roads: What would equality mean for walking and cycling?," *Journal of Transport Geography*, 61:238–248.
Ng, A., Hall, P., and Pallis, A. (2013) "Editorial: Guest editors' introduction: Institutions and the transformation of transport nodes," *Journal of Transport Geography*, 27(1):1–3.
O'Connor, K. (2010) "Global city regions and the location of logistics activity," *Journal of Transport Geography*, 18(3):354–362.
Olvera, L. D., Plat, D., and Pochet, P. (2013) "The puzzle of mobility and access to the city in Sub-Saharan Africa," *Journal of Transport Geography*, 32(1):56–64.
Ong, P., and Houston, D. (2002) "Transit, employment, and women on welfare," *Urban Geography*, 23(4):344–364.
Palm, R. (2002) "International telephone calls: Global and regional patterns," *Urban Geography*, 23(8):750–770.
Patton, J. (2007) "A pedestrian world: Competing rationalities and the calculation of transportation change," *Environment and Planning A*, 39:928–944.

Pedersen, P. O. (2001) "Freight transport under globalisation and its impact on Africa," *Journal of Transport Geography*, 9(2):85–99.

Pred, A. (1984) "Place as historically contingent process: Structuration and the time-geography of becoming places," *Annals of the Association of American Geographers*, 74(2):279–297.

Rain, D. (1999) "Commuting directionality, a functional measure for metropolitan and nonmetropolitan area standards," *Urban Geography*, 20(8):749–767.

Rodrigue, J.-P. (2006) *The Geography of Transport Systems*, London: Routledge.

Rodrigue, J.-P., Noteboom, T., and Shaw, J. (2013) *The SAGE Handbook of Transport Studies*, London: SAGE.

Rodriguez, J. (1999) "Rapid transit and community power: West Oakland residents confront BART," *Antipode*, 31(2):212–228.

Schwanen, T., Banister, D., and Anable, J. (2012) "Rethinking habits and their role in behaviour change: The case of low-carbon mobility," *Journal of Transport Geography*, 24:522–532.

Shaw, J., and Hesse, M. (2010) "Transport, geography, and the 'new' mobilities," *Transactions of the Institute of British Geographers*, 35(3):305–312.

Shaw, J., and Sidaway, J. (2010) "Making links: On (re)engaging with transport and transport geography," *Progress in Human Geography*, 35(4):502–520.

Shearmur, R. (2006) "Travel from home: An economic geography of commuting distances in Montreal," *Cities*, 27(4):330–339.

Sheller, M., and Urry, J. (2006) "The new mobilities paradigm," *Environment and Planning A*, 38(2):207–226.

Shen, Q. (1999) "Transportation, telecommunications, and the changing geography of opportunity," *Urban Geography*, 20(4):334–355.

Shen, Q. (2000) "An approach to representing the spatial structure of the information society," *Urban Geography*, 21(6):543–560.

Sheppard, E. (2001) "Quantitative geography: Representations, practices, and possibilities," *Environment and Planning D*, 19:535–554.

Siemiatycki, M. (2006) "Message in a metro: Building urban rail infrastructure and image in Delhi, India," *International Journal of Urban and Regional Research*, 30(2):277–292.

Siemiatycki, M. (2013) "The global production of transportation public-private partnerships," *International Journal of Urban and Regional Research*, 37(4):1254–1272.

Sigler, T. (2013) "Relational cities: Doha, Panama City, and Dubai as 21st century entrepôts," *Urban Geography*, 34(5):612–633.

Sultana, S. (2002) "Job/housing imbalance and commuting time in the Atlanta metropolitan area: Exploration of causes of longer commute times," *Urban Geography*, 23(8):728–749.

Swyngedouw, E. (2000) "Authoritarian governance, power, and the politics of rescaling," *Environment and Planning D*, 18(1):63–76.

Taaffe, E., and Gauthier, H. (1973) *Geography of Transportation*, New York: Prentice-Hall.

Taaffe, E., Morrill, R., and Gould, P. (1963) "Transport expansion in underdeveloped countries: A comparative analysis," *Geographical Review*, 53(4):503–529.

Taylor, P. J., Catalano, G., and Gane, N. (2003) "Research note: A geography of global change: Cities and services 2000–2001," *Urban Geography*, 24(5):431–441.

Taylor, P. J., Derudder, B., and Witlox, F. (2007) "Comparing airline passenger destinations with global service connectivities: A worldwide empirical study of 214 cities," *Urban Geography*, 28(3):232–248.

Tomaney, J., and Marques, P. (2013) "Evidence, policy, and the politics of regional development: The case of high-speed rail in the United Kingdom," *Environment and Planning C*, 31(3):414–427.

Ullman E. L. (1954) "Geography as spatial interaction," reprinted in 1980 in E. L. Ullman and R. R. Boyce (eds) *Geography as Spatial Interaction*, Seattle, WA: University of Washington Press, pp. 13–27.
Urry, J. (2000) *Sociology Beyond Societies*, London: Routledge.
Urry, J. (2004) "The 'system' of automobility," *Theory, Culture and Society*, 21(4/5):25–39.
Urry, J. (2012) "Social networks, mobile lives and social inequalities," *Journal of Transport Geography*, 21(1):24–30.
van Duppen, J., and Spierings, B. (2013) "Retracing trajectories: The embodied experience of cycling, urban sensescapes and the commute between 'neighbourhood' and 'city' in Utrecht, NL," *Journal of Transport Geography*, 30:234–243.
Vannini, P. (2011) "Mind the gap: The tempo rubato of dwelling in lineups," *Mobilities*, 6(2):273–299.
Vreugdenhil, R., and Williams, S. (2013) "White line fever: A sociotechnical perspective on the contested implementation of an urban bike lane network," *Area*, 45(3):283–291.
Walcott, S. M., and Wheeler, J. O. (2001) "Atlanta in the telecommunications age: The fiber-optic information network," *Urban Geography*, 22(4):316–339.
Weber, J. (2004) "Everyday places on the American freeway system," *Journal of Cultural Geography*, 21(2):1–26.
Weber, J., and Kwan, M.-P. (2002) "Bringing time back in: A study of the influence of travel time variations and facility opening hours on individual accessibility," *Professional Geographer*, 54(2):226–240.
Weber, J., and Kwan, M.-P. (2003) "Evaluating the effects of geographical contexts on individual accessibility: A multilevel approach," *Urban Geography*, 24(8):647–671.
Wheeler, J. O. (2001) "Assessing the role of spatial analysis in urban geography in the 1960s," *Urban Geography*, 22(6):549–558.
Wheeler, J. S. (2002) "Editorial: From urban economic to social/cultural urban geography, 1980–2001," *Urban Geography*, 23(2):97–102.
Whitelegg, J. (1997) *Critical Mass*, London: Pluto.
Wilson, H. (2011) "Passing propinquities in the multicultural city: The everyday encounters of bus passengering," *Environment and Planning A*, 43(3):634–649.
Zavestoski, S., and Agyeman, J. (2015) *Incomplete Streets: Processes, Practices, and Possibilities*, Abingdon: Routledge.

Part I
Intersections
Everyday Places as Nodes

2 Rules of the Road
Choreographing Mobility in the Everyday Intersection
David Prytherch

INTRODUCTION

Every day we traverse intersections, key nodes in the city's infrastructure where streets and personal trajectories converge. Their form is consistent and familiar: curb lines and asphalt engineered to construct the street, lanes and crosswalks marked to guide movements, and signs and signals to direct mobile bodies and vehicles as 'traffic.' But routine as they are, intersections are central arenas for the high-speed and high-stakes ballet of urban life, where social relations and conflict can be a matter of life or death. The design and regulation of everyday intersections—the standardized rules of the road—are thus fundamental to the structured *choreography* of networked circulation and urban space.

A variety of factors shapes the practices and politics of mobility through the public street, which structures social relations in particular ways. To understand and engage with the street thus raises an important overarching question: How, exactly, is the everyday street ordered, socially and spatially? And as spaces in the public sphere, we may ask more specifically: What rule systems shape the design and regulation of the public street as a *place of flows*, and how? This chapter focuses on the discourses of order and control as manifested through the social and material space of transportation infrastructure—of asphalt, concrete, metal, and paint—in the attempt to structure the practice and politics of mobility. To illuminate both the general spatiality of the street and policy arenas for intervention, it broadly overviews the key technical and legal discourses that construct and regulate public roadways, including statutory law, roadway geometry and engineering, traffic control, police enforcement, and judicial adjudication. These standardized rule systems may be familiar to engineers or planners, but are less well understood within transport geography (which has often focused on transport systems) and mobilities scholarship (which often focuses on the experience of mobility). Extending recent writings on the legal and infrastructural politics of urban circulation (see Blomley 2011; Henderson, Culver, this volume), such a broad survey of major rule systems is essential for understanding the spatiality of public streets and flows upon them, a recursive relationship between the social relations and material infrastructure of mobility on the street. This review does not explore how individuals navigate such rule systems and the spaces modeled upon them, but how the public street is materialized and regulated as fundamental context for understanding how mobility is experienced and contested.

This chapter approaches the intersection not as a question of transport policy so much as the basic spatial and temporal fabric—the choreography—of everyday urban life. Drawing upon recent reconsideration of space-time geographies, and recent writings on choreography in dance and beyond, it considers the urban intersection as "movement space" (Thrift 2004), and in terms of "bodies-in-movement-in-spaces" (Merriman 2012). To understand how these are ordered, it focuses on the very specific rules of the road that are physically woven into urban space and enforced upon human and vehicular bodies. Drawing from recent writings, it interprets the intended orderings of such mobility as choreography: "techniques and technologies" from discourses to notational techniques that help direct and "score" bodily movement in space, and can "come to frame how people actually think about and practice these movements" (Merriman 2012, p. 22). Full understanding of mobility on the street would require assessing the experience and subjectivity of the mobile (for an excellent ethnographic approach to mobility as 'performative action,' see Jonasson [2004]). However, this chapter starts by analyzing these rule systems whose purpose is to control and score flows as traffic, which reinforce intended patterns of movement through the durable materiality of infrastructure and power of the law (including bodily arrest). Thus, if the street is among the city's most potentially chaotic spaces, it is also one of its most carefully constructed and rigidly choreographed. Understanding the ballet of urban flows therefore begins with the rule systems that physically and socially channel them.

What's more, given their uniformity, these rules provide geographers and others particularly valuable insights into a vast array of everyday streets and intersections across the United States. This chapter begins with a broad review of applicable national standards. But because most roadways are designed and regulated by states (and localities), one must be cognizant of local standards. This chapter thus looks at both uniform national standards and their translation in the Midwestern, and perhaps stereotypically average, state of Ohio. And to illustrate the local materialization of these rules, it explores them through an Ohio intersection not far from my home, representative in its utter banality. High Street and Martin Luther King Boulevard intersect in Hamilton, a small city located between Cincinnati and Dayton. High St. is both a state route and downtown thoroughfare, while MLK Blvd. is a U.S. route passing adjacent to Hamilton's downtown (Figure 2.1). We have all passed through thousands of such generic intersections. But in their remarkable ubiquity and uniformity, such intersections manifest the power of federal, state, and local rules to structure the city's spatiality and social relations of bodies-in-movement-in-spaces (Merriman 2012). Such an intersection could thus stand in for an untold number of others like it.

Figure 2.1 Intersections like this in Hamilton, OH are ubiquitous and uniform settings for everyday mobility. Photo by author.

Everyday intersections are at once nodes in wider regional transport networks, mundane places, and arenas for the life-and-death practice of mobility. Mobile subjects—their practices, experiences, and creativity—encounter the city and each other within the structured environment of the street. Thus any critical approach to urban transport must begin with powerful rule systems and their manifestation—in Hamilton, or whichever intersection we might select—that seek to create order through the durability of asphalt and the power of the law. How discourses of regulation, design, and enforcement are institutionalized and materialized upon the street illustrates the 'pedestrian' yet powerful way that particular cultures of mobility construct urban life and social relations.

MOBILITY, SPACE/TIME, AND STREET CHOREOGRAPHY

A key challenge of mobilities research is, as Sheller and Urry note, "to open up all sites, places and materialities to the mobilities that are always already coursing through them" (2006, p. 209). Approaching transport spaces requires a thoroughly relational perspective, which sees actions in space in relation to "alignments of actors, entities and resources" (Murdoch 2006, p. 69).

Not surprisingly, such inquiry has prompted renewed interest in space-time geographies, long a theme in geography. Many like Hägerstrand

(1995, 2004) have sought to "rise up from the flat map with its static patterns and think in terms of a world on the move" (Sui 2012, p. 323). As Sui points out, Hägerstrand drew upon and integrated Greek conceptualizations of not only space (*choros*) and time (*chronos*), but also more embodied conceptions of space as *topos* (concrete place/landscape) and time as *kairos* (embedded/event time). *Chronos-kairos-choros-topos*, following Sui, offers both a "holographic" worldview and "broadly based conceptual framework for geographical synthesis" (ibid., p. 13). Jensen et al. also draw upon Greek concepts of "*kinein* 'to move' + *aisthēsis* 'sensation' " (2014, p. 2) to argue that a fully relational view of mobilities starts with our haptic sense of touch and kinaesthetic sense of bodily motion, which "create place (and affect) through the frictions and rhythms of our movement through natural and built environments" (ibid.).

To move still farther in rethinking such categories as "dynamic, open, in process and becoming" Merriman (2012, p. 19) advocates a focus on "movement spaces" (Thrift 2004) and "bodies-in-movement-in-spaces" (Merriman 2011, p. 99). Given the intertwining of spaces and practices, Merriman suggests *choreography* as a way to conceptualize their "affective atmospheres, movements, folding forces and vibrating rhythms" (2012, p. 24). Dance, he notes, is "choreographed, codified, notated, and practiced in relation to a broad array of techniques and technologies," from discourses to notational techniques (ibid.), which can frame how people both practice and think about movement (ibid., p. 22). But the metaphor of dance is more widely applicable. Indeed, Merriman notes how architects like Lawrence Halprin drew inspiration from dance to approach streets,

> not simply as inactive material environments in which driving or walking could be practices, but as active spaces which must be carefully designed, "scored" and choreographed to produce particular movements, experiences, emotions and affects.
> (Merriman 2011, p. 100)

Concepts of movement-space and choreography thus extend naturally to urban streets, which are indeed a meaningful, "kinaesthetic movement-space" rather than empty corridors (Jensen et al. 2014, p. 3).

The kinaesthetics of urban circulation are improvisational and affective, in which every participant in traffic "simultaneously produces spaces and places as a social and physical project" in ways that are "not all about compliance to formal traffic rules" (Jonasson 2004, p. 50). "Seemingly routinized" actions in traffic represent intersecting social practices that produce meaning, order, and space itself, and the practice of socio-spatial abilities (ibid., p. 60). But these relations and alignments are nonetheless subject to "processes of negotiations, techniques and technologies of control and enactments of order" (Ek and Hultman 2008, p. 224). These seek to choreograph mobility to enable fluid mobility or "fix" mobile subjects and their trajectories in "prescribed, 'sticky' places at specific times" (ibid., p. 226), limiting or excluding other orderings in space *and time*, structuring rhythmic movements towards "sculpting the textures of the built environment and its infrastructures for mobility" (Jensen et al. 2014, p. 1).

Thus the intended ordering and choreography of traffic is part of, and simultaneous to, the "construction of culture, space and place" (Jonasson 2004, p. 60). The streetscape is a "zone of circulation" whose use, flow, and blockage is structured through a "complex array of codes and specifications, operative at multiple spatial scales" (Blomley 2011, p. 38). Jensen (2006, p. 152, following the work of Goffman 1963, 1972) highlights the "regulatory regimes of diverse municipal orders" that choreograph urban circulation and intersect with tacit mobility codes and norms. Such disciplinary and normative regulation "cannot be underestimated" (ibid. p. 160) in its power to choreograph individual agents and "mobility cultures" (ibid.). Mobile subjects draw upon "a repertoire of skills and the discovery of ways to flow through diverse urban spaces" (Butcher 2011, p. 251), but do such amidst an ordered and "expected choreography" (ibid., p. 244).

The legal and design ordering of urban circulation—as the choreography of "meaningful everyday life practice" (Jensen 2009)—thus prompts some key questions.

- How do legal and design standards structure street spaces and places towards the ordering of bodies (*choros/topos*) and the sequencing of their embedded movement (*chronos/kairos*)?
- How do such techniques and technologies choreograph mobility by *scoring* or "orchestrating and arranging" (Oxford American Dictionary 2005) street spaces and movements, including by marking ("making a line," [ibid.]) to delineate spaces or direct flows?
- How are intersecting trajectories negotiated or reconciled, whether through social communication or control, physical channeling of flows, or adjudication of competing claims?

Closer consideration of the codes and specifications designed to choreograph movement and "desired forms of behaviour and conduct" (ibid., p. 6)—in the name of social order and the general welfare—is thus a precondition for fully appreciating the street as a place of flows. And since traffic places are constructions that bear "spatio-temporal ideology" (Jonasson 2004, p. 52), the rules behind them merit close critical analysis.

THE RULES AND STANDARDS THAT CHOREOGRAPH MOBILITY IN THE EVERYDAY INTERSECTION

The Law: Statutes and Case Law Define Mobility Spaces in Terms of Rights and Duties

When one steps (or rolls) onto the street, one has entered a space of law. The very existence of any public street—like Main St. and MLK Blvd.

in Hamilton—is inextricable from the ordering system of the law, which enables governments to create and control public spaces for transportation and flow amidst a landscape of private property/rights (otherwise defined by the ability to exclude). In Ohio, the Ohio Revised Code (O.R.C.) Title XII "Municipal Corporations" §719.01 enables

> any municipal corporation (to) appropriate, enter upon, and hold real estate within its corporate limits. . . . For opening, widening, straightening, changing the grade of, and extending streets, and all other public places.

This power extends throughout the public *right-of-way* or "land, property, or the interest therein, usually in the configuration of a strip, acquired for or devoted to transportation purposes" (O.R.C. §4511.01BB). Furthermore, statutes grant municipalities their "care, supervision, and control" (§723.01). Intersections can be publically owned, controlled, or both (as is the case with the intersection of Main/MLK in Hamilton).

Statutes not only enable the creation and control of the streets we circulate upon, but also regulate our movement through them under separate statutory titles named "Motor Vehicles" or "Streets and Highways" or "Traffic Control." These date to the 1920s when the Department of Commerce convened what would become the independent National Committee on Uniform Traffic Laws and Ordinances (NCUTLO), whose motto was "Safety with Freedom through Law" (American Association of State Highway and Transportation Officials 2012b). The NCUTLO developed a *Uniform Vehicle Code* (UVC) and *Model Traffic Ordinance* for states and localities, which became incorporated within federal highway safety laws (NCUTLO 1968, 2000). Chapter 11 of the *Uniform Vehicle Code*, entitled "Rules of the Road," established standardized laws for traffic control, use of the roadway, rights and duties of vehicles and pedestrians, speed restrictions, etc.

Like the *Uniform Vehicle Code* upon which it is based, Ohio Revised Code Chapter 4511 "Operation of Motor Vehicles" defines the space of the street most broadly as right-of-way "open to the use of the public as a thoroughfare for purposes of vehicular travel," including the roadway as "improved, designed, or ordinarily used for vehicular travel" (O.R.C. §4511.01.EE). But the law also envisions distinct users, which are (to some degree) choreographed separately. Within the right-of-way, streets and intersections can often provide separate sidewalks and crosswalks, the latter defined by both the UVC and Ohio code as the prolongation of property/curb lines or roadway edge, whether marked or not (§4511.01.LL). The physically distinct spaces of asphalt roadway, concrete sidewalk, and painted crosswalk one encounters in the everyday intersection—like Main Street and MLK Boulevard—are defined and constructed through statutory law.

Intersections represent particularly important stages for the choreography of multiple bodies-in-movement, where flows intersect and can potentially collide (on an average day nearly 150,000 vehicles flow through the Main/MLK intersection in four different directions [Ohio Department of Transportation 2014]). Thus, the UVC seeks to order movements of diverse bodies by granting particular users, under particular circumstances, a distinct definition of "right-of-way" as

> right of one vehicle or pedestrian to proceed in a lawful manner in preference to another vehicle or pedestrian approaching under such circumstances of direction, speed and proximity as to give rise to the danger of collision unless one grants precedence to the other.
> (National Committee on Uniform Traffic Laws and Ordinances 2000)

Ohio (O.R.C. §4511.01) defines this right still further to mean

> (1) . . . right of a vehicle, streetcar, trackless trolley, or pedestrian to *proceed uninterruptedly* in a lawful manner in the direction in which it or the individual is moving in preference to another (emphasis added).

In very literal terms, statutes choreograph converging movements by ordering mobile relations, defining who has the right to be mobile and who must yield. Such explicit rules can become implicit understandings of where and how we should circulate, to whom we yield, and when we expect our motion to be uninterrupted. Within the roadway, outside crosswalks, pedestrians "shall yield the right of way to all vehicles, trackless trolleys, or streetcars," and may only walk along the margin when sidewalks are absent (§4511.50). On the sidewalk, pedestrians have preferential right, and vehicles must yield. In crosswalks, where pedestrian and vehicular networks overlap, §4511.46 mandates drivers "shall yield the right of way, slowing down or stopping if need be to so yield" to a pedestrian approaching their traffic lane, except when signals indicate otherwise. Enter any intersection in Ohio, and statutes choreograph one's movements through rights-based power relations with others—including the fundamental right to proceed.

The law not only choreographs our mobility by conferring rights, but also by assigning to us responsibility or 'duties.' For example, all roadway users must obey signals at intersections like that controlling Main/MLK. Indeed, the UVC's "Rules of the Road" specify "Pedestrians' Rights and Duties," as well as drivers' duties to "exercise due care to avoid colliding" with pedestrians or bicyclists (National Committee on Uniform Traffic Laws and Ordinances or NCUTLO 2000). Ohio statutes do not specify such duties, but these are elaborated separately in case law of torts (see below). Statutes grant right-of-way to cars in the open roadway or in intersections when they have the signal, and pedestrians must yield. But in crosswalks that balance shifts, vehicles turning into the path of pedestrians create ambiguous situations through which pedestrians and drivers must improvise.

When mobile bodies and vehicles enter the street, they thus enter a legal arena where competition for space and mobility is framed by a clear set of rules establishing and defining streets' spaces, potential users, and rights to proceed or duties to yield. These rules are, in turn, inculcated through education (e.g., driver's education and licensing), public communication (e.g., signs and signals), and enforcement. Main and MLK in Hamilton may be geographically unique in its situation, but the rules that help order movement upon it apply uniformly across Ohio, and with only slight variations across the U.S. The street is a public space of circulation and expression, but one tightly constrained by statute.

Roadway Engineering: The Geometry of the American Intersection

In case users forget such rules, they are built into the roadway. Professional engineers translate statutory principles through development and application of technical discourses and standards of 'roadway geometry' and 'design,' which guide the construction and maintenance of the streets we experience. The power geographies of mobility on the street are, quite literally, by 'geometric design.' Standards for design, construction, materials, etc. are developed primarily by American Association of State Highway and Tranportation Engineers (AASHTO) to "foster the development, operation, and maintenance of an integrated national transportation system" (AASHTO 2004a). These *de facto* rules for street design help ensure that particular intersections like Main/MLK are built to nationwide standards, a uniformity in design that reinforces uniformity in regulation.

Our experience of intersection spaces is particularly shaped by the details of AASHTO's *A Policy on Geometric Design of Highways and Streets*, or 'The Green Book' (and called by many 'The Bible' on roadway design), which provides a "comprehensive reference manual" for planning and administering roadways to provide "operational efficiency, comfort, safety, and convenience for *the motorist*" (AASHTO 2004a, p. xliv, emphasis added). Although the book asserts the "joint use" of transportation corridors (ibid.), its text and standards focus on ordering vehicular flow. It begins by systematizing the street network as a "hierarchy of movements and components," channeled into local, arterial, collector streets, and freeways. Second, it establishes design control elements, or "criteria for the optimization or improvement" of such as vehicle turning radii, driver performance, traffic patterns, and highway capacity (ibid.).[1]

These concepts and measures, and the assumptions underlying them, inform elements of design common to all classes of streets (e.g., sight distance, curvature, grades, widths, topography/vertical alignment, traffic control, etc.), which in turn inform cross sectional elements (e.g., pavement materials and specifications, widths, curbing) to channel flows. The application of AASHTO standards through new construction and retrofits transforms the roadway—and thus choreographs our own mobility—in their image.

How such geometric design criteria are applied depends on the classification of particular roadways within the overall network, which choreographs different kinds of mobility differently. Most streets are local, designed "to provide access to . . . abutting properties," but also provide movement of through traffic (AASHTO 2004a, p. 379). As movement spaces, they are local: designed for lower traffic volumes, narrower travel lanes, slower speeds, and lower levels of service. Conversely, others, like Hamilton's High Street and MLK Boulevard, are designated as arterials to provide a "high-speed, high-volume network for travel between major points" (ibid., p. 443), serving "major traffic movements" and "mobility with limited or restricted service to local development" (ibid., p. 469). As a principal arterial, Hamilton's intersection of High St./MLK Blvd. channels high volumes of cars in four travel lanes and one turning lane in each direction, fulfilling AASHTO standards of wider lanes (12 ft.), greater speeds (30 to 60 mph.), and higher levels of service. Circulate through such different streets, and the contextual cues shift to encourage us to slow down or speed up, access local land uses, or speed by them en route to regional destinations.

Because various movements—at various scales—come together at intersections, nodes pose a major engineering challenge: The entire road network's efficiency, speed, and capacity "depend on their design" (AASHTO 2004a, p. 555). How engineering standards order such potential chaos is central to the practices and experiences of urban mobility. For AASHTO, the main objective of intersection design is to "facilitate the convenience, ease, and comfort and people traversing the intersection while enhancing . . . efficient movement" (ibid.). Depending on the different types of intersection, like the basic 'four-leg' intersection at High St./MLK Blvd. in Hamilton, AAHSTO's 'Green Book' standardizes optimal angles of intersection, continuation of approach pavement, and corner turning radii. Intersections are points of conflict, so AASHTO emphasizes the importance of alignment and grade to allow "users to recognize the intersection and the other vehicles using it, and readily perform the maneuvers needed to pass through the intersection with minimum interference" (p. 579). Together with manuals like AASHTO's *Highway Capacity Manual* and the *Manual of Uniform Traffic Control Devices* (MUTCD), the 'Green Book' provides standardized rule for engineering/designing the socio-space of the roadway.

If intersections seem to facilitate automobility over pedestrianism, that's by design. AASHTO design standards are openly written on the assumption of vehicular flow, prioritizing it over other forms of movement. It may acknowledge the importance of 'accommodating' pedestrians and bicyclists, but rather than incorporating pedestrian or bicycle design within the comprehensive 'Green Book' AASHTO has developed separate—and for many local engineers, optional—pedestrian and bicycle design manuals (AASHTO 2004b, 2012a, respectively).[2] Recognizing the different bodily characteristics and vulnerabilities of distinct modes, these manuals also systematize the design and performance characteristics of nonmotorized bodies (walking speeds, spatial needs, mobility issues, and behavioral patterns) and establish distinct attributes of good design—circulation, balance, connectivity, safety, accessibility, landscaping, etc. For pedestrians, these are manifested in standards for sidewalks (e.g., types, widths, connections, etc.) and crosswalks (e.g., curb radii, crossing distances, crosswalks, etc.). The pedestrian features at High Street/MLK Boulevard include both marked crosswalks and 'walk' signals that both advise pedestrians and "assign the right of way" (AASHTO 2004b, p. 101). Such signals become another essential element in choreographing the intersection's spatial flows and temporal rhythm.

Together, AASHTO manuals and guiding standards—in close conjunction with state statutes and other standards—profoundly influence the design and operation of the streets we circulate upon. They systematically analyze mobility practices and design the roadway and its flows in a language of geometry in three-dimensional space and time, helping translate statutory rules into the materiality of asphalt, concrete, paint, and metal to shape the flow of bodies-in-movement-in-space.

Traffic Control: Uniform Marking and Signing the Street

Statutes and engineering may set the stage for ordered mobility, but choreography in real time depends on 'control.' Such scoring is effectuated in part through traffic control devices like markings and signage that communicate statutory rules and direct movement, techniques which became particularly necessary with the advent of the fast and dangerous automobile. The 1920s saw the creation of the later-named National Committee on Uniform Traffic Control Devices (NCUTCD 2012) to create a set of nationally uniform standards—a *Manual on Uniform Traffic Control Devices* (MUTCD) was first published in 1935—that was later incorporated and administered by the Federal Highway Administration (FHA) after the 1966 National Traffic and Motor Vehicle Safety Act (AASHTO 2012b). According to the national MUTCD, traffic control devices provide for "orderly movement of all road users" by notifying "road users of regulations and provid(ing) warning and guidance needed for the uniform and efficient operation of all elements of the traffic stream" (ibid.). Because these devices are designed to regulate behavior in accordance with the law, states like Ohio have adopted their own MUTCD, modified to reflect local conditions and laws (Ohio Department of Transportation 2012). The uniform look and function of intersections across the U.S. is largely a product of these federal standards.

The first element in controlling traffic is signage, whose need and purpose is at once obvious and profound: to "provide regulations, warnings, and guidance information for road users" conveyed through words, symbols, and arrows. They visibly and legibly inform road users of selected traffic laws or regulations and their applicability in particular contexts

Rules of the Road 55

(FHA 2012, p. 53). Regulatory signs seek to control movements by directing the direction of travel, allowed/prohibited turns, maximum speeds, and when movement should stop. Hamilton's Main/MLK intersection is signed with wayfinding signs to help drivers recognize and navigate local and state routes.

The MUTCD also literally 'scores' the roadway through "pavement and curb markings, delineators, colored pavements, and channelizing devices and islands" (FHA 2012, p. 389). Markings subtly yet powerfully choreograph mobility by providing important information while "allowing minimal diversion of attention from the roadway" (ibid.) (Figure 2.2). The MUTCD (and OMUTCD) prescribe the kinds and qualities of marking materials, as well as their function and application: yellow center lines to "delineate the separation of traffic lanes" (ibid., p. 393), arrows to guide turns, 'transverse' markings to make traffic stop or yield lines, or crosswalks

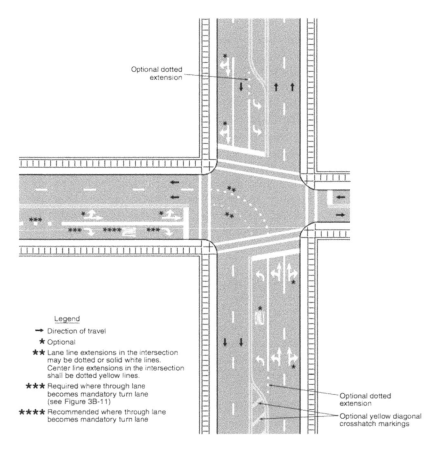

Figure 2.2 Intersections like Main/MLK are 'controlled' through signs, signals, and standard markings (shown here) (Source: FHA 2012).

to help in "defining and delineating paths and alert drivers of designated pedestrian points" (ibid., p. 430). Individuals entering Main/MLK are channeled in their forward movement by all of the above.

The movements and rhythms of users are further controlled through signals, which "attract the attention of a variety of road users" to assign the right-of-way to various traffic movements and "thereby profoundly influence traffic flow" (FHA 2012, p. 489). Engineering criteria are used to determine when there is a "warrant" for a signal, based on factors like vehicular and pedestrian volume, crash experience, and contextual factors (ibid., p. 493). And as a key element in the symbology and choreography of the street, the MUTCD provides very firm standards for the "meaning of vehicular signal indications" (ibid., p. 511), such as steady or flashing green, yellow, and red, and how these lights are arranged on the signal face and positioned within the intersection. The manual also provides guidance for pedestrian signals like those found at High St. and MLK Blvd. in Hamilton to control pedestrian flows.

Kinaesthetics in the everyday intersection—one's movements in time and space—are choreographed via the MUTCD's uniform signs, markings, and signals to order the street as place of flows. Developed by engineers over the past century, these help reinforce sanctioned mobility practices through symbols and material infrastructure, channeling communication and material interactions between users and the street. As users we have freedom of movement to choose our direction and speed, but these are channeled and paced by systems of signs and symbols, which are so ubiquitous and uniform as to operate both consciously and unconsciously. Furthermore, the culture of mobility they represent and reinforce is based on the assumption of circulating bodies as vehicular 'traffic,' helping build the assumption of automobility into everyday intersections like Main Street and MLK. The manuals increasingly envision and design for other users, but the roadway is, for the most part, scored for the longitudinal flow of cars.

Policing: Enforcing the Rules of the Road

Statutory and design rules seek to order our movements through both material channeling and symbolic communication, but such intended orders must still be performed by individuals. Thus, compliance—and where that fails, police enforcement—becomes another essential element in the street's "regulatory regimes" (Jensen 2006, p. 152). To reinforce the choreographic techniques of the law and roadway design, states empower state or local "peace officers" (O.R.C. §2935.01) and create penalties "imposed upon an offender . . . as punishment for the offense" (O.R.C. §2929.01). Police departments like that in Hamilton typically issue citations that inform alleged violators of the offense (§2935.27) or may exercise the power of arrest to literally stop and jail violators in more serious offenses.

Just as statutory rules order mobility, sanctions also help categorize and order types and degrees of noncompliance. According to O.R.C. Title 45 "Motor Vehicles," most traffic violations are misdemeanors penalized ranging from minor (up to $1,000 fine but no jail time) up to first degree (up to $5,000 and/or 180 days in jail) (O.R.C. §2929.24). Some of these crimes address how motor vehicles relate to traffic controls upon the roadway, like minor misdemeanors for speeding beyond established limits (§4511.21) or disobeying traffic control devices (§4511.12), or how vehicles relate to each other, like failing to yield right-of-way to other vehicles (§4511.41) or pedestrians in crosswalks (§4511.46). Perhaps surprisingly, in Ohio it is only a misdemeanor to operate a vehicle "in willful and wanton disregard of the safety of persons and property" (§4511.20) or "without being in reasonable control" (§4511.202). Pedestrians have legal responsibilities too, so it is a minor misdemeanor to fail to yield to vehicles in the roadway (§4511.50), "jaywalk" outside of crosswalks (§4511.48), or fail to obey traffic control devices (§4511.12).

The mechanisms of legal enforcement are latent, but they can become effective when one runs afoul of the police and legal system. One can discern the power relations and cultures of mobility in the proportionality of how the conduct—and crimes—of the mobile public is penalized. Many vehicular collisions are considered 'accidents' and minor misdemeanors, however, so the maximum fine cannot exceed $1,000 for the first offense in Ohio. In Hamilton, the maximum fine for such minor misdemeanors is $150, with no possibility of imprisonment (Hamilton Codified Ordinances 303.99). Penalties for pedestrian violations are still less: in Hamilton they include $10 to $15 fines for failing to yield or obey signals (Hamilton Codified Ordinances 371.99). So while policing and judicial penalties help choreograph life and death in the intersection, they do so with a relatively light touch. Many collisions are 'accidents' and entail no penalty, and even operating a vehicle recklessly (O.R.C. §4511.20) is penalized with only a $150 fine (!).

Mobility is a potentially deadly practice, however, so an accident can become a felony when "the death of another or the unlawful termination of another's pregnancy" results from operating a vehicle "recklessly" or "negligently," or when "the offender used a motor vehicle as the means to commit the violation" (O.R.C. §2903.06). Felony penalties use "maximum sanctions" (§2929.11) to protect the public from future crime and punish the offender. In Ohio, fatal accidents can become aggravated vehicular homicide—a felony—when drivers drive drunk, recklessly, or with past convictions. The "degree of culpability" is attached to mental state: A person acts "recklessly when, with heedless indifference to the consequences, he perversely disregards a known risk" and "negligently" when, "because of a substantial lapse of due care, he fails to perceive or avoid a risk that his conduct may cause a certain result" (O.R.C. §2901.22). Mandatory sentences for aggravated vehicular homicide in the first degree in Ohio can include 10 years in prison and a $20,000 fine. But absent the determination of recklessness or negligence, charges may not always result when vehicular bodies harm others, including the killing of vulnerable pedestrians.

As mobile bodies, our conduct on the street is thus under close scrutiny, perhaps closer than any other aspect of public life. Break the rules in Hamilton, and you may be cited or arrested and prosecuted by local courts of law. When fatalities result, and can be attributed to recklessness or negligence, even more severe penalties are possible. But while strong rule and enforcement systems are in place to maintain order, the kind of order they produce is—like other rules of the road—ideological. Most forms of illegal vehicular conduct are mere misdemeanors, despite the potentially catastrophic dangers everyday behaviors—like running a red light—can have for others. Driving recklessly through this intersection in Hamilton, OH only risks a $150 fine. Such processes of enforcement, though indirect, help reinforce the high-speed and risky cultures and choreographies of (auto)mobility.

Judicial Adjudication: Weighing Rights, Responsibility, and Liability

The final, major set of rules that helps choreograph mobility on the street does not emanate from legislators or engineers, but from lawsuits among civil litigants. When accidents occur, criminal enforcement can be paired with civil litigation, the latter of which is resolved by judicial/common-law rulings, *post hoc* mechanisms that impact the social and material space of the roadway directly by negatively reinforcing misconduct. Where a breach of a legal duty results in "injury, death, or loss to person or property" the pursuit of compensatory damages can lead to "tort actions" (O.R.C. §2307.311). In adjudicating such actions, judges consider not only the rights of different parties but also their duties or "obligation to do or forbear from doing something" (Calnan 2009, p. 8). Damage can be the result of intentional conduct, but traffic fault and liability is usually attributed to failures of "reasonable care" when doing otherwise legal things, known as negligence or "behavior that creates unreasonable foreseeable risks of injury" (Wright 1995, p. 250). Judicial rule systems may not play a direct or conspicuous role in choreographing mobile practices, but they are illustrative both for the assumptions they reveal and the risks they pose for the negligent. The scope of potential civil damages is so much greater than statutory sanctions—driving recklessly may be a $150 fine in Hamilton, but damages for noneconomic loss after an accident can reach $350,000 in Ohio, and greater where there is permanent injury—that tort law is thus potentially a significant force in structuring practices.

Innumerable and diverse accidents produce varied legal precedents. In adjudicating subsequent claims, judges seek to determine degree of negligence or "contributory fault" (O.R.C. §2315.33), in which damages should be "proportionately equal to the degree of tortuous conduct" (ibid.). Sometimes negligence is established because a statute is broken; such negligence *per se* represents an "active violation of a standard of care established by a safety legislative enactment adopted for the protection of the public" (Bush v. Harvey Transfer Co. et al. 1946). Fail to obey traffic signals and markings in Hamilton, or fail to properly yield, and you can be held negligent *per se* and liable for damages. In other cases, the violation is not of a statute, such as when parties are conducting themselves legally but "fail to observe the duties imposed on them in the exercise of ordinary care" (ibid.). Hit someone as the result of inattention or distraction and you also can be held liable.

Among these many cases, perhaps the most interesting are where different modes—and their distinct trajectories—intersect. Ohio's clear statutory assumption is that the roadway is a vehicular space where cars have right-of-way (Betras v. McKelvey Co. 1947; State of Ohio v. Goldie L. Huston 2004). Judges have affirmed that pedestrians must yield right-of-way to all vehicles outside crosswalks (Saade v. Westley 1986). While motorists must "use ordinary care to avoid colliding with any pedestrian upon the street," pedestrians must yield and can be held negligent if struck in the roadway (ibid.). The court in Wall v. Sprague (2008) has reiterated that motorists have a "preferential right of way" under §4511.01, and "a driver of a vehicle owes no duty of care to a pedestrian who walks into the path of the vehicle in an area not marked by a crosswalk." If a pedestrian is hit whilst 'jaywalking' across Main Street in Hamilton, she can be both cited *and* held negligent in a civil lawsuit.

Courts affirm the pedestrian's statutory right to proceed uninterruptedly within the crosswalk (e.g., Martinovich v. Jones Co. 1939; Grass v. Ake 1950; Ballash v. Ohio Department of Transportation 2002), but clarify this right to be *preferential* and not absolute. A pedestrian shares with drivers a duty of care when crossing the street (Molodylo v. Kent 1979) and "must use his or her sense, hearing or must exercise such care for his or her safety" (ibid.) across the entire crosswalk. As a pedestrian, if you enter the Main/MLK intersection with a 'walk' signal you may have the right-of-way, but must remain aware of various turning movements. If hit by a turning car, you could be negligent for failing to maintain continuous duty of care.

Awareness of one's legal responsibilities thus becomes part of the everyday performance of mobility. Tort law, like the enactment and enforcement of statutory law, seeks to standardize otherwise implicit rules about rights and duties among the mobile through the public street. If one is not careful, informal breaches of mobility practices (like yielding the right-of-way) can result in collision, legal citations, civil lawsuits, and potentially severe civil penalties. Whether these *post hoc* mechanisms modify behavior is a separate question, but the financial consequences of being held negligent—whether through civil damages or increased insurance costs—can be significant. These may not choreograph mobility directly, but individuals, corporations, and cities become cognizant of liability through the actuarial and policy interpretation of insurers, made manifest in insurance rates and occasional settlement. Thus, any mobile practices to avoid liability (and resulting penalties) are not only the enactment of statutory regulation and roadway design, but also the penalizing standards of tort law.

CONCLUSION

This chapter diagrammed the basic legal and design standards—the rules of the road—that help choreograph spatial relations within the street as place of flows. By examining statutory laws, engineering/geometric design, uniform traffic control, enforcement, and judicial adjudication, we may better understand the complex *choros/topos/chronos/kairos* of everyday bodies-in-movement-in-space. These rule systems seem banal: While some are highly visible (e.g., traffic lights), others are more subtle (e.g., traffic lane markings), and others functionally invisible most of the time (e.g., enforcement sanctions, civil penalties). They are everywhere, however, and in their ubiquity become naturalized elements in the time-space of the street. A cursory review of these powerful techniques and technologies reveals the very intentional ways urban movement is orchestrated, arranged, and scored in and through regulation and design. These help design the stage upon which the ballet of urban circulation unfolds, 'block' the movement and positioning of mobile actors, and otherwise direct the city streets' complex and contested sequence of movements. And they do so in highly standardized and institutionalized ways, replicated across the innumerable intersections central to urban life.

The choreography of the intersection is thus a recursive process by which social norms about mobility are codified as law, materialized in structured street spaces, signed and marked through the control of traffic patterns and rhythms, enforced through policing and judicial sanction of misconduct, and adjudicated through tort law. How such systems conceive of the street and mobility thus reflects dominant cultures of mobility, and reinforces them through an array of symbolic meanings, material infrastructure, and social controls. Translating statutory norms into material spaces, the street is physically engineered to channel and structure flows. But these cannot guarantee 'appropriate' mobile conduct, nor entirely prevent conflict and collision among mobile bodies, so additional systems of enforcement and adjudication become necessary. Together these help negotiate and reconcile the trajectories of different mobile bodies where they come into most direct conflict: the everyday street intersections we pass every day.

These rule systems are not all encompassing, and individual experiences and action within the street may not conform to governmental intentions. Jonasson (2004, p. 55) has argued that the inability to completely control humans produces a gap between formal and informal structures of mobility. Rules and regulations are not determinative in shaping mobility on the street, and people improvise in innumerable and creative ways. But at the same time, one cannot ignore the impact of institutionalized systems designed for scoring bodies-in-movement-in-space to produce particular orderings, enabling or constraining flow towards broader public ends through the power of law and materiality of concrete. Even this brief survey of these different rules highlights how elaborately these systems are "designed, 'scored' and choreographed to produce particular movements, experiences, emotions and affects" (Merriman 2011, p. 100) in ways that ultimately define urban places. Their uniformity and ubiquity across the U.S. suggest something of the power of certain ideologically laden techniques to shape urban places of flows.

In Ohio, as across the U.S., the dominant culture of automobility—and the diverse symbolic and material practices that constitute it—is not only produced through legal bias, but these uneven power relations are literally materialized through geometric design and socially reinforced through enforcement mechanisms (which view vehicular movement along the roadway as a dominant right and reckless misconduct as a minor offense). The rules of the road, like the roadway itself, are socially constructed around the primary object of vehicular speed and flow. Improvisation on the roadway is, of course, easier when the stage and scoring are built for your particular form of mobility. Even in a downtown urban place like Hamilton, Ohio, a key intersection like Main Street and MLK Boulevard has been designed to prioritize high-speed, high-volume vehicular flow over the safety of pedestrians or the character of place. Its choreography—both intended and improvised—is defined by speed, mass, and vehicular flow. This downtown place and its flows, like so many others in American cities, belongs more to the circulatory network than to the community that surrounds it.

But attention to standardized rule systems reveals their potential as objects of progressive social change. The landscape and its sense of place of the everyday intersections are products of particular rules, applied in particular ways, to produce particular urban outcomes. When everyone obeys those rules, and follows the signs, a social order results. Such is the choreography of which we, as kinaesthetic bodies-in-movement-in-spaces, all take an active part the minute we enter the everyday intersection, and accede to its complex rules. The kind of automobile-oriented spatial order we find in places like Main Street and MLK Boulevard may not be a social order everyone appreciates, however. Differently worded and more inclusive rules—statutory, design, enforcement, and common law—could manifest quite different orderings, which could produce more justice among the diverse users of the street, not all of whom are (or should be) in cars. Challenging such orderings begins by appreciation of the rule systems already at work, and the kinds of spaces they already produce. It's not hard, since they're all around us. The challenge is envisioning different rules and spaces. It's time for a new choreography of the street.

NOTES

1 Among these is the concept of level of service, an important measurement of roadway and intersection performance, including "speed and travel time,

freedom to maneuver, traffic interruptions, and comfort and convenience" (AASHTO 2004a, p. 84), addressed in other chapters (Culver; Henderson, this volume).
2 The *Guide for the Development of Bicycle Facilities* (AASHTO 2012b) argues that *all* streets, "except where bicyclists are legally prohibited, should be designed and constructed under the assumption they will be used by bicyclists" (pp. 1).

WORKS CITED

American Association of State Highway and Transportation Officials (AASHTO) (2004a) *A Policy on Geometric Design of Highways and Streets* (5th ed.), Washington, DC.
American Association of State Highway and Transportation Officials (AASHTO) (2004b) *Guide for the Planning, Design, and Operation of Pedestrian Facilities*, Washington, DC.
American Association of State Highway and Transportation Officials (AASHTO) (2012a) *Guide for the Development of Bicycle Facilities*, Washington, DC.
American Association of State Highway and Transportation Officials (AASHTO) (2012b) "History of the National Committee on Uniform Traffic Control Devices (NCUTCD)." Retrieved July 23, 2014 from http://www.ncutcd.org.
Ballash, Tina M. et al., Plaintiffs-Appellants, v. Ohio Department of Transportation et al., Defendants-Appellees. Court of Appeals of Ohio, Tenth Appellate District, Franklin County. 2002-Ohio-620; 2002 Ohio App. LEXIS 602.
Betras, Admr., Appellee, v. G.M. McKelvey Co., Appellant. No. 30970. Supreme Court of Ohio. 148 Ohio St. 523; 76 N.E.2d 280; 1947 Ohio LEXIS 377; 36 Ohio Op. 173.
Blomley, N. (2011) *Rights of Passage: Sidewalks and the Regulation of Public Flow*, New York: Routledge.
Bush, Admr., Appellant, v. Harvey Transfer Co. et al., Appellees. Supreme Court of Ohio. 146 Ohio St. 657; 67 N.E.2d 851; 1946 Ohio LEXIS 366; 33 Ohio Op. 154.
Butcher, M. (2011) "Cultures of community: The mobile negotiation of space and subjectivity on Delhi's metro," *Mobilities*, 6(2):237–254.
Calanan, A. (2009) *Duty and Integrity in Tort Law*, Durham, NC: Carolina Academic Press.
Ek, R., and Hultman, J. (2008) "Sticky landscapes and smooth experiences: The biopower of tourism mobilities in the Öresund region," *Mobilities*, 3(2): 223–242.
Federal Highway Administration (FHA) (2012) *Manual on Uniform Traffic Control Devices* (2009 ed., including revisions), Washington, DC.
Goffman, E. (1963) *Behaviour in Public Places. Notes on the Social Organization of Gatherings*, New York: The Free Press.
Goffman, E. (1972) *Relations in Public. Micro Studies of the Public Order*, New York: Harper & Row.
Grass Appellant, v. Ake, Exr., Appellee. Supreme Court of Ohio. 154 Ohio St. 84; 93 N.E.2d 590; 1950 Ohio LEXIS 388; 42 Ohio Op. 151.
Hägerstrand, T. (1995) "Action in the physical everyday world," in A. D. Cliff, P. R. Gould, A. G. Hoare, and N. J. Thrift (eds) *Diffusing Geography: Essays for Peter Haggett*, Cambridge, MA: Blackwell, pp. 35–45.
Hägerstrand, T. (2004) "The two vistas," *Geografiska Annaler*, 86B(4):315–323.
"Hamilton Codified Ordinances." (2014) Hamilton, OH. Retrieved June 15, 2014 from http://www.conwaygreene.com/hamilton.htm.

Jensen, O., Sheller, M., and Wind, S. (2014) "Together and apart: Affective ambiances and negotiation in families' everyday life and mobility," *Mobilities*. doi:10.1080/17450101.2013.868158
Jensen, O. B. (2006) " 'Facework,' flow and the city: Simmel, Goffman, and mobility in the contemporary city," *Mobilities*, 1(2):143–165.
Jensen, O. B. (2009) "Flows of meaning, cultures of movements—Urban mobility as meaningful everyday life practice," *Mobilities*, 4(1):139–158.
Jonasson, M. (2004) "The performance of improvisation: Traffic practice and the production of space," *Acme: International E-Journal for Critical Geographies*, 3(3):41–62.
Martinovich, a minor, Appellee v. The E.R. Jones Co., Appellant. Supreme Court of Ohio. 135 Ohio St. 137; 19 N.E.2d 952; 1939 Ohio LEXIS 357; 13 Ohio Op. 529.
Merriman, P. (2011) "Roads: Lawrence Halprin, modern dance and the American freeway landscape," in T. Cresswell and P. Merriman (eds) *Geographies of Mobilities: Practices, Spaces, Subjects*, Burlington, VT: Ashgate Publishing Company, pp. 99–118.
Merriman, P. (2012) "Human geography without time-space," *Transactions of the Institute of British Geographers*, NS 37:13–27.
Mologdylo, Wasylina, Appellee v. Dean C. Kent, Appellant. Court of Appeals of Ohio, Eighth Appellate District, Cuyahoga County. 1979 Ohio App. LEXIS 12209.
Murdoch, J. (2006) *Post-Structuralist Geography: A Guide to Relational Space*, London: Sage.
National Committee on Uniform Traffic Control Devices (NCUTCD) (2012) *History of the National Committee on Uniform Traffic Control Devices*, Washington, DC: American Association of State Highway and Transportation Officials.
National Committee on Uniform Traffic Laws and Ordinances (NCUTLO) (1968) *Uniform Vehicle Code and Model Traffic Ordinance*, Washington, DC. Retrieved March 10, 2014, from https://law.resource.org/pub/us/cfr/ibr/004/ncutlo.vehicle.1969.pdf.
National Committee on Uniform Traffic Laws and Ordinances (NCUTLO) (2000) *Uniform Vehicle Code and Model Traffic Ordinance*, Washington, DC.
New Oxford American Dictionary (2nd ed.) 2005 Oxford: Oxford University Press.
Ohio Department of Transportation (2012) *Ohio Manual of Uniform Traffic Control Devices*, Columbus, OH.
Ohio Department of Transportation (2014) *Traffic Counts*, Columbus, OH: Division of Planning, Office of Technical Services. Retrieved July 22, 2014 from http://www.dot.state.oh.us/.
Ohio Revised Code (O.R.C.) "LAWriter Ohio laws and rules." Retrieved July 23, 2014 from http://codes.ohio.gov/orc/.
Saade, Zahra Plaintiff-Appellant v. William K. Westley Defendant-Appellee. Court of Appeals, Eighth Appellate District of Ohio, Cuyahoga County, Ohio. 1986 Ohio App. LEXIS 6776.
Sheller, M., and Urry, J. (2006) "The new mobilities paradigm," *Environment and Planning A*, 38:207–226.
State of Ohio, Plaintiff-Appellee, v. Goldie L. Huston, Defendant-Appellant. Court of Appeals of Ohio, Tenth Appellate District, Franklin County. 2004-Ohio-6069; 2004 Ohio App. LEXIS 5531.
Sui, D. (2012) "Looking through Hägerstrand's dual vistas: Towards a unifying framework for time geography," *Journal of Transport Geography*, 23:5–16.
Thrift N. (2004) "Movement-space," *Economy and Society*, 33:582–604.
Wall, Meghann D., et al., Plaintiffs-Appellants, v. Adam C. Sprague, Defendant-Appellee. Court of Appeals of Ohio, Twelfth Appellate District, Clermont County. 2008-Ohio-3384; 2008 Ohio App. LEXIS 2867.
Wright, R. (1995) "Right, justice and tort law" in D. Owen (ed) *Philosophical Foundations of Tort Law*, Oxford: Clarendon Press, pp. 1–28.

3 Concrete Politics and Subversive Drivers on the Roads of Hyderabad, India

Bascom Guffin

INTRODUCTION

I sat in the passenger seat as Ram steered his new Tata Manza out of our gated community and onto the roads of 'Cyberabad,' the rapidly urbanizing western periphery of Hyderabad's urban agglomeration. Infotech corporate campuses, malls, and clusters of residential high-rises have been cropping up here amongst the erstwhile villages, farms, and granite scrublands that once defined the boundary around the city. And all of it is connected by an expanding, densifying network of roads that themselves carry a burgeoning fleet of motor vehicles spurred by a conglomeration of neoliberal government policies that have grown a new middle class both able and willing to engage in an expanded consumer economy.[1] The bullock carts and bicycles of an earlier era are steadily fading from public view, and pedestrians have been progressively pushed to the edges of the city's thoroughfares. Even smaller neighborhood side streets are becoming increasingly populated by motorized transport, especially auto-rickshaws, private cars, radio cabs, and motorcycles and scooters.

It was into this cauldron of moving metal that Ram pulled his sedan, and we headed toward the new Cyberabad outpost of Paradise, a restaurant so iconic the original lends its name to the neighborhood surrounding it. We eventually encountered one of the few traffic lights scattered about the area, and as it had just turned red, Ram stopped and waited. This did not suit the driver of the car behind us at all, who honked his horn incessantly at us to move. But Ram, who had grown up in Bombay and learned his driving as a postdoctoral researcher in Maryland, kept his car resolutely still, seething with righteous anger, until the light turned green and we proceeded along our way. To his mind, Ram was a disciplined citizen of the road and would not be bullied into violating the law, whether or not there was any cross traffic.

But the driver behind us considered Ram to be in the wrong, and felt entitled to honk at us for being stopped at a red light. As a taxi driver once told me, "in Hyderabad, drivers don't use their brakes, only their horns." I repeated this quote to many of my friends living there, and not once did anyone disagree with that driver's assessment. Because in Hyderabad, people indeed do not brake unless they absolutely must, and the honking horn has become the preferred substitute for slowing down or even stopping. The impetus to forward motion among drivers is not particularly unique to Hyderabad (cf. Laurier 2004; Taylor 2003). But when you combine this with the fact the city's drivers tend to see abstract regulatory traffic symbols—a system of colored lights, or two-dimensional words and pictures on signs, or lines painted onto roadways—as mere suggestions, you end up with a situation that diverges from what you might expect in many other parts of the world, and even in other Indian cities. Simply put, traffic signals and other abstract enforcement symbols in Hyderabad are very often either ignored or interpreted to a driver's advantage.

As far back as 1976, Hiebert documented this phenomenon in Hyderabad, contrasting it with Seattle. Drivers in Seattle, he noted, tend to rely heavily on *mediate* (i.e., mediated) means of control—that is, those lines and signs that create an order "in the geographic space of the highway" (p. 327). In Hyderabad, by contrast, drivers tend to make their calculations in light of the *immediate* (i.e., nonmediated) situation they are faced with, whether this be figuring out what other drivers will do based on status, inertia, and the like, or considering how different landscape features might impact their vehicle's integrity. Even today, Hyderabad's drivers seem only to respond to three-dimensional material objects placed directly in their way. These forms are both mobile and immobile. Mobile concrete forms include the various kinds of vehicles on the road, as well as bodies in locomotion, whether humans or other animals. Immobile, concrete forms include buildings, pillars, occasional raised walkways, trees, religious shrines, potholes, parked vehicles, and a plethora of others. Some of these are manifestations of state authority, such as Jersey barriers, speed bumps, blacktopped roads with their contrasting dirt shoulders, or the occasional raised walkway. The state projection of power also sometimes takes the fleshy form of traffic police physically standing in the middle of intersections directing the flow of vehicles.

In this chapter, I present an ethnographic look at how authorities in Hyderabad have responded to the intransigence of the city's drivers regarding mediated or abstract signaling systems by constructing immediate, concrete regulatory measures—and the ways that many drivers have managed to subvert even these.[2] What I recount is based on observations and interviews I conducted while living in Hyderabad for roughly 18 months, regularly riding with male and female drivers of private cars, motorcycles, and scooters while they commuted to work, went shopping, picked children up from school, and pursued leisure interests. I conducted interviews during ride-alongs and in more formal settings. These drivers tended to be middle-class people with extensive formal educations, and the large majority either worked in Hyderabad's burgeoning information technology sector or were married to these 'techies.' As such, they were mostly in their twenties and thirties, had come to Hyderabad from just about every region of India, and some, like Ram, had experience driving in other countries. I also talked with auto-rickshaw and radio cab drivers, traffic police, and planners. In addition, I spent time at a number of major traffic junctions and stretches of roadway, monitoring traffic interactions as both stationary observer and vehicle passenger, mostly in the Cyberabad area of the city. Finally, I have continually monitored for five years the highly active email list for the residential township where I stayed—paying particular attention to threads addressing issues of mobility, traffic, and road infrastructure—as well as the Hyderabad Traffic Police's Facebook page, its main venue for online interactions with citizens.

On a day-to-day basis, the fraught relationship between the city's drivers and planners takes the form of interactions between drivers, the physical artifacts that make up Hyderabad's road system, and the occasional traffic cop. I address these players in reverse order. First, I consider the embodiment of state authority that traffic police represent. I then examine the (often literally) concrete measures that road planners put in place, focusing on the shifting microlandscape of a single intersection, as well as discussing general road features and their purposes. Finally, I explore some of the ways drivers subvert the intentions traffic planners have built into the city's roadscape. In doing so, I am highlighting the material basis of state power in constructing regimes of mobility, and the limits of that power. While in many driving contexts it may seem that mediated forms of control are all it takes to govern traffic behavior, the situation in Hyderabad calls attention to the important, foundational role that concrete infrastructures necessarily play in shaping the behaviors of urban mobile agents. At the same time, as scholars such as Jonasson (2004) and Hall (this volume) point out, and this chapter affirms, road transportation spaces are made through complex negotiations amongst and between drivers as mobile citizens and the physical landscape as shaped by traffic planners—as well as, at times, the simultaneously mediate and immediate presence of traffic police.

FLESHY STATE POWER

To briefly consider the power of the embodied presence of police on Hyderabad's roadways is to glimpse some of the dynamics that have led to the government's reliance on concrete, immediate measures of traffic control over abstract, mediate ones. To wit: While traffic police rely on mediate, symbolic gestures to guide traffic, their presence is a physical manifestation of state authority.

For instance, during rush hour at the JNTU junction in Kukatpally, named for the Jawaharlal Nehru Technological University that occupies one corner of the T-junction, a white-clad traffic officer stands in the center of the intersection, using hand signals to direct traffic as a supplement to the traffic lights. In the face of the general chaos at this crossroads between a national highway and the key spinal road leading from Kukatpally to Hyderabad Information Technology and Engineering Consultancy (HITEC) City—and notwithstanding the obvious limitations of the human body versus the inertial might of thousands of motor vehicles—drivers overwhelmingly follow the messages conveyed by the officer's outstretched arms. Despite this traffic officer's general lack of recourse to technologies such as chase cars or cameras, he (almost always he) demonstrably garners more respect than the automated traffic signal with which he shares the intersection. Indeed, drivers' eyes are on him rather than on those lights. This is not much different from what happens on roadways around the world, where if a human (whether police officer or road crew) is present to direct traffic, his or her signals often take precedence over traffic lights, lane lines, and other abstract signaling systems. However, unlike in many other places, without the officer in place, drivers in Hyderabad simply ignore those other signals. The situation is not unique to the present time, nor to Hyderabad. Arnold documents the 1941 statement of a district traffic board in Madras: "[N]ot only were the public 'sadly wanting in road sense' but they took 'perverse delight in disregarding all traffic rules.'" They obey the rules "only so long as the policeman is present and break them immediately his back is turned" (2012, p. 136).

Because he is, in a fairly literal sense, a personification of state power, the officer's uniformed body is not only recognized as giving him license to be out in the middle of the wheels and fumes, but it gives him an inherent authority to direct drivers traversing his intersection. He can even get people to pull to the side just with hand signals, despite the fact that this often means bad news, including tickets written for rule infractions and even the occasional (or notsooccasional, according to the stories of some friends) demand for a bribe.

That there is an actual human being present takes state power out of the abstract and, appropriately, humanizes it. By 'humanize' I do not mean that it necessarily inspires in anyone warm sentiments about the government, its laws, or its enforcement practices and technologies. Rather, the presence of the traffic cop serves as a clear reminder that the government works based on human actions and agency, which enhances the state's potency in the minds of its subjects (perhaps part of a general tendency of humans to privilege their fellow species-members over other animate and inanimate entities). And though he functions as a mediate form of control, the officer is also, through his humanity, a more concrete manifestation than the symbolic order of squiggles on two-dimensional planes and blinking, colored lights—an order that Hyderabad's drivers understand even as they very often honor it in the breach.

But as agents of state power, Hyderabad's traffic police are in a fraught relationship with the city's drivers, who, like many other Indians, view the agents and institutions of governmental power with distrust, opposition, and general antipathy. As expressed repeatedly by many of my friends, as well as by members of my township's email list—and confirmed in personal observations of friends' dealings with government agencies—state bureaucracies are generally perceived as incompetent and inefficient at best, and very often as simply corrupt. They are to be endured and not relied upon for proficient service. Dealing with these structures can involve frustration, large investments of time, and cultivation of personal relationships, often via the passage of money from your own hand to that of the relevant bureaucrat. Or, for those who can afford it, payment to third-party brokers who have already cultivated those relationships can smooth your way as you attempt to obtain a passport or transfer your car's title.

This general mistrust and animus toward government instruments extends to traffic police, who are notorious for arbitrary rule enforcement and shaking drivers down for bribes. Indeed, it may in part be a wish to remain outside a traffic cop's notice that prompts drivers to follow his hand signals in an intersection. However, a sense that traffic rules are capriciously applied may play a role in drivers' trying to get around them or, if they seem to be unavoidable, to follow them only reluctantly. For instance, during a citywide enforcement drive to get two-wheeler riders to don helmets, my friend Matthew would either hand his helmet to me or keep it on his handlebars and only put it on when he spotted a police checkpoint up ahead. Matthew did this in large part because he knew, based on long experience, that the enforcement drive would end after a few weeks, and the police would more than likely once again ignore violations of the helmet law. Enforcement initiatives tend to be the result of the interest of a particular, powerful individual, as an interviewee in an article in the English-language daily *The Hindu* makes clear: "Focus on traffic management changes whenever top level officer is shifted [sic]. When Ms. Tejdeep Kaur Menon was traffic police chief, helmet rule enforcement was vigorously pursued. It is everybody's experience that traffic police ignore this rule now" (Ramu 2013).

At the same time drivers scoff at arbitrarily enforced rules, they are forced to reckon with them at the occasional police checkpoint. These tend to be a series of wood, plastic, and metal barricades to slow and then stop traffic, allowing a group of police to go through and inspect vehicles and drivers at their leisure. My friends found these checkpoints to be especially anxious spaces because they were rendered symbolically and concretely vulnerable in the face of this manifestation of state power, and thus exposed to the danger of what they generally thought of as the arbitrary and often unfair application of that power—particularly demands for bribes. My friends' unease was strong enough that on more than one occasion someone I was riding with would see a checkpoint up ahead and make the first turn possible—even a U-turn—and take a sometimes substantially longer route to our destination to avoid dealing with the police. Nonetheless, these checkpoints made from physical barricades are generally quite effective at immediately modifying driver behavior, by making one slow down, stop, or divert the vehicle. This brings me to concrete politics.

THE CONCRETE ORDER

Even if the relationship between Hyderabadis and their government was a harmonious one—and remember, drivers do accede to the demands of the officers' hand signals—traffic cops are not in every intersection of this sprawling city, nor does any given intersection enjoy continuous police presence. Police are human, with human needs such as rest and a regular paycheck, and are thus precluded from being permanently stationed along all of Hyderabad's roads. And so, in the face of motorists' intransigence regarding more abstract forms of regulatory power, Hyderabad's planners and police have resorted to (often literally) concrete measures. These carry both symbolic and material power, but I am most interested here in the material aspects of built forms of control, as their symbolic power is most often wrapped up in the possibilities and limitations afforded by the relation of a physical object to an active agent (for more on affordance, see Gaver 1991; Norman 1999; Dant 2004).

Michel de Certeau, a foundational theorist for the argument that urban actors enact their own, sometimes subversive, agency in relationship to space on a day-to-day basis (a point I shall return to), is also clear to note the "possibilities fixed by the constructed order" (1984, p. 98). That is, the concrete has a role in governing human actions. It may theoretically be most convenient for a person to cut straight through the middle of a city block to walk home from the bus stop, but if she is going to have to contend with an uninterrupted row of three-story buildings to do so, she will probably have an easier time just going around. Likewise, Michel Foucault, in mapping the genealogy of the rise of the modern disciplinary order, famously describes Jeremy Bentham's panopticon (1979), whose power (at least in theory) relies on the very concreteness and materiality of the built form. The material properties, technologies, and arrangements of Bentham's prison create a disciplinary order by giving its inmates the sense that they may be under surveillance at any time. The material technologies built into the panopticon are key to its governing effects.

Traffic planners also rely on built forms to regulate the individual and collective actions of drivers, as part of a web of concreteness, materiality, and immediacy that work to enmesh drivers in a regulatory regime. This starts with the relationship between driver and vehicle, in which the two combine into a hybrid (Beckmann 2001) or assemblage (Dant 2004) with its own set of characteristics. And while the human driver part of the assemblage has, as this chapter demonstrates, a great deal of agency, so too does the vehicle (Thrift 2004). Cars, for instance, have antilock brakes to control stops and, as famously described by Latour (1992), bells and lights to nag a driver into putting on her seat belt. A driver's very body becomes trained to the demands of the vehicle, "with eyes, ears, hands and feet all trained to respond instantaneously, while the desire to stretch, to change position, or to look around must be suppressed" (Sheller and Urry 2000, p. 747; also see Mauss 1973).

Thus disciplined, the "car-driver hybrid" (Beckmann 2004) becomes an agent in the landscape. Particularly in a large, populous city such as Hyderabad, this agent quickly becomes involved in a series of negotiations with other mobile agents. Tim Ingold states, "walking down a city street is an intrinsically social activity," and one must be "continually responsive to the movements of others in the immediate environment" (2004, p. 328). This is no less the case for motorists. The underlying principle to their immediate (that is, unmediated) negotiations is, as Jonasson puts it, *continuity*, meaning "that a traffic participant who is believed to have a continual movement in space imposes the right to go first" (2004, p. 44). In Hyderabad, negotiations are slightly more complex and follow a sort of 'law of traffic inertia.' The equation here is: mass × speed = right-of-way. If an auto-rickshaw calculates it can get across an intersection before a slow-moving truck crosses its path, it will probably make a dash. But a car that sees a bus bearing down with a full head of steam will generally let that bus go by. (Also see Edensor 2004, who has similar observations regarding Indian cities in general.)

These negotiations, of course, take place in time and space. Huijbens and Benediktosson point out that the car-driver hybrid itself "enters into a combination with the terrain traversed; a combination that is unique in each and every moment of travel" (2007, p. 155). At this landscape-level interface, traffic planners have the opportunity to regulate the behavior of drivers by quite literally *producing space* (see Lefebvre 1991), creating a concrete system that fundamentally underlies negotiations involving any number of variables between throngs of car-driver hybrids and other mobile (and non-mobile!) entities in a cityscape. Of course (and especially as I shall discuss later), concrete infrastructure is not purely determinative of driver behavior, and there arises "a recursive relationship between the social relations and material infrastructure of mobility on the street" (Prytherch, this volume,. But it does provide a delimiting foundation upon which relations of mobility can take place.

As in most cities, arguably the most important regulators of traffic in Hyderabad are the roads themselves. These produce order, as Cresswell points out, by "channeling motion . . . [and] producing correct mobilities through the designation of routes" (2010, p. 24). The paved, relatively flat surfaces of roads carve long corridors through the urban landscape, enabling vehicles to traverse substantial distances with a minimum of jarring. Paired with the many walls and building façades that line them, they are effective at channeling traffic through specified spaces in a fairly linear manner. Even when they are laid down in unbuilt areas in the periphery, they can provide a much smoother driving experience than the uneven dirt, boulders, plants, and mud of the wilderness or cultivated lands they cut through. And, as I shall discuss later, motorists in Hyderabad, being human, have an appreciation for paths of least resistance. So they stick to the roads rather than clatter across rice paddies and snake burrows.

Traffic planners also place other concrete forms on the traffic landscape to supplement the paved road itself. These tend to rely on verticality to physically prevent vehicles from crossing over particular lines in space, and include permanent medians and roundabouts, as well as more modular barriers made from cement, metal, or even an impromptu line of small boulders. Collectively, they are meant to keep traffic flowing in specific directions; although especially in the case of roundabouts, navigating the spaces they create can take some learning. It is not inherently apparent, for instance, which way to go around, or how to deal with rights-of-way—the latter of which, in Hyderabad, has more to do with the traffic law of inertia than anything else. Thus, while a roundabout's concrete presence generally dictates where a driver may *not* go, there is some social learning needed so that drivers are all on the same page about where *to* go. The concrete order interacts with social understanding to create a space for a relatively orderly flow of traffic (cf. Jonasson 2004). These forms are thus a physical manifestation of traffic planners' efficiency goals, which are to move as many vehicles along a linear stretch as quickly and safely as possible over any given period of time.

INTERSECTION AT A 'SPACE STATION'

That is not, of course, the limit of traffic planners' toolset. They can erect traffic lights, paint lane lines, and put up regulatory signs. But in Hyderabad, as I have noted, these mediate forms of control are largely ineffective. The example of one particular intersection should be instructive, and might be seen as a counterpoint to Prytherch's well-regulated intersection in this volume. During my fieldwork in the Cyberabad area of Hyderabad, I lived in a large township (gated housing colony) of around 2,000 units that I shall call Space Station.[3] The township's main entrance opens onto a busy six-lane road, forming a T-junction. In part because the main road does not encounter any major intersections for hundreds of meters at a time, the traffic along it tends to go fast.

When I first moved into Space Station and began observing the intersection, the junction had no traffic controls here other than the blacktop, raised roadside curbs, and meter-high brick medians separating the north- and southbound lanes from each other. There were also short, curb-separated pullout lanes on either side of the road, presumably to allow for pickups and drop-offs without disturbing the normal flow of traffic. The intersection itself was an empty field. There was nothing to interrupt traffic so that pedestrians could cross the road or vehicles could make wide right turns or U-turns. Because the township is so heavily populated, with a regular flow of vehicles and pedestrians coming in and out, and because the main road is itself so busy, the lack of traffic controls led to situations that were inconvenient at best, and frequently dangerous (Figure 3.1). Pedestrians and motorists would often have to wait several minutes at a time to find a break

Figure 3.1 An attempt to calm traffic near Space Station's intersection via signage is unsuccessful. On the right, a median keeps northbound and southbound traffic separated. Photo by author.

in traffic in order to cross or turn; I regularly experienced this situation as a pedestrian or passenger. Many times people would become impatient and simply head into the oncoming traffic, relying on drivers to be willing and able to slow down or stop for them. This would generally result in honking vehicles (rather than the application of brakes, as I've noted). The situation could also sometimes lead to shouted imprecations and, occasionally, to collisions.

Traffic authorities finally decided to do something about this state of affairs at the Space Station junction and installed traffic signals. These operated for about a month, but no matter whether they were set to cycle between green, amber, and red, or with lights in all directions blinking red, drivers largely ignored them. It is possible that this could be at least somewhat attributable to drivers simply being accustomed to there being no traffic controls at this particular intersection. But during the month the lights operated, whenever a police officer was present to direct traffic, drivers obeyed his signals, which were generally timed to what the lights were indicating. When an officer was not standing there, drivers reverted to ignoring the lights. And so one day the lights were unceremoniously turned off, though they were never removed—they simply hung there, forlorn and abandoned.

Not long after this failed experiment, speed bumps (or speed breakers, as they are called in India) were installed in either direction of the main road for traffic entering the intersection. Even without the presence of a cop, traffic through the Space Station junction calmed immediately. The colorful British term for a speed bump, "sleeping policeman," might give satisfaction to disgruntled motorists who can feel as if they are symbolically running over an oft-reviled agent of the state, but is also appropriate to its function as a passive regulator of traffic. The jarring you experience if you go over one too quickly is not particularly pleasant, and gives the impression that you could be damaging your car. And because there is little signage or contrasting paint to highlight the bumps, it is up to motorists to either spot new speed bumps when they sprout up or learn about them by having their bones rattled (Space Station's email list was abuzz one day with news of a motorbike unwittingly hitting one of the new speed bumps at pace, launching it and its two passengers into the air and scattering them across the road). But as drivers learned, one way or another, of their existence, the speed bumps did have the intended effect of slowing traffic enough that pedestrians and drivers entering and exiting Space Station felt optimistic about their chances of getting across the road unscathed. Road-crossers still had to gauge the speed and deceleration of oncoming traffic to make sure drivers knew about and were taking the speed breakers seriously. But even the relatively slow, fragile pedestrians no longer had to wait long for a break in traffic to get across—the break was pretty much always there.

The dynamics of regulating the Space Station junction speak to the overall approach that traffic planners have largely taken with Hyderabad's roads. To recall: Abstract symbols are generally ignored while police officers' bodies and material emplacements are not. So the solution to regulating the wild rumpus that is Hyderabad's traffic situation would seem to be equipping the road system with concrete limiters and the occasional traffic cop stationed in an intersection. But if concrete measures by their very nature do affect how people drive in this city, people also find ways around them and through them. And sometimes the concrete controls can themselves encourage unintended behaviors that may even directly contradict their purpose. This is the phenomenon to which I now turn my attention.

SUBVERSIVE DRIVERS

One of the key points of de Certeau's essay, "Walking in the City," is that despite the best efforts of planners, people will tend to make their own uses of urban spaces and carve their own routes through them. His clearest passage on this bears repeating:

> And if on the one hand he [the walker] actualizes only a few of the possibilities fixed by the constructed order (he goes only here and not there), on the other he increases the number of possibilities (for example, by creating shortcuts and detours) and prohibitions (for example, he forbids himself to take paths generally considered accessible or even obligatory). He thus makes a selection.
>
> (de Certeau 1984, p. 98)

As explicated by anthropologists such as Holston (1989), Scott (1998), and Ghannam (2002), this can run to out-and-out subversion of designs of control that urban planners construct into cityscapes. Holston, for instance, outlines the ways in which residents of Brasília undermined the intentions of planners by, among other things, turning what was meant to be the utilitarian backs of shops, dominated by the ugliness of parking spaces, into the *de facto* shop fronts that the overwhelming majority of patrons traversed when doing their shopping. Meanwhile, the clean, orderly front façades remained barren of foot traffic.

Thrift (2004) extends de Certeau's arguments regarding walkers to consider *drivers* in the city, and how they can at times engage in tactics that run counter to the expectations of planners and even other drivers, while Jonasson (2004) considers how drivers engage in place-making through their improvisational activities. And Hall, in this volume, describes how truckers in Vancouver often follow routing logics that subvert transport analysts' models.

Similarly, drivers in Hyderabad constantly make choices that mutate or even thwart the purposes of traffic planners as laid out in the concrete landscape. In designing separated pickup/drop-off lanes approaching intersections, planners have inadvertently allowed traffic to drive straight through while avoiding any speed breakers. What was meant as a space for vehicles to slow and stop to allow people on and off has now become a thoroughfare of its own, with a steady stream of cars speeding along its hundred-meter length. Even this subversive flow of traffic is interrupted a couple of afternoons a week when middle-class parents from Space Station and nearby neighborhoods set orange cones on either end of the lane and transform it into a *de facto* roller rink. For a couple of hours, cars are replaced by children with polystyrene shells strapped to their heads and wheels tied to their feet, careening up and down the blacktop with varying levels of skill.

Hyderabadis similarly repurpose other spaces of transit, sometimes in less ephemeral fashion. For instance, the few sidewalks that do exist often get taken over for parking. With the dramatic rise of private vehicles in the city, there is a general shortage of parking, even for motorcyclists, as this exchange with Ashok, a young infotech worker, makes clear:

Ashok: Sometimes you won't find parking at all. You go to Building D, and the attendant says, "no sir, please sir go to C, you go to C." And the next says, "please sir go to B, you go to B," and, "please go to A." And when you land up in A, "sorry sir, there's no parking, you've got to park it outside."

Bascom: So then do you just park it right outside the gates then?

Ashok: Uh, no, actually, along the wall of the building, there's some space where you can put your bikes . . . It's the whole tech area where all the techies sit and work, so all of them, they end up parking bikes on the roads; along where actually the pedestrians should walk, they end up parking. We changed that into a parking lot and everybody keeps parking bikes over there.

Bascom: So then pedestrians are just on the road.

Ashok: Yeah, that's the sad part.

Due to a lack of designated parking spaces, motorcycle commuters along this stretch of the HITEC City area of Hyderabad have created their own unsanctioned infrastructure simply by all placing their bikes in a concentrated agglomeration. This is not a unique phenomenon; you can find other unofficial parking agglomerations throughout the city. The general situation causes some friction in the Hyderabadi press, which features constant complaints about shops and malls opening without adequate parking spaces—in contravention of zoning rules—and vehicles subsequently taking over nearby roadsides and sidewalks. This and other examples—the repurposing of roadsides and empty lots as auto-rickshaw stands, the takeover of pedestrian spaces by small-scale vendors—demonstrate Hyderabadis' willingness to create their own infrastructure through collective daily action.

Coming back to the driving experience itself, aside from driving faster than allowed by law (but as fast as they feel the physicality of the roads allows them to—though see Vanderbilt [2008] on how humans are notoriously bad on judging road conditions), drivers in Hyderabad subvert the intentions of traffic planners in two main ways: driving on the dirt shoulders of roads and driving on the wrong side of roads. Dirt shoulders are common along the major roads of Cyberabad, although some properties, such as Space Station, have been built to the edge of the road. Motorists sometimes resort to briefly driving on the shoulder to avoid an obstacle such as a shrine, cow, giant pothole, or speed bump. Planners would probably view this sort of deviation as relatively inoffensive (with the exception of avoiding that speed bump). But particularly on busy thoroughfares when traffic gets congested, the shoulder becomes an unofficial extra lane. Auto-rickshaw drivers and motorcyclists are most likely to take this option, partly because their nimble forms are most able to maneuver along the narrow, oftenpitted dirt track, but I have been in the car with more than one techie friend who has also opted to steer a hatchback or small sedan along the side of the road. This phenomenon is similar to that of unauthorized parking agglomerations, in that motorists take over currently unoccupied spaces and, through their actions, turn them into new or expanded forms of infrastructure to suit their needs at the moment. As large numbers of drivers make use of the shoulders this way, it becomes a normally accepted behavior. While you might not normally want to drive on the dirt shoulder since the paved road is smoother, faster, and easier on you and the car, drivers come to view these shoulders as almost designed to be available to help ease congestion, rather than as buffers between the road and the non-transit landscape.

Driving on the wrong side of the road even more clearly subverts the intentions of traffic planners. There are two main varieties of this wrong-side driving in Hyderabad. During rush hours, when roads become particularly congested in a single direction, a stream of drivers may begin to overflow onto the other side of the road, forcing oncoming traffic to become restricted to fewer lanes. An ebb and flow of this kind of encroachment is particularly common on roads that do not feature raised medians. But while traveling from Kukatpally to HITEC City in the early evening it was not uncommon to be confronted with a wave of oncoming two-wheelers, auto-rickshaws, and cars that should have been on the other side of the concrete walls of the flyover that separates the two directions. And because the flyover prevented them from moving back over into their designated lanes even if they wanted to, they were committed to their transgression for at least half a kilometer. Since there were large numbers of motorists doing the same thing, and no police along that stretch of the road to stop them, this became a regular feature of the traffic-scape during particular times of the day. It can, of course, be argued that this behavior actually increased the efficiency of traffic flows during times when vehicles going one direction greatly outnumbered those going the other way. And traffic planners in cities around the world have taken measures to designate more lanes for one direction or another to accommodate rush hours, including lane designation lights and shifting concrete barriers along highways to create more lanes in either direction. But absent these measures, and in the face of a road infrastructure that has quickly become overwhelmed by the rapid proliferation of vehicles in the city, Hyderabad's drivers have once again taken matters into their own steering wheel–clutching hands and adapted the infrastructure to better suit their own immediate needs.

The other major reason to drive against the stream has ironically to do with medians, which are meant to separate traffic but end up encouraging many drivers to improvise in order to make turns. Along the city's busier stretches, Hyderabad's traffic planners have often installed medians as uninterrupted walls, even when a thoroughfare encounters cross streets. As a result, if you are driving on the left side of the road, you cannot make a right turn onto the cross street you might want to drive into, and you cannot make an immediate right turn to get to a shop on the other side of the road. Most motorists would rather not have to drive on the left side of the street for several hundred meters, make a U-turn, and head back another few hundred meters just to get to the sweets shop they are aiming for. And so many of them jog over to the other side at an opportune intersection and drive, salmon-like, up the opposite shoulder for 10 or 50 or a few hundred meters to their destination. In other words, when you also take into account Hyderabadi drivers' overall relationship with the city's traffic planning and enforcement, the long, linear nature of medians actually *encourages* some of them to drive on the wrong side of the road. Since they feel reasonably assured they will not be prosecuted for doing so, they create their own immediate efficiencies to get around what, in the moment, seems to be not much more than an infrastructural annoyance.

These are but a few examples that, while planners in Hyderabad have built what they see as a particular set of limitations and capabilities—affordances— into their concrete order, the rampant imaginations and experimentations of Hyderabadi drivers leads to whole other sets of relations between drivers (and others) and these built forms.

CONCLUSION

As a fellow anthropologist, Nick Seaver not long ago explained to me, two key phrases in an anthropologist's toolbox are, "it's complicated," and "it depends." This of course applies quite well to the relationship between drivers, traffic planners, and police in Hyderabad. There is, for instance, a cadre of drivers in Hyderabad that has extensive experience driving in North America, Europe, and even other Indian cities and who, like my friend Ram, insist on following the law as written and hold in contempt that majority of the city's drivers who have a more flexible view of traffic rules. Some, like Ram, will shout imprecations from behind the wheel or simply insist on driving in what they view is the proper way. Many drivers also participate on the Hyderabad Traffic Police's remarkably active Facebook page (facebook.com/HYDTP), lodging complaints about driving behaviors they have witnessed or the perceived failings of the police and posting jokes about the general lawlessness of the city's drivers.[4]

The number of 'lawful' drivers, however, has not yet reached critical mass, as the incessant complaints of these drivers in various forums attest. This is ultimately a demonstration of the limits of state power on the roads of Hyderabad. Mediate, symbolic projections of authority such as traffic lights and painted lane lines have been largely ineffectual, and enforcement officers are only effective when they are bodily present. What Hyderabad's government is left with are immediate, concrete measures that physically limit what drivers can do. Even here, motorists and pedestrians find ways to subvert the order that planners have implied through their constructions, sometimes in direct response to the material landscape planners have erected.

State power is neither an absolute nor a given, and there is a constant negotiation between governing regimes and those being governed. As the relationship between drivers and traffic planners in Hyderabad demonstrates, this pertains just as much to daily regimes of mobility as it does to explicitly political realms of struggle. A number of writers on mobile agents in the city have considered the improvisational nature of traveling through spaces of transit (de Certeau 1984; Ingold 2004; Jonasson 2004; Thrift 2004; Hall, this volume), and some of these have contemplated ways that mobile agents might subvert planners' intentions. The situation in Hyderabad brings this fraught dynamic into sharp relief and highlights the material foundation underlying regimes of mobility. The fact that driving experiences can be radically different from city to city or country to country (cf. Edensor 2004) of course indicates that there is a strong sociocultural element shaping how motorists behave. But even in places where drivers largely follow mediate signs of authority such as lane lines and traffic lights, there is still at least the implication that there is a fleshy, concrete human apparatus on the other side of that mediation. And notwithstanding a proliferation of signaling systems in a given jurisdiction, traffic engineers continue to turn to concrete measures, shaping roads, intersections, and crash barriers to both take account of and try to influence driver behavior (Vanderbilt 2008). Though, as I have shown, this does not always lead to predictable results.

Nonetheless, as scholars such as Latour (1992) and Law (1992) have pointed out, we humans and our systems are inherently entangled with non-human materialities. The realm of human mobility is a place where this reality becomes quite evident—particularly in a place like Hyderabad, where traffic planners overwhelmingly rely on concrete measures to try and bring intransigent drivers into line.

NOTES

1 From 1981 to 2005, the number of vehicles per 1,000 people has gone from 35 to 221, for a total of 1.43 million vehicles (Sudhakara Reddy and Balachandra 2012, p. 154; also see Agarwal and Zimmerman 2008, p. 1).
2 The fieldwork upon which his chapter is based was funded by a Wenner-Gren Foundation Dissertation Fieldwork Grant. He would like to thank the

following for their valuable feedback: Jonathan Echeverri, Timothy Murphy, Kari Britt Schroeder, Smriti Srinivas, Li Zhang, and this book's editors, Julie Cidell and David Prytherch.
3 There is an actual Space Station built by Aliens Group in Hyderabad, and I had hoped to stay there and be the first anthropologist to ever conduct an ethnography of an Aliens Space Station, but alas, perpetual delays in funding and construction prevented me from doing so. I have appropriated its name as a consolation.
4 This type of discourse about traffic behavior in India is not new, as Arnold's (2012) account of motorized street transportation in late-colonial India demonstrates.

WORKS CITED

Agarwal, O. P., and Zimmerman, S. L. (2008) "Toward sustainable mobility in urban India," *Transportation Research Record: Journal of the Transportation Research Board*, 2048:1–7.
Arnold, D. (2012) "The problem of traffic: The street-life of modernity in late-colonial India," *Modern Asian Studies*, 46(01):119–141.
Beckmann, J. (2001) "Automobility—A social problem and theoretical concept," *Environment and Planning D: Society and Space*, 19(5):593–607.
Beckmann, J. (2004) "Mobility and safety," *Theory, Culture & Society*, 21(4–5): 81–100.
Cresswell, T. (2010) "Towards a politics of mobility," *Environment and Planning D: Society and Space*, 28(1):17–31.
Dant, T. (2004) "The driver-car," *Theory, Culture & Society*, 21(4–5):61–79.
de Certeau, M. (1984) *The Practice of Everyday Life*, Berkeley: University of California Press.
Edensor, T. (2004) "Automobility and national identity: Representation, geography and driving practice," *Theory, Culture & Society*, 21(4–5):101–120.
Foucault, M. (1979) *Discipline and Punish: The Birth of the Prison*, New York: Vintage Books.
Gaver, W. W. (1991) "Technology affordances," *Proceedings of the SIGCHI Conference on Human Factors in Computing Systems*.
Ghannam, F. (2002) *Remaking the Modern: Space, Relocation, and the Politics of Identity in a Global Cairo*, Berkeley: University of California Press.
Hiebert, P. G. (1976) "Traffic patterns in Seattle and Hyderabad: Immediate and mediate transactions," *Journal of Anthropological Research*, 32(4):326–336.
Holston, J. (1989) *The Modernist City: An Anthropological Critique of Brasília*, Chicago: University of Chicago Press.
Huijbens, E. H., and Benediktsson, K. (2007) "Practising highland heterotopias: Automobility in the interior of Iceland," *Mobilities*, 2(1):143–165.
Ingold, T. (2004) "Culture on the ground: The world perceived through the feet," *Journal of Material Culture*, 9(3):315–340.
Jonasson, M. (2004) "The performance of improvisation: Traffic practice and the production of space," *ACME: An International E-Journal for Critical Geographies*, 3(1):41–62.
Latour, B. (1992) "Where are the missing masses? The sociology of a few mundane artifacts," in E. B. Wiebe and J. Law (eds) *Shaping Technology/building Society: Studies in Sociotechnical Change*, Cambridge, MA: MIT Press, pp. 151–180.
Laurier, E. (2004) "Doing office work on the motorway," *Theory, Culture & Society*, 21(4–5):261–277.

Law, J. (1992) "Notes on the theory of the actor-network: Ordering, strategy, and heterogeneity," *Systems Practice*, 5(4):379–393.
Lefebvre, H. (1991) *The Production of Space*, Oxford, UK: Blackwell.
Mauss, M. (1973) "Techniques of the body," *Economy and Society*, 2(1):70–88.
Norman, D. A. (1999) "Affordance, conventions, and design," *Interactions*, 6(3):38–43.
Ramu, M. (2013, January 26) "Road accidents and the unabated dance of death," *The Hindu*, Cities.
Scott, J. C. (1998) *Seeing Like a State: How Certain Schemes to Improve the Human Condition Have Failed* [Yale Agrarian Studies Series], New Haven, CT: Yale University Press.
Sheller, M., and Urry, J. (2000) "The city and the car," *International Journal of Urban and Regional Research*, 24(4):737–757.
Sudhakara Reddy, B., and Balachandra, P. (2012) "Urban mobility: A comparative analysis of megacities of India," *Transport Policy*, 21:152–164.
Taylor, N. (2003) "The aesthetic experience of traffic in the modern city," *Urban Studies*, 40(8):1609–1625.
Thrift, N. (2004) "Driving in the city," *Theory, Culture & Society*, 21(4–5):41–59.
Vanderbilt, T. (2008) *Traffic: Why We Drive the Way We Do (and What It Says About Us)*, New York: Alfred A. Knopf.

4 A Bridge Too Far
Traffic Engineering Science and the Politics of Rebuilding Milwaukee's Hoan Bridge

Gregg Culver

It is easy to take for granted the impact of the technical aspects of traffic engineering on our daily lives. However, as both Prytherch and Henderson (this volume) have highlighted, the tools and design standards of traffic engineering are crucial for better understanding the (re)production and dominance of automobility in American society. Indeed, perhaps traffic engineering fails to attract attention because it seems so straightforward, apolitical, and objective. On the contrary, however, the tools and design standards of traffic engineering are quite normative, and the practice of implementing them is ultimately subjective and political. In this chapter, I argue that critically engaging traffic engineering—as both a discursive and spatial practice—can provide an important perspective on the politics of mobility, contributing to transportation and mobilities scholarship alike. In particular, I hope to contribute to making sense of two interrelated processes: first, how the ultimately value-laden tools and concepts of traffic engineering help turn normative spatial visions of mobility, or people's visions about how mobility and its spaces ought to be, into a material reality (Henderson 2004, 2011b); and second, how the values embedded in such tools can be obscured and partly naturalized through a discourse of scientific objectivity.

I use a politics of mobility framework (which I take to refer to the inherently normative and value-laden struggles over how transportation and its spaces ought to be [Henderson 2013]) to interrogate one specific traffic engineering concept called 'level of service' (LOS), using case study research of a 2011 controversy over whether a bicycle and pedestrian path (or simply bike path) should have been added to Milwaukee's iconic Hoan Bridge during its reconstruction—a debate in which LOS came to be the core point of contention. As part of a larger research project, in this chapter I specifically draw on qualitative analysis of official project documents, video and audio recordings and my personal notes from public meetings on the project, and 16 semi-structured interviews with key engineers, planners, politicians, and activists.

I begin the next section with a review of the critical geographies of transport and mobilities on the significance of traffic engineering and its historical construction as an apolitical and objective science, after which I outline

LOS as a significant theme in the controversy. Then, I briefly describe the background of the case study before presenting the research results. I conclude by considering some insights gained from examining the role of LOS as exemplar of a tool of ostensibly objective traffic engineering.

FORGING VISIONS OF MOBILITY THROUGH TRAFFIC ENGINEERING

There is a general consensus among critical transport, mobilities, and urban geographers that a deeply institutionalized 'traffic logic' has been established through the technical concepts and design standards of traffic engineering, as well as through the traffic code (Blickstein 2010; Blomley 2007, 2012; Carr 2010; Cresswell 2006, 2010; Henderson 2004, 2006, 2011b; Hess 2009; Patton 2007; Prytherch 2012, Chapter 2 in this volume; Simons 2009). This traffic logic has been a predominant discourse surrounding urban public space within the North American city and constitutes such spaces as transport corridors and the people and objects within these spaces as blockages and flows (Blomley 2007, p. 64). These scholars have also underlined the clear bias within North American cities toward the circulation and flow of automobiles over 'otherly mobile' bodies. In sum, the "uneven power geometries" among motorists, pedestrians, cyclists, and so on are materialized at least in part through the interrelated technical and regulatory aspects of transport (Prytherch 2012, p. 295). Consequently, the tools and discourses of traffic engineering should be considered a central site of the production of the material bias toward automobility.

Traffic engineering has long been understood as crucial for the (re)production and hegemony of automobility in the United States (Barret and Rose 1999; Duany, Plater-Zyberk, and Speck 2001; Henderson 2006; Jackson 1985; Jacobs 1961; Kunstler 1994; Norton 2008; Shoup 2011). And while certain scholars, cultural critics, and alternative mobility activists are aware of this bias, to the general public automobility has generally appeared unproblematic (Böhm et al. 2006, p. 4)—as the natural product of consumer demand and technological progress (Bruegmann 2005).

Significant to this teleological perspective of automobility is the historical construction of traffic engineering as an objective and apolitical science. During the early 20th-century Progressive Era of scientific positivism (whose proponents viewed most problems as being solvable using scientific methods), the road engineer's professional responsibility shifted away from building and maintaining infrastructure to performing scientific research focused on the human component of traffic flow (Barret and Rose 1999; Brown 2006; Brown, Morris, and Taylor 2009). This new paradigm of traffic engineering took the unimpeded flow of automobiles as an unquestioned core value and a predict-and-provide ideology as its central policy orientation, meaning that the necessity to anticipate and accommodate future automobile traffic growth was uncritically accepted as the norm (Barret and Rose 1999; Hebbert 2005; Henderson 2009; Hess 2009; Østby 2004; Norton 2008).

While questions of mobility had formerly been resolved through contentious politics involving various stakeholders, the creation of a traffic science from the Progressive Era onward "technif[ied]" and "transform[ed]" these fraught politics "into concerns for the objective arena of science and technology" (Blomkvist 2004, p. 300). This had the effect of making traffic engineers appear to be value free, objective, and apolitical, imparting them with a voice of authority that was perceived to transcend and supersede other voices in the contentious politics of mobility (Brown 2006; Østby 2004). In other words, such an appeal to science has since functioned to "render ordinary political life impotent through the threat of an incontestable nature" (Latour 2004, p. 10), as the traffic engineer's ostensibly objective assessment trumps laypeople's value-based judgments. Yet despite this objective and apolitical appearance, traffic engineering has historically been closely aligned with particular business, political, and bureaucratic interests, as a general consensus developed that easing conditions for motorists was good for business (Brown 2006; Weinstein 2006).

However, since the latter part of the 20th century, this consensus has increasingly broken down. Planning professionals in particular have moved beyond viewing streets strictly as transport corridors, to also thinking of streets as potential spaces of place-making (Hess 2009). But despite planners' desire to move beyond traffic logic, at least two major points act as barriers to significant change away from the bias toward automobility. First, automobile-biased traffic engineering standards have become institutionalized as considerable hurdles in the urban design process (Henderson 2011a; Hess 2009). Second, and at least partially through the discourse of scientific objectivity assumed by traffic engineering, the specialist knowledge of engineers is privileged over that of planners. On the one hand, engineers' tools are able to provide highly precise quantitative figures, typically in favor of the (re)production of automobility, while alternative mobility proponents are often forced to rely on imprecise figures and qualitative arguments (Patton 2007). On the other hand, these quantitative figures are favored by elected officials, particularly as their relative specificity and historical precedence is often implicated in transportation safety concerns and legal liability issues (Hess 2009; Patton 2007).

Alongside this breakdown of consensus on traffic logic, a number of noteworthy changes have taken place since the Progressive Era, and thus a few qualifications are in order. First, the bias of traffic engineering toward automobility has been critiqued, while new road construction has been deemphasized somewhat (Brown, Morris, and Taylor 2009; Duany, Plater-Zyberk, and Speck 2001; Hess 2009). Second, traffic engineers are not an unchanging, ideologically coherent group, and, generally speaking, city traffic engineers have been more attuned to local concerns than federal and state traffic engineers (Brown, Morris, and Taylor 2009). Third, there are engineers who are eager to encourage nonmotorized modes but face the aforementioned firmly institutionalized standards, which pose considerable political and legal hindrances to making change (Hess 2009). Finally, there are differences both historically and geographically in the U.S. among states' departments of transportation and their approach to mobility. Despite these qualifications, on the whole, traffic engineering in the U.S. continues to function largely in terms of its traditional paradigm, which prioritizes the high-speed flow of cars over other modes (Brown 2006; Henderson 2011a, 2011b; Patton 2007).

LEVEL OF SERVICE

Simultaneously embodying both unambiguous automobile bias and discourses of apolitical scientific objectivity is the traffic engineering concept of LOS, which is used virtually universally throughout the United States. Henderson (this volume) explores the application of LOS to urban intersections; however, it is also commonly applied to freeways, which is the central concern of this chapter. In this subsection, I describe and critically examine LOS using the 2010 edition of the U.S. Transportation Research Board's *Highway Capacity Manual* (henceforth, IICM [TRB 2010])—a definitive source of traffic engineering in the U.S.—and the source used and cited by the Wisconsin Department of Transportation (WisDOT) in this case.

Freeway LOS is the simplification of operating conditions (the speed and density of automobile traffic) for a given stretch of freeway into six levels of service, represented by letters ranging from A to F, with LOS A representing "the best operating conditions from the traveler's [i.e., motorist's] perspective and LOS F the worst" (TRB 2010, p. 5-3). LOS A thus represents free-flow traffic with a high degree of driver maneuverability, meaning there is enough space for individual drivers to remain relatively unaffected by other drivers. As the density of traffic increases with more cars on the roadway, driver maneuverability and speed decrease until a breakdown in the steady flow of traffic, or LOS F, is reached. The calculation of freeway LOS begins with an assumption of drivers of personal cars (especially regular commuters) as the default, optimal road users. Then, trucks and recreational vehicles (RVs) are factored in as complicating factors, while nonmotorists remain entirely excluded (Mannering, Kilareski, and Washburn 2005). LOS is calculated somewhat differently depending on specific circumstances, but generally (and as was applied for the Hoan Bridge) this involves: first, using several years' worth of traffic counts on a given freeway segment to establish a historical trend in automobile traffic flow; second, extrapolating the future growth rate of automobile traffic from this trend; and third, based on the amount of predicted traffic relative to the available road space, determining the operating conditions for both the design year of the project and for 20 years in the future (WisDOT 2011c).

According to the HCM, LOS is a simplified metric that offers guidance to decision makers regarding whether predicted freeway performance will remain acceptable, and whether changes to this performance would "be perceived as significant by the general public" (TRB 2010, p. 5-3). If operating conditions are deemed unacceptable, the results of LOS analysis guide engineers and decision makers in deciding how much additional roadway/infrastructure to build to ensure that the predicted flow of automobile traffic remains within the range deemed acceptable. Of course, the flow of traffic (and thus the LOS) fluctuates significantly, with the lowest LOS typically occurring during the weekday rush hours. So, LOS during these peak periods is given the greatest weight in the analysis (WisDOT 2011c).

As the HCM notes, "roadways are not typically designed to provide LOS A conditions during peak periods, but rather some lower LOS that reflects a balance between individual travelers' desires and society's desires and financial resources," and "it is up to local policy makers to decide the appropriate LOS for a given system element in their community" (TRB 2010, p. 5-3). So, LOS standards should theoretically be the product of a negotiation between these often competing interests, making the notion of acceptable LOS political and subject to judgment. However, although there is no universal LOS standard, freeways in "heavily developed sections of metropolitan areas" are generally given a target of LOS D during peak periods—the standard used by WisDOT in this case study (AASHTO 2011, p. 8–2; WisDOT 2011a).

While the letter scheme of LOS provides an easy to understand rubric for describing the flow of automobile traffic, due to numerous assumptions and uncertainties, the HCM cautions that "one should also be mindful of [the weaknesses of LOS]" (TRB 2010, p. 5-3). First, LOS only accounts for the experience of individual motorists, and not for other types of road user. Second, individual motorists may perceive the LOS differently under different conditions, differently than other motorists, and differently than the HCM method would predict (TRB 2010). Third, the step nature of the model (from one letter to the next) is a "particularly sensitive issue," because a relatively small change in operating conditions could lead to a change in one or even two levels of service (TRB 2010, p. 5-4). This means that when fixed standards of LOS are used (for instance, local authorities determine LOS on Freeway X must not drop below LOS D), a small change in the flow of traffic predicted could require local authorities to build costly infrastructure. Fourth, due to the uncertainties in data collection and analysis that accompany making predictions about the future, "the 'true' LOS value may be different from the one predicted" (TRB 2010, p. 5-4). In light of these weaknesses, the HCM states that LOS results can be thought of as "*statistical 'best estimators'*" of conditions and aggregate traveler perception" (TRB 2010, p. 5-4, emphasis added). From a more critical perspective, the purpose of LOS can be summed up as a tool that allows predict-and-provide ideology to be operationalized. In particular, the emphasis on striving to accommodate the peak period of traffic reflects a vision embedded within LOS of a minimally restrained automobility—one in which the growth of automobile traffic must be anticipated and accommodated, while nonmotorists can only act as obstacles to the flow of automobiles (Henderson 2011a, Chapter 5 in this volume; Patton 2007). LOS is a goal-oriented tool, not an objective, apolitical one. It is inherently normative, as the system of A–F letters establishes what is generally desirable (LOS A, or unrestrained automobility) and undesirable (LOS F, or restricted automobility) for the future (WisDOT 2011c). Meanwhile, the establishment of acceptable LOS standards is directly political, as it is achieved through a process of negotiation between competing interests.

The institutionalized commitment to LOS encourages the fragmentation of the built environment, stifles attempts at multimodality, and helps to create the conditions for the (re)production of automobility (Henderson 2011a, Chapter 5 in this volume; Hess 2009; Patton 2007). However, tools such as LOS are not only hurdles to multimodality institutionalized in the urban roadway design process *per se*. In what follows, I argue that the interpretation and representation of tools such as LOS/LOS analysis are themselves also active sites of struggle in the politics of mobility. Further, I argue that a discourse of apolitical, scientific objectivity allows traffic engineers to conceal their values and judgment and simultaneously grants traffic engineers an authoritative voice in the politics of mobility.

ENGINEERING FOR MORE THAN CARS?

A recent controversy over whether to include a bike path on a major bridge reconstruction project in Milwaukee sheds light on the discourses of apolitical and scientific objectivity within the politics of mobility, particularly as this plays out in debates over LOS. Milwaukee's iconic Hoan Bridge, the most direct connection between downtown and the bicycle-friendly Bay View neighborhood, has long appeared as offering the best potential bike route between two areas separated by industrial zones and the Port of Milwaukee (Figure 4.1). Moreover, this section was long viewed as a major gap in Milwaukee County's extensive Oak Leaf Trail system, as well as in the 162-mile network of bicycle trails along the western shores of Lake Michigan from Chicago to Oostburg, Wisconsin. Consequently, bike path advocates considered this gap a problem impacting local bicycle commuting, tourism, and quality of life.

Consideration of a Hoan Bridge bike path began in the early 1990s, when Milwaukee County was awarded federal funds to create a Bay View-Downtown bike connection. Despite its location at the center of the city, the bridge is a state facility and thus WisDOT, not the city or county, has

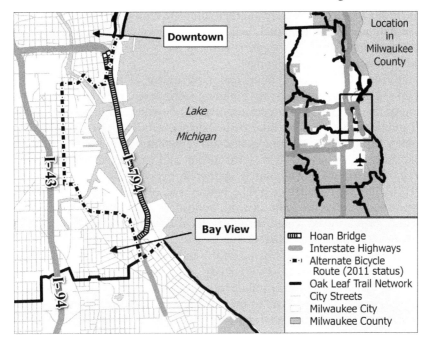

Figure 4.1 Case study location, Milwaukee's Hoan Bridge. Map by author.

authority over the bridge. As part of the 3.5 mile-long freeway spur Interstate 794 (I-794), the 2.5 mile-long bridge connects downtown Milwaukee with lakeshore neighborhoods and suburbs to the south. Running parallel just west is the combined I-43 and I-94, which serves as the main freeway connection between Milwaukee and major destinations south, west, and north. So, despite its centrality, I-794 is one of the least congested freeways in Milwaukee County (Milwaukee Department of Public Works 2011).

An initial study of a bike path ended in 2002, with WisDOT deciding against a bike path over the bridge, despite city and public support for it. However, WisDOT promised local authorities it would revisit the issue during the future reconstruction of the bridge, and in the meantime, some of the funds were to be used toward an alternate bicycle route (see Figure 4.1). Though this mostly on-street route was in place by 2011 and would see improvements in the future—with over 20 intersections on city streets—it remained a complicated and potentially daunting route for many bicyclists.

Due to WisDOT's earlier promise, as well as federal and state law requiring consideration of bicycles and pedestrians on projects using federal and state funds, WisDOT announced in 2011 that it was beginning the Hoan Bridge Bicycle/Pedestrian Feasibility Study (henceforth, simply feasibility study), and that the bike path cost could be covered by the overall estimated bridge reconstruction budget of $275 to $350 million (Held 2011a). Because the bridge would not require another complete renovation for 40–50 years, bike path advocates reignited their lobbying efforts, viewing this as a 'generational opportunity' to secure a path on the bridge (personal interviews, local politicians, bicycle planner).

In time, Milwaukee's municipal government, Department of Public Works, Long Range Lakefront Planning Committee, the Harbor Commission of the Port of Milwaukee, and over 40 downtown business executives all publicly endorsed the bike path. Further, a number of city aldermen, county supervisors, and state politicians representing the area led advocacy bike rides, gave interviews to local media, and supported a long-standing petition drive which ultimately gathered over 5,000 signatures (Held 2011b). In contrast, fewer public figures and organizations spoke out in opposition to the bike path. Instead, opposition to the bike path was more diffuse, primarily expressed within the local news media, blogs, and public comments submitted to WisDOT.

After months of local lobbying efforts and debate, WisDOT officially presented the results of its feasibility study at a packed, town hall-style public meeting (henceforth, simply public meeting) in November 2011. The public meeting was a key moment of the debate, as it was the only opportunity for the public to discuss the feasibility study results directly with WisDOT representatives, who had informed the audience that they would use the feasibility study results and public comments to inform their decision about the bike path. Due to its significance to this controversy, I organize the following themes around key moments at the public meeting.

OBSCURING BIAS THROUGH APOLITICAL OBJECTIVITY

From early on in the public meeting, the theme of objectivity became increasingly apparent. WisDOT described itself as an "objective regional [expert]" that had "45+ years experience" (WisDOT 2011b), and which used "objective study criteria" (Sell 2011). In introducing the source of their LOS methodology to the audience, the 2010 HCM was described as "the latest edition of a manual that was first published back in 1950, so about 60 years ago" (WisDOT 2011b). Immediately after this, the representative gave a personal aside, noting that he had "been using and applying this methodology since grad school at Purdue in the late 80s," making him "certainly very familiar . . . with its application" and allowing him to "provide good guidance to it as [WisDOT was] moving forward" (WisDOT 2011b). At the same time, WisDOT did not mention any of the weaknesses of LOS identified by the HCM. Consequently, the language used and information given to the audience helped to ground WisDOT and its methods in the realm of objective science and tried-and-true expertise.

A further theme was WisDOT's framing of itself as apolitical and external to the politics of the debate, and in a position of information gatherer and arbiter between value-laden, normative positions. Early on, a WisDOT representative informed the audience that there were "people on many different sides on the particular issue," and those in attendance were asked to respect each other's opinions (WisDOT 2011a). Meanwhile, WisDOT was presented as an apolitical authority that hosted meetings in order to gather information: "[W]e have a very inclusive public involvement process. That's why we're here this evening, to try and be open and answer questions and listen to what people have to say. Very, very good process" (WisDOT 2011a). Together, such rhetoric helped to frame WisDOT as an unbiased arbiter who deserves to have the authoritative voice in how the bridge ought to be rebuilt and whose mobility the bridge should facilitate.

Despite discursively positioning itself as objective, apolitical, and unbiased, upon more critical examination it is clear that WisDOT was committed to a particular mobility vision for the bridge through its very approach in ascertaining the engineering feasibility of the project. In the feasibility study, WisDOT explicitly defined five criteria with which to evaluate the 'engineering feasibility' of the bike path. The *first* of these criteria involved "[t]he ability of the facility to maintain acceptable operating conditions for both existing and future [automobile] traffic growth" (WisDOT 2011c, p. 10). In other words, WisDOT committed itself to predicting and providing for automobile traffic, which was operationalized through the use of LOS. With this, the feasibility of a bike path went beyond questions of the physical and financial limitations of engineering to become inextricably tied to the operating conditions of automobiles. Therefore, in order to be 'feasible,' a bike path could not negatively impact, or even partially substitute, current and future automobile traffic by some other mode, such as bicycling. This criterion hierarchized mobility, placing car driving at the pinnacle and subordinating bicycling and walking as modes that are allowed only insofar as they do not impede automobility, and embedded a vision of unrestrained automobility for the bridge in the study from its inception. Consequently, despite its claims to value-free objectivity, WisDOT had already taken a normative position on whose mobility the bridge should facilitate, and thus fundamentally could not be an unbiased moderator in the politics of the bike path.

ELEVATING EXPERT ASSESSMENT OVER PERSONAL JUDGMENT

Having introduced both itself and the Hoan Bridge project, WisDOT presented the LOS analysis results of five bike path alternatives to the public (WisDOT 2011c). Both Alternative 1A at $9.4 million and Alternative 1B at $27.5 million would repurpose one of the three northbound lanes to accommodate the bike path, thereby impacting LOS. The remaining three alternatives—2A, 3B, and 4—would all add a structure to the bridge, and thus would not impact LOS. These three were far more expensive—$76.4, $95.5, and $84.4 million, respectively. Of these five alternatives, then, the focus quickly centered on Alternative 1A—Alternative 1B would have the same LOS impact as 1A, but at approximately three times the cost, while the remaining three alternatives were so expensive that they seemed obviously publically and politically untenable. Consequently, the LOS results for Alternative 1A immediately appeared crucial for the fate of a bike path.

In sum, LOS under Alternative 1A was deemed acceptable for the design year and near future. However, for 2035, it was predicted to reach LOS E to F during the morning peak hour in the northbound lanes, when the flow of automobile traffic was predicted to drop to between 47.6 and 51.9 miles per hour (mph), or close to the posted legal speed limit of 50 mph (WisDOT 2011c). For all other periods of the day, and for all times in the southbound lanes in 2035, LOS was predicted to be within the acceptable range. Simply put, LOS for Alternative 1A was labeled with a 'red flag' and identified as a 'major concern,' because the northbound peak morning rush hour traffic was predicted to travel at the legal speed limit in the year 2035 (see Figure 4.2).

Confused by WisDOT's determination of the 'unacceptability' marked by LOS E to F, bike path advocates contested the accuracy and significance of this result during the question and answer session:

> Questioner: I think the speed limit should be considered because 47 mph seems reasonable to me and you're flagging Alternative 1A

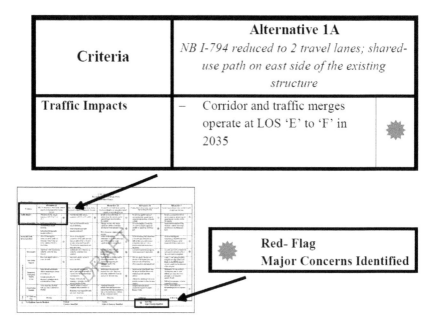

Figure 4.2 Excerpts from WisDOT document depicting LOS impact of Alternative 1A as a "red flag" and "major concern" (adapted from WisDOT 2010c).

> as a traffic impact as a red flag and major concern, going 47 mph I don't think that's, that 60 mph, that's a reasonable . . .
>
> WisDOT: [interjects] I think that part of that is a *personal judgment*. What we looked at is the *department's objective criteria*.
>
> Questioner: [in an assertive tone] It's actually the law. It's a speed limit.
>
> WisDOT: I appreciate that but I think that you understand, that if I came in and used a free flow speed of 50 miles an hour the same individuals would criticize, different ones would criticize and say that "oh no, I've been out there and I go like 55, 60 over the Hoan"—which I don't by the way. Sheriff, Lieutenant if you're listening [laughter].
>
> (Sell 2011, emphasis added)

Despite a serious disagreement over what constituted unacceptable operating conditions on the bridge, throughout the politics of the bike path, WisDOT (through a standard practice of traffic engineering derived from the empirically observed, 85th percentile operating speed of automobiles on the bridge [personal interview, local transportation planner]) had the sole ability to define the parameters of acceptability, and thus was the only 'objective' voice in the production of knowledge on the bridge. So, any assessment that did not follow WisDOT's parameters of acceptability could not be 'objective' but instead was, by definition, 'personal judgment.' Thus, even while bike path advocates argued that, from their perspective, the predicted LOS was entirely acceptable—a nonissue, even—their own local knowledge and assessments of acceptability could not be integrated into the feasibility study.

REPRESENTING RESULTS, LIMITING INTERPRETATIONS

Still dissatisfied by the representation of the 2035 operating conditions as unacceptable, another person asked,

> Questioner: Let's assume I live in the south side now, and I drive on the Hoan Bridge and I actually obey the law, I drive over there at . . . 50 miles per hour. . . . If I still live there in 2035 and I still drive over the Hoan Bridge at 50 miles per hour, my commute time is exactly the same thing. Right? It hasn't changed. And even if I drive it at 60 miles an hour and 20, 30 years from now I'm driving at ten miles less, at 50 miles an hour, that's a total increase in my commute time over the Hoan Bridge of less than 30 seconds. Right? [Applause from audience]
>
> WisDOT: So, let me see if I can capture the question correctly [Laughter; audience member calls out "there was no question"].

> The question was, with regard to the change in speed that is associated with 2035 whether you drive the speed limit or whether you drive above the speed limit *does it represent a modest reduction in speed? The answer is: yeah. The numbers are what the numbers are.*
>
> (Sell 2011, emphasis added)

This was the only instance in which WisDOT publically suggested that the bike path would have had a modest impact on automobile traffic, and contrasts with WisDOT's general assessment of the LOS impact as plainly unacceptable, marked by a failing 'grade' of LOS E to F.

In a letter from the Milwaukee Department of Public Works (2011) to WisDOT, the city traffic engineers disputed this assessment of unacceptability as represented to the public, expressing to WisDOT that

> LOS should not be represented or perceived as a "grade." To minimize the confusion the LOS should be referenced as an index representative of a range of typical operating conditions. . . . [G]reater emphasis should be made to communicate the short period of time that motorists would encounter . . . LOS E or F. . . . [E]ven during [these] short periods, traffic continues to travel at speeds near the speed limit (50 miles per hour). [Even under Alternative 1A] motorists traveling over the Hoan Bridge will continue to enjoy some of the highest levels of service available anywhere on the freeway system in Milwaukee County.

The salient issue for city engineers was not the LOS results *per se*. Rather, at issue was that the short and 'modest reduction in speed' on the bridge in 2035 was represented in WisDOT's presentation and documents in such a way that limited the interpretation of the results: WisDOT's red-flagging of Alternative 1A as a 'major concern,' and its characterization of the LOS as a static 'failing grade,' rather than as the predicted low point of a fluctuating range, suggested to the public the LOS impact would be thoroughly, obviously, and 'objectively' unacceptable. Whatever the 'real' reason was that the results were represented to the public in a way that unambiguously suggested the bike path would have an unacceptable impact on the bridge, it reveals the subjectivity and judgment that are involved in an ostensibly objective process. Of course, as WisDOT had the full authority to determine the parameters of the study as well as the 'correct' interpretation of the LOS results, the voices of city engineers, local politicians, and bike advocates could not effectively challenge WisDOT's interpretation.

DETERMINING WHOSE MOBILITY COUNTS

While WisDOT's broad vision of unrestrained automobility was built into the study itself through the required criterion of LOS, supporters of the bike path challenged this with alternative visions in which the automobile has no exclusive claim to the bridge; in which congestion is not inherently negative, but a marker of a dynamic, vibrant urban place; and in which nonmotorists must, at the very minimum, be accommodated. As one person at the public meeting commented,

> what I found disturbing about your analysis is that it's all based upon sort of engineering, but have you considered at all the benefit of alternative transportation, encouraging the use of bikes? Not only that, but what it does to the county of Milwaukee? Intangible things like that don't seem to be a part of your calculations.
>
> (Sell 2011)

These more qualitative aspects, as the WisDOT representative responded, were "beyond the engineering feasibility analysis" of the study. Significantly, though, hard data on bicycling and walking were also not considered in the study. When asked whether WisDOT, an organization otherwise armed with precise data for automobile traffic, had data on predicted bicycle traffic on the bridge, the answer was "no. Honestly, with being able to predict bicycle demand for using a given facility, really, it's a very, you know, it's not something that we really have the ability to do" (Sell 2011).
Another person asked:

> Since one of the concerns about this . . . will be safety issues for vehicles and for bicyclists, . . . is there any sort of information base that we could go to that would be able to track bicycle incidents, injuries, accidents, and stuff like that within the Milwaukee area?

And the response was: "As part of our feasibility study, we did not look at specific safety analysis, and I'm personally not aware of [any such database]" (Sell 2011). The hesitation and uncertainty with which these questions were answered in the meeting only helped to reveal a stark, built-in bias toward automobility to the detriment of other modes. Of course, tools do indeed exist and are being refined to capture performance measures for nonmotorists, as evidenced by the existence of a multimodal LOS tool in the 2010 HCM, the very manual referenced by WisDOT. However, WisDOT neither acknowledged nor used tools to collect and analyze such data in a meaningful way, signaling that these non-automobile data were not considered worthy of counting. And, by excluding aspects (such as improved safety of, the demand for, and potential economic impact of a bike path) from the scope of the study, alternatives to unrestrained automobility were significantly undermined, as they could not be expressed with any authority or legitimacy. In contrast, automobile data truly counted, as evidenced by WisDOT's final decision that a bike path "would impair [their] ability to provide safe, efficient travel and deliver less value than other possible department investments in economic development in Milwaukee" (WisDOT

Secretary Mark Gottlieb, quoted in Glauber 2011). This announcement brought closure to the controversy. LOS F allowed WisDOT's decision against the bike path to appear unbiased to the general public, on the one hand, while also undermining the possibility of legal challenge by the city and bike path advocates, on the other hand.

CONCLUSION

LOS analysis is but one tool—albeit a significant one—that helps to structure mobility and its spaces. The example of LOS in the politics of the Hoan Bridge bike path helps to demonstrate two main points. First, it highlights how the very tools with which traffic engineering is able to measure and conceive of mobility can encourage certain forms of mobility and uses of space while inhibiting others. Particular values are inscribed in such tools and enacted through them, materializing unequal power relations between 'differently mobile' bodies. LOS is an easy to overlook factor in the larger politics of (auto)mobility, yet as a standard criterion in the roadway design process, it is a powerful means by which automobility comes to be (re)produced. Second, and more broadly, this case highlights how a discourse of dispassionate, scientifically objective traffic engineering can wield great power in the politics of mobility, as it obscures the embedded values and the power relations at play. Such an authority stifles politics by not allowing other voices to articulate competing visions of mobility with the same legitimacy as that of traffic engineers. In this case, WisDOT engineers had the authority to construct the parameters of the debate, parameters which, from their inception, privileged a particular mobility vision and associated set of values. However, WisDOT's discursive construction of itself and its tools as scientifically objective, apolitical, and evenhanded obscured its bias toward unrestrained automobility. In contrast, essentially the opposite occurred for bike path advocates, whose bias toward alternative mobility visions was highlighted through the process—drawn out and exposed as 'personal judgment'—and therefore excluded from producing knowledge toward determining an acceptable and desirable vision of mobility for the bridge. This experience mirrors that of multimodal advocates in similar research (Cidell 2008; Henderson 2011a; Patton 2007), where local knowledge and judgments are devalued through a process that elevates the quantitative, expert knowledge of traffic engineers, positioning it as irrefutably correct.

In drawing less attention in the politics of mobility literature, there is a risk that many of the technical aspects of traffic engineering remain black boxed, meaning that the work that goes into them remains hidden, while the products themselves appear as a given (Latour 1999). Such tools are not only significant for the politics of mobility in the sense that they act as institutionalized barriers to alternatives. In the case of the Hoan Bridge, the result of the LOS analysis in and of itself did not lead to a simple, straightforward

A Bridge Too Far 95

decision against the bike path. Indeed, this is evidenced by the fact that the city traffic engineers did not dispute the LOS results themselves, but how WisDOT represented and interpreted these results. Aside from serving as a caution against treating traffic engineers as an ideologically coherent block, this disagreement between city and state engineers over WisDOT's characterization of the bike path's LOS implications suggests that the representation and interpretation of such tools and the results they produce can be as subjective and politically fraught as the bias embedded in the tools themselves.

The tools and concepts of traffic engineering are not simply a given, and thus they have a much more extensive impact on the politics of mobility than simply acting as stumbling blocks to alternatives to automobility. Increasing attention to the technical aspects of engineering throughout the entirety of controversies could help to better understand how they come to matter in the politics of mobility. Toward this end, one might ask: Which tools and concepts are deemed essential, and for what purpose were they created? How is the work of carrying out, interpreting, and representing the tools and their results done, and whose voices and knowledge are included/excluded in this process? What aspects of (whose) mobility are counted/taken into account, and which are left out?

The case of the Hoan Bridge helps to underline that how we choose to structure our mobility is not a matter of fact, but of judgment, and that traffic engineers play a crucial role in turning particular mobility visions into material reality. Judgment and bias toward particular outcomes are inherent in the arrangement of mobility and its spaces precisely because producing mobility is an inherently value-laden process. Consequently, exposing biases embedded in the tools and concepts of traffic engineering will not necessarily make transportation projects less contentious, but doing so may help lead to a shift in the starkly unequal power relations characteristic of the production of contemporary automobility.

WORKS CITED

American Association of State Highway and Transportation Officials (AASHTO) (2011) *A Policy on Geometric Design of Highways and Streets* (6th ed.), Washington, DC.

Barret, P., and Rose, M. (1999) "Street smarts: The politics of transportation statistics in the American City 1900–1990," *Journal of Urban History*, 25(3): 405–433.

Blickstein, S. G. (2010) "Automobility and the politics of bicycling in New York City," *International Journal of Urban and Regional Research*, 34(4):886–905.

Blomkvist, P. (2004) "Transferring technology—Shaping ideology: American traffic engineering and commercial interests in the establishment of a Swedish car society, 1945–1965," *Comparative Technology Transfer and Society*, 2(3):273–302.

Blomley, N. (2007) "Civil rights meet civil engineering: Urban public space and traffic logic," *Canadian Journal of Law and Society*, 22(2):55–72.

Blomley, N. (2012) "2011 urban geography plenary lecture—Colored rabbits, dangerous trees, and public sitting: Sidewalks, police, and the city," *Urban Geography*, 33(7):917–935.
Böhm, S., Jones, C., Land, C., and Paterson, M. (2006) "Introduction: Impossibilities of automobility," in S. Böhm, C. Jones, C. Land, and M. Paterson (eds) *Against Automobility*, Malden, MA: Blackwell, pp. 3–16.
Brown, J. (2006) "From traffic regulation to limited ways: The effort to build a science of transportation planning," *Journal of Planning History*, 5(1):3–34.
Brown, J., Morris, E. A., and Taylor, B. D. (2009) "Planning for cars in cities," *Journal of the American Planning Association*, 75(2):161–177.
Bruegmann, R. (2005) *Sprawl: A Compact History*, Chicago: University of Chicago Press.
Carr, J. (2010) "Skating around the edges of the law: Urban skateboarding and the role of law in determining young peoples' place in the city," *Urban Geography*, 31(7):988–1003.
Cidell, J. (2008) "Challenging the contours: Critical cartography, local knowledge, and the public," *Environment and Planning A*, 40:1202–1218.
Cresswell, T. (2006) "The right to mobility: The production of mobility in the courtroom," *Antipode*, 38(4):735–754.
Cresswell, T. (2010) "Towards a politics of mobility," *Environment and Planning D: Society and Space*, 28:17–31.
Duany, A., Plater-Zyberk, E., and Speck, J. (2001) *Suburban Nation: The Rise of Sprawl and the Decline of the American Dream*, New York: North Point Press.
Glauber, B. (2011, December 16) "DOT kills Hoan Bridge bike lane proposal," *Milwaukee Journal Sentinel*. Retrieved September 15, 2013 from http://www.jsonline.com/news/milwaukee/dot-kills-hoan-bridge-bike-lane-proposal-7e3fk0g-135758273.html.
Hebbert, M. (2005) "Engineering, urbanism and the struggle for street design," *Journal of Urban Design*, 10(1):39–59.
Held, T. (2011a, August 23) "Advocates for bike/pedestrian lane on the Hoan show momentum in Bay View," *Milwaukee Journal Sentinel*. Retrieved September 15, 2013 from http://www.jsonline.com/blogs/lifestyle/128289773.html.
Held, T. (2011b, November 29) "Sen. Larson set to present petitions supporting Hoan Bridge bike path," *Milwaukee Journal Sentinel*. Retrieved September 15, 2013 from http://www.jsonline.com/blogs/lifestyle/134673163.html.
Henderson, J. (2004) "The politics of mobility and business elites in Atlanta, Georgia," *Urban Geography*, 25(3):193–216.
Henderson, J. (2006) "Secessionist automobility: Racism, anti-urbanism, and the politics of automobility in Atlanta, Georgia," *International Journal of Urban and Regional Research*, 30(2):293–307.
Henderson, J. (2009) "The politics of mobility: De-essentializing automobility and contesting urban space," in J. Conley and A. T. McLaren (eds) *Car Troubles: Critical Studies of Automobility and Auto-Mobility*, Farnham, UK: Ashgate, pp. 147–164.
Henderson, J. (2011a) "Level of service: The politics of reconfiguring urban streets in San Francisco, CA," *Journal of Transport Geography*, 19:1138–1144.
Henderson, J. (2011b) "The politics of mobility in the south: A commentary on sprawl, automobility and the Gulf oil spill," *Southeastern Geographer*, 51(4): 641–649.
Henderson, J. (2013) *Street Fight: The Politics of Mobility in San Francisco*, Amherst: University of Massachusetts Press.
Hess, P. M. (2009) "Avenues or arterials: The struggle to change street building practices in Toronto, Canada," *Journal of Urban Design*, 14(1):1–28.

Jackson, K. T. (1985) *Crabgrass Frontier: The Suburbanization of the United States*, Oxford: Oxford University Press.
Jacobs, J. (1961) *The Death and Life of Great American Cities*, New York: Modern Library.
Kunstler, J. H. (1994) *The Geography of Nowhere*, New York: Touchstone.
Latour, B. (1999) *Pandora's Hope: Essays on the Reality of Science Studies*, Cambridge, MA: Harvard University Press.
Latour, B. (2004) *The Politics of Nature*, Cambridge, MA: Harvard University Press.
Mannering, F. L., Kilareski, W. P., and Washburn, S. S. (2005) *Principles of Highway Engineering and Traffic Analysis* (3rd ed.), Hoboken, NJ: John Wiley.
Milwaukee Department of Public Works (2011, November 30) "Public letter to Governor Scott Walker and Secretary Mark Gottlieb, WisDOT."
Norton, P. D. (2008) *Fighting Traffic: The Dawn of the Motor Age in the American City*, Cambridge, MA: MIT Press.
Østby, P. (2004) "Educating the Norwegian nation: Traffic engineering and technological diffusion," *Comparative Technology Transfer and Society*, 2(3):247–272.
Patton, J. W. (2007) "A pedestrian world: Competing rationalities and the calculation of transportation change," *Environment and Planning A*, 39:928–944.
Prytherch, D. L. (2012) "Codifying the right-of-way: Statutory geographies of urban mobility and the street," *Urban Geography*, 33(2):295–314.
Sell, B. (2011, November 14) "Public hearing on draft feasibility study." Audio file. Retrieved August 2, 2014 from http://www.bikethehoan.com/9PublicHearing.mp3http://www.bikethehoan.com/9PublicHearing.mp3.
Shoup, D. (2011) *The High Cost of Free Parking* (updated ed.), Washington, DC: American Planning Association.
Simons, D. (2009) "Bad impressions: The will to concrete and the projectile economy of cities," in J. Conley and A. T. McLaren (eds) *Car Troubles: Critical Studies of Automobility and Auto-Mobility*, Farnham, UK: Ashgate, pp. 77–93.
Transportation Research Board (TRB) (2010) *Highway Capacity Manual* (5th ed.), Washington, DC: National Research Council.
Weinstein, A. (2006) "Congestion as a cultural construct: The 'congestion evil' in Boston in the 1890s and 1920s," *The Journal of Transport History*, 27(2): 97–115.
Wisconsin Department of Transportation (WisDOT) (2011a) "1/5 Hoan Bridge Mtg—Nov 14—Opening of bike/ped study." Video file. Retrieved August 2, 2014 from http://www.youtube.com/watch?v=nMxmGgaAjFk&lr=1http://www.youtube.com/watch?v=nMxmGgaAjFk&lr=1.
Wisconsin Department of Transportation (WisDOT) (2011b) "5/5 Hoan Bridge Mtg—Nov 14—Traffic, structures, more and next steps of bike/ped study." Video file. Retrieved August 2, 2014 from http://www.youtube.com/watch?v=cOOhdFTBCrY.
Wisconsin Department of Transportation (WisDOT) (2011c) "I-794 Lake Freeway bicycle/pedestrian feasibility study draft." Retrieved March 11, 2012 from http://www.dot.wisconsin.gov/projects/seregion/794hoan/docs/feis-complete.pdf http://www.dot.wisconsin.gov/projects/seregion/794hoan/docs/feis-complete.pdf http://www.dot.wisconsin.gov/projects/seregion/794hoan/docs/feis-complete.pdf.

Part II
Corridors
Links in the Network

5 From Climate Fight to Street Fight
The Politics of Mobility and the Right to the City

Jason Henderson

INTRODUCTION

Recent science suggests that 2°C above the 1900 global mean is a tolerable temperature window, or threshold, for averting social and economic disruption to human systems due to climate change (Hanson et al. 2008). This target could be achieved by reducing current global carbon emissions by 2–3% per year, thus avoiding a future "trillionth ton" of cumulative carbon emissions (Foster 2013). But this requires keeping a substantial portion of known fossil fuels in the ground. Meanwhile, transportation emissions make up 22% of current global carbon emissions, and transportation is almost entirely fossil fuel dependent. With the number of passenger cars forecast to triple to 1.7 billion by 2035, transportation is also the fastest growing sector of emissions and by most accounts will remain a voracious consumer of fossil fuels even with electrification or hybrid vehicles (International Energy Agency 2012). This in turn suggests that expedient mitigation of greenhouse gases (GHGs) must come in the form of reducing driving—or reduced vehicle miles traveled (VMT) in the transportation planning lexicon.

At the urban scale, reducing VMT is a vexing and contentious affair because it means reallocating street space by prioritizing transit, pedestrians, and bicycles while limiting the space of private automobiles. Systemic interface between transportation emissions and climate change must be brought down to street intersections and bus stops. Scaling down to local conflicts over street and bus stops may seem mundane, but the stakes are high. These hyperlocal nodes in urban transportation networks cumulatively structure the flows within cities, and ultimately the amount of GHGs released in the atmosphere. The stakes are also ideological, with competing progressive, neoliberal, and conservative political viewpoints shaping how streets are conceptualized and contested, which then scale back up to how modern society reacts to the urgency of climate change.

To better understand how climate politics unfold through street fights (and the ideologies that underpin them), I look to San Francisco, California, because it is at the cutting edge in reducing transportation GHG emissions by reallocating street space, but also has a car density of almost 10,000 motor vehicles per square mile, one of the highest in the United States. This means every inch of street space is coveted by incumbent motorists in ways more pronounced than in many other American cities with lower automobile densities, and San Francisco's debates over street space blend into sharp ideological debates over mobility and the right to the city.

Briefly, San Francisco has an ideologically progressive politics of mobility, invoking concerns about the environment and social justice, and seeking to use government to limit the overall amount of automobility in the city through reallocating street space. This differs from the logic of neoliberalism, which, while sometimes pursuing the use of state power to maximize private profit, is fundamentally about the allocation of urban space as a scarce resource, determined by pricing and markets rather than by regulation and collective action. Rounding out San Francisco's politics of mobility is a conservative ideology of mobility which posits that unfettered movement is a prerequisite of individual liberty and freedom and that government should proactively accommodate uninhibited movement mainly by car, even when that requires generous subsidy or undermines broader, collective environmental and social goals (Henderson 2013).

In the following chapter I describe how San Francisco progressives, while looking for ways to reallocate street space, come up against entrenched conservative frameworks used to evaluate and design streets, such as traffic engineering metrics that now privilege cars over other modes. Changing these frameworks is politically charged and means confronting long-standing conservative politics towards roadway engineering and circulation systems. Meanwhile, reallocating streets towards a progressive form of livability, while a laudable goal, can come with vexing contradictions, such as housing displacement near bus stops that are used by a new, *ad hoc*, private bus system (often given the 'Google bus' moniker), largely the result of neoliberal and conservative politics towards transit and failures of regional planning.[1] A new stratum of affluent technology workers, many commuting in these new private buses, have engendered a divisive backlash about livability—and whom it is for—at a time when livability is most needed to mitigate climate change. In that vein, the climate fight is not just a street fight, but also a struggle over the right to the city.

FROM CLIMATE FIGHT TO STREET FIGHT

The time for rethinking streets and automobility apropos to GHG emissions is urgent. Global greenhouse gas emissions need to peak in 2020 (some say 2016) and then decline if a safe target of 2°C warming above preindustrial times is to be met (IEA 2013). Current voluntary emissions reduction pledges will not achieve this, and the World Bank (2012) frets that the lack of a universal, cooperative global climate policy will result in temperature rises exceeding a disastrous 4°C within this century—perhaps as early as

2060. Meanwhile, transportation is not only 22% of the global total, but is also the fastest growing sector of global GHG emissions, forecasted to grow by 40% by 2035 (IEA 2013). Such growth largely results from the expansion of global automobility; presently 500 million passenger cars are in use, but by 2030 this figure is expected to reach 1 billion, accompanied by another billion trucks, motorcycles, and other motorized vehicles (including electric bikes).

Global automobility and transport GHG emissions have been highly uneven, considering that the U.S. has 4% of the world's population but 21% of the world's cars, produces 45% of the global carbon emissions that come from cars, and overall produces 25% of the total global GHG emissions while consuming 23% of the world's oil annually (United States Department of Energy 2011). If China had the same per capita car ownership rate as the U.S., there would be more than one billion cars in China today—double the current worldwide rate.[2]

'Sustainable motorization' or an 'enlightened car policy' of electric or hybrid cars might contribute to reducing some emissions but will not contribute to climate stabilization. Given the host of negative effects from mass automobility, one energy scholar asks that if doctors should not encourage low-tar cigarettes, why do environmentalists encourage hybrid (or electric) cars (Zehner 2012)? Consider that if the world's fleet of gasoline-powered automobiles shifts to electric, hydrogen fuel cells, or biofuels, the change will also draw resources away from industrial, residential, and food systems, or it will have to involve an entirely new layer of energy production. Massive quantities of petroleum will be needed to scale up to wind turbines, solar panels, and other, cleaner energy sources. Untenable amounts of GHGs will continue to be emitted just to replace the existing petroleum-based automobile palimpsest, and the world will still hit atmospheric carbon levels of 600 to 1,000 parts per million (ppm) (and maybe even higher) within this century (exceeding 420 ppm is considered unsafe for the planet) (Intergovernmental Panel on Climate Change 2013). Moreover, future emissions from automobiles do not include the full life cycle of automobiles, which itself contributes to substantial emissions and fossil fuel consumption (Chester and Horvath 2009).

As the U.S. and other developed nations expect China, India, and other developing nations to realistically participate in climate change mitigation, America will need not only to provide leadership, but also to decrease its appetite for excessive, on-demand, high-speed automobility. An incremental adaptation of allegedly clean automobility is not enough. Mitigation must include scaling down, away from large-scale energy system approaches as a transport solution to considering mundane urban street configurations and bus lines.

Reflecting this necessity to scale downward, people and organizations throughout the world are rethinking the connection between mobility and cities. In the U.S., the livability movement seeks to reduce car use by reconfiguring urban space into denser, transit-oriented, walkable built forms, a development pattern also associated with smart growth or new urbanism. The movement argues that Americans must undertake a considerable restructuring and rescaling of how they organize transportation in cities, and calls for reductions in annual VMT—that is, less driving in order to address global-scale GHG emissions, reduce oil consumption, and address a plethora of other social and environmental problems stemming from high VMT. Some suggest a need to redesign American cities to reduce driving by 25–50% of present levels by 2050, an enterprise that will require urban freeway removal, reduced parking, and replacement of road space with other uses (Newman et al. 2009).

A full reorientation away from private vehicle use must include prioritizing transit and bicycles over automobiles through the deployment of exclusive transit and bicycle lanes, the removal of lanes and curbside parking available to cars, signal prioritization for transit and bicycles at intersections, queue jumping so that transit can bypasses traffic stalled at intersections, restrictions on turns for automobiles, and transit stop improvements including bus stop bulb-outs and amenities. Thus, while global warming must be addressed as a global problem, much of its mitigation, at least in terms of mobility emissions, comes in the form of local street regulation. This is no small task, as current patterns of automobility and associated emissions are the direct product of historical decisions about transportation technology, including the prioritization of vehicular flow in the space of the street. I therefore ask, by way of case study in San Francisco, what can be done to transform street spaces, and what political challenges stand in the way of more progressive change?

MITIGATION AND MOBILITY IN SAN FRANCISCO AND THE BAY AREA

In San Francisco, city and regional planners and sustainable transportation advocates have begun to imagine what is needed to substantially reduce driving, reduce emissions, and allow livable infill. As a desirable and livable city, planners expect that 92,000 housing units and 190,000 new jobs will be added to San Francisco by 2040, increasing the city's current population from 800,000 to over 1 million (Association of Bay Area Governments and Metropolitan Transportation Commission 2013).

In anticipation, large swaths of former industrial areas have been rezoned for intensive residential and commercial infill development centered on transit and walkability, including rules that restrict the amount of parking made available to residents and visitors. Parking caps (known as maximums) and allowing zero parking (zero minimums) make these rezoning efforts some of the most ambitious examples in the nation of the implementation of the tenets of the livability agenda (Henderson 2009, 2013). Planners estimate that if San Francisco is to reduce emissions by 50% below 1990 levels by 2035, driving must be reduced from 62% to 30% of all daily trips in the city. Concomitantly, transit must increase from 17% to 30% of all daily trips by 2035, while increasing walking and bicycling to 40% of all daily trips (SFMTA 2010, 2011). These figures reflect a '30-30-40' split that transportation planners envision for future mode shares in the city.

San Francisco's Municipal Transportation Agency (SFMTA) has a strategy, known as 'Muni Forward,' and a bicycle plan that jumpstarts the '30-30-40' vision.[3] The transit and bicycle plans are programs of transportation infrastructure projects and service improvements that would be implemented in phases over a decade with a goal of enabling anyone in San Francisco to contemplate living in the city without a car (SFMTA 2006). A modest start compared to the broader 30-30-40 ambition, Muni Forward would reallocate street space for higher transit reliability, attracting more ridership and enabling San Franciscans to conveniently reduce driving to half of all trips by 2018 (from 62% in 2012). Complementing this, the City also has ambitions to increase the mode share of bicycling from 3.5% to 9% by 2018, and eventually 20% by the 2020s (SFMTA 2012).

Yet juxtaposed against San Francisco's transportation goals is a very high density of automobiles—9,853 per square mile—one of the highest in the U.S. This makes it more challenging to usurp the spaces of automobiles for transit and bicycling, and ignites political backlashes. The fact that so many people still own and drive cars in San Francisco, despite the existing high population density, land use intensity, and excellent transit coverage, is a profound issue to consider with respect to GHG mitigation efforts. But more significantly, it also shows that the fight over street space—just as with the climate fight—is not just physical and technical, but also ideological.

THE POLITICS OF STREET SPACE

That mobility is ideological is not a novel observation. As mentioned elsewhere in this volume, a "new mobilities" paradigm in the social sciences and humanities considers mobility to be culturally nuanced but also a fundamentally important factor underpinning material, social, political, and economic processes (Sheller and Urry 2006). The new mobility paradigm compels one to look at the wider ideological assumptions about mobility to understand how and why decisions involving that phenomenon are made. Just as the social philosopher Henri Lefebvre (1991) theorized that the character and nature of urban space reflect dominant modes of production and social relations (and power) within a given society, so, too, does mobility contain embedded social relations. Understanding this notion requires deconstructing the assumptions that lead to a particular ideology of mobility. This includes reconsidering the claim that movement and flows can be decoupled from ideology, or claims that quantitative, data-driven methods are apolitical, dispassionate, objective, and unbiased examples of professionalism in transportation analysis (see Culver, this volume).

In San Francisco and elsewhere, mitigation of transportation GHGs is hindered by a legacy of ideologically conservative visions of using government to actively preserve automobility as a set of social relations. The city's approach to analyzing streets, despite the erstwhile progressive transit and bicycle ambition, continues to assume an inevitability of automobility, conservative visions that people need cars and will continue to need them. For example, before official adoption, San Francisco's transit and bicycle plans underwent extensive environmental analysis required by state law. The multi-year studies each devoted hundreds of pages, thousands of hours of staff time, and millions of dollars to analyzing intersection level of service (LOS), a traffic engineering metric that assesses the delay motorists experience at street intersections. In a seeming contradiction to modern understandings of the environment and mobility, California's environmental analysis framework, established in the 1970s and 1980s, included delay to incumbent car traffic as change, or impact, on the physical environment that had to be mitigated. While the impact of idling vehicles on local air pollution was part of the rationale, the broader rationale was simply that free-flow traffic, once impeded, was a negative impact on quality of life for drivers.

One of the most widely used tools of traffic analysis in the U.S. (and exemplified in the Milwaukee bicycle debates described by Culver in this volume), LOS, in simplified terms, is part of a logic of flows that considers the actual time it takes for a vehicle to move through a specific local intersection compared to the theoretical optimum time it would take with no interference from other vehicles or impediments. The optimal conditions ascribed to 'good' LOS are 12-foot-wide travel lanes at level grade with no curb parking on approaches, no pedestrians or bicycles, no buses stopping in lanes, and only passenger cars in the vehicle mix (Transportation Research Board 2000). Delay is described by a six-letter grading scale, or range, which is similar to a school report card. LOS A, B, and C indicate reasonable traffic flow but with steadily increasing delay. LOS D, which means a 35-second delay, is considered a point at which an intersection is approaching capacity and should be expanded or modified to avoid 'bad' LOS E or F conditions, F being an extreme delay—that is, one of 80 seconds or more.

Because of this practice of prioritizing automobile LOS in local transportation analysis and environmental review, the findings of the environmental reports for transit and bicycling contradict the goals and ambition of the city outlined above. For example, in July 2013 the environmental study for transit concluded that doing nothing ('No Project Alternative') to improve San Francisco's transit system, known as Muni, was "environmentally superior" because this did not delay cars by reallocating street space (San Francisco Planning Department 2013). Furthermore, in ranking a range of alternatives for prioritizing transit on streets, the environmental impact report (EIR) then concluded the 'moderate' alternative, which minimized street reallocation, was environmentally superior to an 'expanded' alternative that included reallocating street space in ways that reduce capacity for private automobiles in order to prioritize transit operations on city streets and in the public right-of-way.

The 'expanded' alternative of the transit plan would use traffic calming and queue jumping with more frequency, resulting in faster, more reliable and efficient transit operations, compared to the moderate alternative. Yet the expanded alternative would have 'unacceptable' delay at five key city intersections and, cumulatively, 13 intersections by 2035 (the moderate alternative creates no significant delay, but does inconvenience motorists with curbside parking removal). Ironically, despite benefit to transit riders and pedestrians, the expanded alternative was considered more harmful to the environment because, invoking the bias towards the free flow of automobiles inherent to LOS analysis, it increased delay and inconvenience to motorists.

To be sure, San Francisco is far from a bastion of the conservative ideology found in American politics, especially regarding social issues such as gay rights, religion, and immigration policy. Yet there is a pronounced conservative discourse in San Francisco that protests efforts to reallocate street space toward other modes, opposes efforts to densify the city around public transit, and advocates that new development in the city should have abundant and inexpensive parking. Indeed, in July 2014 a Republican-led, *ad hoc* coalition of motorists placed on that year's ballot a 'Restore Transportation Balance' initiative that sought to block the city's progressive, transit-first policies, specifically limiting parking removal and other hyperlocalized street reallocations. This *ad hoc* coalition was made up of disparate neighborhood and merchant organizations that came together over erstwhile isolated and very localized, neighborhood-scale political conflicts over bike lanes, transit lanes, and curbside parking reform. Central to their platform was engineering streets to achieve smoother flowing traffic for car drivers (the objective of LOS) (see http://www.restorebalance14.org/).

Ideologically, in analysis like LOS, momentum for a progressive transit vision is subsequently dampened and confused by what is fundamentally a conservative analysis. By invoking LOS in decisions about streets, localities like San Francisco unwittingly accommodate a conservative vision of using government to actively preserve automobility as a set of social relations. Assuming an inevitability of automobility, LOS is steeped in a vision that people need cars and will continue to need them—that is, in the essentialization of automobility—and assumes there are naturally inherent cultural characteristics tied to national and regional geographies (see Sayer 1997, on essentialism as a concept). For example, it is assumed that Californian or American identity is inextricably bound with automobility, that California is a car culture, and that the car is natural, inevitable, and embodied in state and national identity. In that vein, San Francisco's planning assumptions have for decades included a 1% annual increase in driving, and therefore the city's planning models show more driving in the future regardless of parking restrictions or improved bicycling and transit.

LOS becomes a metric that can be deployed as unbiased scientific proof that reallocating street space would be disastrous for drivers. Further, LOS is wrapped in the critical infrastructure of the production and circulation of automobiles, and undergirds a historically contingent form of capitalism centered on automobility (see Huber 2013). As a metric, LOS assumes as a goal the reduction of circulation time for automobiles, thereby increasing speeds of flows and spatial range and hence quicker accumulation for businesses that perceive their viability to be reliant on automobility. Prioritizing high LOS preserves the incumbency and exchange value of the system and flows of automobility, which is a fixed capital investment threatened by congestion, rising fuel prices, and the entire concept of livability (if implemented). The networks of automobility require metrics like LOS to maintain functionality as a system, but so does the individual motorist, who has sunk costs in a car, insurance, garage, perhaps a home easily accessible by car, and a retail system structured by the car. To ignore, eliminate, or reform LOS in a way that privileges other modes will bluntly devalue the spaces of the car.

The use of LOS has vexed progressive planning for proposals such as Muni Forward, as well as bicycle and pedestrian planning, while also frustrating progressive and neoliberal desires to reurbanize development. This has led to a decades-long effort to dispense with and replace LOS with a new metric that measures the number of automobile trips caused by new development (auto trip generation, or ATG) or a VMT per capita metric that measures how new development or transportation investment reduces and shortens car trips (OPR 2013). Progressives, and especially San Francisco's bicycle advocates, had struggled for almost 20 years to discontinue the use of LOS in that city (Henderson 2013). For a variety of politically frustrating reasons, the city's conservative traffic engineering apparatus prevailed.

At the state level, however, reform—albeit slow—finally seems to have traction. Though not yet adopted as of October 2014, proposals to replace LOS with a new metric have been introduced by Governor Jerry Brown's administration in response to legislation adopted in 2013 (Governor's Office of Planning and Research 2013). This legislation, known as SB 743, mandates that California's environmental review process discontinue using LOS in environmental analysis of development or infrastructure once a new metric is approved. SB 743 states that automobile delay, as described by level of service, shall not be considered a significant impact on the environment (notably, the same legislation also declared that parking was not to be analyzed in environmental review, further easing transit and bicycle plans). Stating that the criteria in environmental analysis must promote the reduction of greenhouse gas emissions, the development of multimodal transportation networks, and a diversity of land uses, SB 743 aligned environmental analysis with the state's GHG mitigation goals (SB 375) and Bay Area planning goals. It effectively scaled up the solution to San Francisco's political gridlock by linking the state's GHG emissions goals to the elimination of hyperlocal street analysis favoring automobiles. Politically and ideologically, it also signified the triumph of neoliberalism, with a progressive edge, over a half century of conservative, automobile-oriented planning.

California's neoliberal development class (which is also linked to global capital) is cognizant of the role mobility has in maintaining the exchange value of cities, their property, and fixed capital investments that enhance profit. As the system of automobility reaches its limits in terms of environmental pollution, congestion, resource unavailability, and practicality for sustained profit, neoliberals have actively sought to steer new mobility investments to bicycling and transit as part of the broader commodification of livability and in order to maintain the exchange value of their investments (San Francisco Planning and Research 2009).

The particularly neoliberal slant for dispensing with the LOS metric was born out of compromise in an 'impossible coalition' of developers, local government officials, and environmentalists willing to make concessions in order to link state GHG caps with land use and transportation change (NRDC 2009; Nagourney 2013). Progressives envision the reforms to LOS as achieving spatial planning goals that mitigate climate change and the environment, while neoliberals envision the reforms as a streamlining of the production of highly profitable infill development. To be sure, some progressive environmentalists have suggested the efforts have been too generous to developers, as the state legislation included exemptions for infill development projects and also did not address economic displacement generated by infill development in cities like San Francisco. Nonetheless, many progressives have cheered the reforms, and moreover, the reforms (especially GHG emissions mandates) show that neoliberals not only see necessity in the new mobility armature for the circulation of capital, but also need government to steer it.

The state mandate to rethink LOS is still underway (as of October 2014), but will likely result in a significant change with regards to how streets are analyzed. However, there are limits, and the outcome is not certain. Local city and county transportation departments, many still displaying conservative viewpoints towards street space, will still be able to use the LOS metric in projects that fall outside the jurisdiction of state environmental code. Bicycle lanes or transit priority projects can still be stymied by a conservative local government that privileges LOS in basic traffic analysis separate from environmental review. Nonetheless, removing LOS from the environmental review process will be a vast improvement for the pace of future transit and bicycling planning in the Bay Area and California. Going forward, the real political battle will be between progressives and neoliberals over the exaction fees for the metric replacing LOS, since for the most part the mitigation fee will be levied on private real estate development at the local scale. Political conflict over the mitigation fees will still involve ideological battles over who the street is for, whose circulation and flows are privileged, and who pays for mitigation.

THE 'GOOGLE BUS' CONUNDRUM

As San Francisco's ideologically laden street fight over LOS approaches some semblance of reform (albeit with progressive/neoliberal tension over fee rates), a new street fight over the city's curbside bus stops engendered parallel debates over whose mobility is privileged in urban space. Briefly, throughout late 2013 and 2014, progressive housing activists in San Francisco decried and protested the proliferation of private, corporate shuttles, or 'Google buses,' as symbols of gentrification and inequity (Solnit 2013; Gumbler 2014). Public transit passengers also complained the private buses blocked Muni bus stops and slowed public transit, because these private buses utilized public bus stops in an *ad hoc* and uncoordinated manner.

This "shadow industry solution" to declining public transit is an expression of neoliberal and conservative inclinations to abandon the public realm, in this case public transit (Baum 2010). These private, employer-provided regional buses reflect a neoliberal model shaped by pricing and markets rather than by regulation and collective action, consistent with the broader agenda of the privatization of space and market-based pricing of public access to space. Such privatized transit constitutes a neoliberal resolution to the muddled labor union work rules like those found in Bay Area Rapid Transit (BART) and Muni. In these new systems drivers are flexible, can work part time, and are largely unorganized compared to the public sector transit unions (Singa and Margulici 2010). The private buses are also conservative in that they reflect the reluctance of wealthier people to ride crowded, slow buses, but who support commuter rail and express buses if they are of high quality and exclude undesirables and the poor. With some exceptions, ordinary people cannot ride the private buses, and the routing and scheduling of these buses are not oriented towards the low-wage, low-skilled workers that clean and cook for the tech firms. Mobility is segregated into a premium and basic class issue, invoking Graham and Marvin's (2001) *Splintering Urbanism*.

To be sure, the private commuter buses show that there is great potential for a public, regional, express bus system. Many employees in the technology sector shun living in suburbia and prefer the city of San Francisco. Although most reverse commuters drive, increasingly thousands are using luxury buses provided by third-party contractors for Google, Yahoo, Facebook, Apple, Genentech, E-Bay, and an expanding array of Silicon Valley firms. These luxury buses are reducing the amount of VMT that would occur if these employees drove to the sprawling office parks. In one survey, 63%

of passengers said they would drive alone were it not for the private bus service (SFCTA 2011). This equated to 375,000 round-trip, solo-driving commutes avoided that year. Moreover, 28% of the respondents in the survey did not own a car, a figure on par with San Francisco's rate of 30% car-free household ownership in 2010. But while they do substitute for some car trips, an *ad hoc* private transit system does not reflect the kind of thoughtful, progressive regional planning that will sustain beyond the boom cycles of private companies. At street level, the buses are oversized, have a difficult time navigating the tight street grid of San Francisco, and, most poignantly, usurped public space by illegally using bus stops reserved for public transit.

The city responded to the latter complaint with a pilot fee of $3.55 per bus stop, per bus to cover the costs of coordinating the private buses with Muni and to underwrite data collection for a study that would consider how to make this emerging private system work better towards the city's sustainability and transit goals. This pilot, initiated in August 2014 to conclude in early 2016, was roundly criticized by housing and environmental advocates. During several public hearings, including a hearing on environmental impacts in early 2014, they argued the scheme failed to take into consideration equity, gentrification, and displacement. This claim is especially relevant to the climate fight, in that gentrification and displacement may be undermining the sustainable transportation goals of the city and region.

Simply put, at a statewide and regional planning level, LOS reform enables neoliberal developers to more easily reorient towards urban cores and transit nodes; however, lower-income households are displaced from these urban and transit nodes to car-oriented sprawl, whereas elite tech workers, benefitting from LOS reforms, purchase livability near a private shuttle line that also potentially disrupts Muni service, all the while redefining who has the right to the city.[4] Here the street fight over LOS intersects with bus stops, and to date no policy makers have addressed the broader structural problem of large, wealthy corporations and their employees' sidestepping public, regional planning solutions and instead creating an inequitable, two-tiered, private system.

This is where the issue of LOS reform enters back into the conversation as a potential tool to enable regional public transit on a different scale. As the LOS metric is replaced with a fee-based ATG or VMT assessment on new development, a regional or semi-regional fund could be established to create public express buses using exclusive lanes on freeways, perhaps operated by BART or Caltrain (commuter rail between San Francisco and Silicon Valley) as part of the next iteration of regional planning. Mitigation fees could supplement other progressive revenue measures, such as a property assessment on the corporations and developers, in part already possible within the existing BART district, as well as other proposals such as vehicle license fees or congestion charging on regional freeways.

Thus LOS reform and the proposed fees may also be intertwined with resolution of the *ad hoc* private commuter bus system. With 660,000 new residences and hundreds of millions of square feet of new commercial and office space expected to be built between 2014 and 2040, ATG or VMT mitigation fees will be critical for financing transit capacity expansion, the lynchpin for reducing regional per capita VMT and GHGs. As the schedule of fees is vetted and progressives contest neoliberal fee proposals, attention will no doubt turn to the reality that current available financing for transit is inadequate. Regional planners identify $289 billion in future funds from existing revenue streams between now and 2040 for roads, bridges, and transit, but upwards of 87% of this is already committed to maintenance of existing roads and transit—not transit capacity expansion for the massive, new reurbanization expected (ABAG and MTC 2013). New homes and jobs might be focused around existing BART and Caltrain stations (enabled by LOS reform), but transit capacity issues will limit effectiveness in reducing GHGs.

In 2013 and 2014, both BART and Caltrain were regularly breaking ridership records, but were near capacity. Meanwhile, most mainline Muni buses and railcars are currently jam-packed, yet the city of San Francisco is somehow expected to absorb 92,000 housing units. Because there's no real capacity expansion in the region's long-range plan, few new residents will be able to squeeze onto the already crowded transit systems. As such, the current iteration of Bay Area planning won't reach its own modest goal of 74% of trips by car in 2040 (it is currently 84%) (ABAG and MTC 2013). With 2 million more people, cumulative emissions from driving will actually increase by 18%. Per capita VMT might decline, but the absolute growth in driving that will accompany population growth will negate regional GHG emissions reduction goals.

Without new revenues, all indications suggest premium, private transit will expand. As Bay Area transit agencies strain under declining revenues, deferred maintenance, and deep federal and state cuts, one Bay Area private transit contractor has rapidly expanded to over 200 vehicles and more than 100 drivers, increased daily ridership to more than 6,000 people a day, and grown 30% a year between 2005 and 2010. The SFCTA (2011) reports that Google had doubled its shuttle bus commuters, and in 2014 it was estimated that well over 35,000 workers commuted on these buses, equaling the ridership of Caltrain commuter rail service. In 2011 real estate listings for some neighborhoods in San Francisco began to mention proximity to private shuttle bus routes, and anecdotally some observers suggested that such proximity increases rents by up to $400, while the *Wall Street Journal* reports a 20% premium on new condominium sales (Keats and Fowler 2012; Said 2011). All of this foreshadows a potential transit future in which a premium system serves the wealthy in first-class coaches—and in premium livable neighborhoods—and a dilapidated, economy-class system serves the lower classes that are gentrified out of the core.

Housing and social justice activists may find it convenient (and in some cases justifiable) to target protests against gentrification at tech workers on private buses, but it might be more constructive over the long term to direct political energy into shaping the debate over impact fees and directing fee revenue to regional transit capacity. With reform of LOS and replacement with a new, fee-generating metric (ATG or VMT), this could not only steer regional planning in a progressive direction, but also result in a public transit system that more efficiently serves elite and non-elite users alike. In this sense, the two street fights—LOS and Google bus—are crosscutting struggles intersecting in the goals of reducing VMT and accessible housing, all of which cumulatively scale back up to the same climate fight.

CONCLUSION

With debates over street space and the right to the city increasing throughout the U.S. and beyond, the degree to which this politics of mobility is accentuated in San Francisco makes it a compelling case study and bellwether for rethinking automobility in an era when climate change mitigation is imperative. From contesting how intersections are analyzed in traffic studies to debates about public bus stops usurped by private luxury coaches, these mundane nodes in cities make up the wider urban flows that produce significant levels of GHG emissions. They are front lines in the climate fight.

Given the pressing need for expedient climate change mitigation in the form of reducing VMT, it is fitting to consider what is unfolding in San Francisco. San Francisco's transportation ambitions are relatively impressive compared to most of the U.S., and its emerging policies in regard to the car are probably closer to where other cities will really need to be focused in terms of addressing GHG emissions, energy policy, and social concerns related to mobility. It is also an exceptionally livable city by many indicators; indeed, it is consistently ranked as one of the most livable in the U.S. and has a very high, arguably insatiable, demand for new housing (which brings some uncomfortable affordability issues). Through efforts to reform traffic-engineering metrics (at least in environmental review) and replace LOS with a new metric that acknowledges the nexus between driving and GHGs, San Francisco would be primed to fund its 30-30-40 split goal. But the emerging reliance on private transit embodies what can go wrong when inadequate regional transit, coupled with poor land use planning, intersects with new strata of workers who seek livability in the urban core.

If climate change mitigation through transportation policy is to be fair, it must avoid the secessionist tendencies emerging with San Francisco's private bus system. Walkable, bicycle- and transit-oriented gentrification with premium, private commuter transit is an inequitable livability for the elite, especially when public bus stops are used for private buses. This is a reality that sustainable transport advocates in San Francisco and elsewhere must judiciously navigate. Livability boosters who crow that livability is legitimized because walkable places have high real estate values undermine the right to the city. This only enables environmental-friendly lifestyles for those who can afford it. If there is to be a serious reorientation of cities that reduce transportation GHGs, it cannot rely on market-based systems that consistently put aside collective interests in the face of potential profit. Linking the much needed LOS reform to a future revenue generating metric (ATG or VMT fees) could help orient regional planning in a more progressive direction and tackle GHGs at the same time.

These means to an end do require careful consideration of the ideological alignments around mobility. While San Francisco has a historically conservative approach to analyzing streets, there is readiness among progressives and neoliberals to dispense with it. In other localities in the U.S. and abroad, the conservative bent may be more pronounced and entrenched, making comparison a matter of accentuation. Furthermore, San Francisco's flirtation with neoliberal, regional transit occurs in a bastion of high-tech industries that lack historical ties to organized labor but have strong connections to environmental concerns. This begs caution for progressives who might appreciate the role private transit has in removing cars, but that also further aggravates broader programs focused on equity in housing, labor organizing, and the right to the city.

NOTES

1 The Google bus moniker is due to Google having the single largest fleet, an outsized number of employees in the city, and global recognition in the mainstream media.
2 In 2009 the U.S. had 828 vehicles per 1,000 people. Given its population in 2012 of 1.343 billion and that rate of car ownership, China would have over 1.112 billion vehicles. Data on vehicles are from USDOE, *Transportation Energy Data Book 30*, table 3–5; China population figure is from "US Census International Population Database," www.census.gov/population/international/data/idb/country.php.
3 Until Autumn 2014, *Moving Muni Forward* was called the Transit Effectiveness Project (TEP).
4 It should also be noted that by defining the program as a pilot, the city was able to skirt detailed environmental analysis that would have, ironically, included the impact these buses may have on intersection LOS. This further reflects neoliberal logic of avoiding LOS analysis.

WORKS CITED

Association of Bay Area Governments (ABAG) and Metropolitan Transportation Commission (MTC) (2013) *Plan Bay Area: Regional Transportation Plan and Sustainable Communities Strategy for the San Francisco Bay Area 2013–2040*, Oakland: ABAG/MTC.
Baume, M. (2010, July 26) "New study recommends augmenting the benefits of private shuttle service," *StreetsblogSF*. Retrieved July 27, 2010 from http://sf.

streetsblog.org/2010/07/26/new-study-recommends-augmenting-the-benefits-of-private-shuttle-service/.
Chester, M., and Horvath, A. (2009) "Environmental assessment of passenger transportation should include infrastructure and supply chains," *Environmental Research Letters*, 4:1–8.
Foster, J. B. (2013) "James Hansen and the climate-change exit strategy," *Monthly Review*, 64(9):1–19.
Governor's Office of Planning and Research (OPR) (2013) *Preliminary Evaluation of Alternative Methods of Transportation Analysis*, Sacramento, CA: Governor's Office of Planning and Research.
Graham, S., and Marvin, S. (2001) *Splintering Urbanism: Networked Infrastructures, Technological Mobilities, and the Urban Condition*, New York: Routledge.
Gumble, A. (2014) "San Francisco's guerrilla protest at Google buses swells into revolt," *The Guardian London*. Retrieved February 1, 2014 from http://www.theguardian.com/world/2014/jan/2025/google-bus-protest-swells-to-revolt-san-francisco.
Hansen, J.M.S., Kharecha, P., Beerling, D., Berner, R., Masson-Delmotte, V., Pagani, M., . . . and Zachos, J. C. (2008) "Target atmospheric CO2: Where should humanity aim?," *The Open Atmospheric Science Journal*, 2:217–231.
Henderson, J. (2009) "The spaces of parking: Mapping the politics of mobility in San Francisco," *Antipode*, 41(1):70–91.
Henderson, J. (2013) *Street Fight: The Politics of Mobility in San Francisco*, Amherst, MA: University of Massachusetts Press.
Huber, M. T. (2013) *Lifeblood: Oil, Freedom, and the Forces of Capital*, Minneapolis, MN: University of Minnesota Press.
Intergovernmental Panel on Climate Change (IPCC) (2013) *Climate Change 2013: The Physical Science Basis, Summary for Policymakers*, Geneva, CH: United Nations Environment Program and the World Meteorological Society, p. 33.
International Energy Agency (IEA) (2012) *2012 World Energy Outlook: Executive Summary*, Paris: International Energy Agency.
International Energy Agency (IEA) (2013) *CO2 Emissions From Fuel Combustion Highlights 2013*, Paris: International Energy Agency.
Keates, N., and Fowler, G. (2012, March 16) "The hot spot for the rising tech generation," *Wall Street Journal*: D1.
Lefebvre, H. (1991) *The Production of Space*, Malden, MA: Blackwell.
Nagourney, A. (2013, September 10) "California takes steps to ease landmark law protecting environment," *New York Times*: A1.
Natural Resources Defense Council and California League of Conservation Voters. (2009) *Communities Tackle Global Warming: A Guide to SB 375*.
Newman, P., Beatley, T., and Boyer, H. (2009) *Resilient Cities: Responding to Peak Oil and Climate Change*, Washington, DC: Island Press.
Said, C. (2011, August 11) "S.F. apartment rent rises as vacancy rates fall," *San Francisco Chronicle*: A-1.
San Francisco County Transportation Authority (SFCTA) (2011) *The Role of Shuttle Services in San Francisco's Transportation System*, San Francisco: SFCTA.
San Francisco Municipal Transportation Agency (SFMTA) (2006) *Transit Effectiveness Project Briefing Book*, San Francisco: SFCTA.
San Francisco Municipal Transportation Agency (SFMTA) (2010) *State of the SFMTA: Presentation to SFTA Board of Directors Workshop, September 21, 2010*, San Francisco: SFCTA.
San Francisco Municipal Transportation Agency (SFMTA) (2011) *2011 Climate Action Strategy for San Francisco's Transportation System*, San Francisco: SFCTA.
San Francisco Municipal Transportation Agency (SFMTA) (2012) *SFMTA Strategic Plan: Fiscal Year 2013–2018*, San Francisco: SFCTA.

San Francisco Planning and Urban Research Association (2009, May) "Critical cooling: San Francisco can fight global warming through smart changes to local policy. What can we do to lead the way?," *Urbanist*, 482:6–19.

San Francisco Planning Department (2013) *Transit Effectiveness Project Draft Environmental Impact Report Volume 1—Chapters 1 to 7*, San Francisco: San Francisco Planning Department.

Sayer, A. (1997) "Essentialism, social constructionism, and beyond," *Sociological Review*, 45(3):453–487.

Sheller, M., and Urry, J. (2006) "The new mobilities paradigm," *Environment and Planning A*, 38:207–226.

Singa, K., and Margulici, D. (2010) *Privately-Provided Commuter Bus Services: Role in the San Francisco Bay Area Regional Transportation Network*, Berkeley: California Center for Innovative Transportation, University of California, Berkeley.

Solnit, R. (2013) "Diary," *London Review of Books*, 35(16). Retrieved February 20, 2015 from http://www.lrb.co.uk/v35/n16/rebecca-solnit/diary.

Transportation Research Board (2000) *Highway Capacity Manual 2000*, Washington, DC: National Research Council, Transportation Research Board.

USDOE (2011) *Transportation Energy Data Book: Edition 30*, Oak Ridge, TN: Center for Transportation Analysis, Engineering Science and Technology Division, Oak Ridge National Laboratory.

World Bank (2012) *Turn Down the Heat: Why a 4 Degree Celsius Warmer World Must Be Avoided*, Washington, DC: The World Bank.

Zehner, O. (2012) *Green Illusions: The Dirty Secret of Clean Energy and the Future of Environmentalism*, Lincoln, NE: University of Nebraska Press.

6 The Social Life of Truck Routes

Peter V. Hall

INTRODUCTION

However much we might wish to believe in the dematerialization of the postindustrial economy, it is difficult to imagine a contemporary, metropolitan landscape in the developed world (or increasingly the developing world) without the movement of some goods by truck. Trucks and urban communities have long had a complex and contradictory relationship; we want the goods, but not their movement. Truck routes promise to reduce these tensions by codifying the conception that certain roads and routes are 'for trucks,' and that others are 'not for trucks.' Indeed, truck routes have become a taken-for-granted regulation of movement in urban space (Blomley 2010), but behind the technical and rational mask they are a result of multiple social and political processes. In this chapter I explore why and how certain city roads become accepted as routes for truck movement (while others are not) or, more generally, how the mobility of goods in urban space "is 'channeled' into acceptable conduits" (Cresswell 2010, p. 24), and as conduits for the acceptable movement of goods, how truck routes are implicated in the dialectical relationship between urban places and transport flows (Hall and Hesse 2013).

Truck route networks generally share basic characteristics: They clarify what constitutes a truck, permit them on certain roads, and often draw a distinction between local (i.e., intraplace) versus through (i.e., interplace) movements. The New York City Truck Route Network, for example, is defined as

> a set of roads that commercial vehicles must use in New York City. This network is comprised of two distinct classes of roadways, Local Truck Routes and Through Truck Routes. The network is defined in Section 4–13 of the New York City Traffic Rules. All vehicles defined as a truck (two axles and six tires, or three or more axles) are required to follow the Truck Route Network. Commercial vehicles that do not meet the definition of a truck are not required to follow this network,

but must follow all posted signage regarding the operation of commercial vehicles.

(NYC 2014)

Publics are more likely to accept commercial goods movements that are small and local, occur during business hours, and are kept out of residential areas. Large, nonlocal, and frequent movements are less likely to be okay. Designation as a truck route implies something about the possibilities for the nonmotorized use of road space, and it informs adjacent land values and uses. And, despite some recent experiments in dedicated truck-only lanes (see Fischer et al. 2003), designated truck routes do not share the exclusivity of freight rail corridors. Rather, in order to preserve a deeper assumption of urban mobility that is expressed in the mostly unrestricted right of access to public road space, they regulate rather than separate the movement of trucks (goods) and automobiles (people). At the same time, they contribute to the expansion and maintenance of the urban road network.

All urban regions are today confronted with intense and dynamic inter- and intra-metropolitan flows, although these mobility pressures, and the responses to them, vary from place to place. Greater Vancouver is home to Canada's largest seaport complex. The truck routes that connect port terminals with associated, yet widely dispersed, logistics facilities—and major infrastructure development programs like the Asia-Pacific Gateway and Corridor Initiative and the Pacific Gateway, or known together as the 'Gateway'— form the empirical case of this chapter. I draw inductively upon my ongoing research into the multidimensional and multiscalar urban entanglements of the region's port-logistics industry. The evidence presented here comes from mixed methods: in-depth interviews with industry managers, workers, elected officials, and public servants; a survey of municipalities; review of planning and policy documents; and analysis of census data.[1]

We already know that transport and land use have a dynamic interrelationship (Allen et al. 2012), yet we know less about how the creation, designation, and use of truck routes are implicated in mediating this recursive relationship. Through related vignettes, I trace how infrastructure investment decisions, successful and unsuccessful community resistance to truck routing, the social organization of trucking, and deeply contested imaginaries of urban space shape actually existing truck routes and the city-region they traverse. The Vancouver case shows that truck route construction, selection, and use display remarkable social regularities that reinscribe differentiation between places within the metropolitan area.

DEPARTURES

The idea that trucks are part of the proper use of particular roadways, and equally that they are not part of the proper use of most others, constitutes a social fact that may or may not be codified in regulation and fixed in place through public infrastructure investments. In many places, including Vancouver, British Columbia, truck routes are municipally regulated and formalized pathways. Investments in such roadways—pavement strength, turning lanes, road signs, over- and underpasses—lock these social choices into the physical landscape.

Truck route designations 'make sense' as a public policy. Roadways that are safe for heavy truck movements have additional technical requirements, from sturdier paving and bridgework to additional engineering requirements for turning lanes and sightlines, as well as signage (see Culver, this volume). And the designation of roads as (non-)truck routes makes a difference to traffic patterns. For example, a simulation exercise for the Chicago region in which trucks were restricted from driving on just 562 of 39,018 links in the regional road network, including such important roads as Lake Shore Drive, and some boulevards and arterial roads showed that substantial differences in road allocation result from truck restrictions (Boyce 2012). Hence the efficiency of the transportation system may be altered by influencing the routing choices of such vehicles (van de Riet, de Jong, and Walker 2008).

The existing transportation literature on truck routes typically seeks to understand patterns of truck movement within the given spatial structure. Transportation geographers think about truck routes through origins and destinations, accounting for the truck movements between them in terms of the cost, techno-engineering, and legal regulation of movement. Truckers choose the most financially and temporally 'efficient' route that can accommodate their vehicles as permitted (Rodrigue et al. 2009). Engineering and operations perspectives approaches formalize these assumptions into models addressing such matters as scheduling optimization (Ronen 1988; Boerkamps et al. 2000). Other approaches are oriented towards supporting planning and public decision-making, for example around planning for hazardous truck routes (Saccomanno and Chan 1985; Frank et al. 2000). And a few empirical modeling approaches point to the recursive nature of the relationship between goods movement and land use, even if they cannot 'explain' the human behavior that underpins them. For example, in a logistics-oriented transportation land use model developed for Calgary, Woudsma et al. (2008) showed that variations in accessibility and congestion patterns today are likely to result in shifts in land use by the logistics industry that will be measureable in five to 10 years.

In this chapter I seek to complement these transportation approaches with a mobility perspective that emphasizes the context in which truck routes are constructed as social facts (Urry 2007; Creswell 2010). Indeed, trucks and truck routes would not be such taken-for-granted dimensions of contemporary urban life if they were not so deeply embedded in urban economic and social structures that allow for the provision (and hence consumption) of just about anything at any time, without the unpleasantness of having to live close to points of production. Truck routes represent a compromise between convenience and ubiquity, on the one hand, and scale and efficiency, on the other. Like railroad lines and shipping lanes they concentrate high volume flows between points of production and distribution. But trucks are quite unlike railcars and ships in that they provide convenience and flexibility. Trucks are not nearly as dependent on the schedules of others as are trains and ships, and they can reach consumers just about anywhere.

Truck routes exist in an urban, spatial, economic context that has unfolded for more than a century. As Marshall observed in industrial England at the end of the 19th century, "factories now congregate in the outskirts of larger towns and in manufacturing districts in their neighbourhood rather than in the towns themselves" (Marshall 1890, p. 226). In his 1960 study of the future New York metropolitan economy, Vernon wrote of the relocation of manufacturing activity away from the region's core: "Other conditions emerged to make the change possible. One of these was the truck, which allowed manufacturers to shuttle cloth and garments between New York and the country districts" (p. 44). Recent scholarship has noted the importance of polycentric metropolitan areas in the organization of global goods movements (see Hall and Pain 2006; O'Connor 2010; Hall and Hesse 2013). In a fragmented, specialized, economic landscape, effective connections within and between metropolitan places are essential. Given the paucity of data on urban goods movement, it is probably unwise to claim that the number of goods movement trips within city-regions has everywhere increased more rapidly than the population or final demand. But what is surely new is the scale and intensity of the flows that originate in and are destined for, or are routed through, urban spaces. Furthermore, the organization of these flows has become more complex and dynamic, and they appear to be subject to more rapid reorganization.

Industry advocates have a strong desire to maintain the potential for movement through the system of truck routes and through the rules and regulations governing their formation. But these advocates do not necessarily get all that they wish for. Environmental health experts have helped the wider public to become aware of the health hazards resulting from exposure to diesel particulate matter and other emissions of truck traffic (Lena et al. 2002; Kozawaa et al. 2009; Perez et al. 2009). Among the most important sites for revealing the public health impacts and their unequal social distribution are the urban areas closest to the ports of Long Beach and Los Angeles, California (Houston et al. 2008). The negative impacts of living, playing, attending school, and working in close proximity to truck routes are large and known, even if they are too often met with callous indifference from the goods movement industry.

Freight and truck route designations and routing choices are hence intensely political (see Cidell, this volume), and the ways particular roads come to be understood as truck routes may be viewed as elements of distinct urban assemblages of production, distribution, and consumption. This idea connects to the work of Deleuze and Guattari (1987), and others who have emphasized the "indeterminacy, emergence, becoming, processuality, turbulence and the sociomateriality of phenomena" (McFarlane 2011, p. 206). Assemblage theory thus views cities as constantly emerging at the intersection of global and local flows of ideas, data, finance, people, and goods. However, this may have an implication of impermanence that is in tension with the empirical case discussed here, and indeed with the observation that there are similarities between places that are confronting increasing goods movement.

In particular, what the Vancouver case discussed below suggests is that actually existing truck routes display more stability than might be expected given the increasingly dynamic and complex patterns of flow they must channel. This observation calls for greater attention to the way existing built forms, land values, and plans exert a continuing influence on the routes that are built, designated, and actually chosen for truck movements. In this regard, the work of Molotoch et al. (2000) is helpful because it shifts our attention from the differences and similarities between the elements that constitute places to analyzing how those elements combine in distinct ways. In their seminal paper, Molotch et al. (2000) argued that despite being confronted by similar, 'homogenizing' forces, communities can maintain their differences, "because what is distinctive is not a list of attributes but the way these attributes lash-up and how the structuration process moves the resulting conjunctures forward through time"(p. 816). By extension, I argue that while all developed urban places today have to deal with heightened intra-metropolitan truck movement, the mobility practices that shape how truck routes are built, designated, and enacted both create and perpetuate distinct and uneven urban geographies.

VANCOUVER AND BRITISH COLUMBIA LOWER MAINLAND

The greater Vancouver region is relatively small in global terms; it is home to about 2.5 million people, with a mostly service-oriented economy. As in most other metropolitan areas in the developed world, governance is complex and fragmented. This creates openings for a politics of scale in which higher levels of government seek, under pressure from industry actors, to facilitate the 'free' flow of goods, while lower levels of government look to protect place-based quality of life (see Van Neste, this volume). There are 24 local governments that together comprise a regional government federation with some bulk service delivery functions, but little controlling authority. Local governments retain power over land use, although they have consented to cede some authority over industrial and green land uses to the regional government (Metro Vancouver 2011). As some have noted, fragmented metropolitan spaces exhibit unevenness in truck routes and standards (see Dablanc and Fremont 2013). In Metro Vancouver, municipalities have varying definitions of a truck (based on different gross vehicle weights, the number of axles, length, tractors, etc.), which are differentially applied through route, time of day, and other restrictions.

A regional agency known as TransLink is responsible for transit and regional transport planning for Metro Vancouver. Most roads in the region are either owned by the municipalities themselves or by the Province, though TransLink owns a few roads and bridges. The agency plays a vital role with respect to truck routes in at least two ways. First, TransLink is responsible for defining and cofunding "an integrated system of highways throughout the transportation service region" (British Columbia 1998), known as the Major Route Network (MRN). Created in 1998, at the same time that TransLink was established, the MRN consists of approximately 2,300 km of roads considered to be important for regional movement of people and goods (see Figure 6.1). Municipalities receive funding from TransLink—on a shared cost basis—for maintenance and improvements to these roads. Designation of the MRN is made by TransLink, with consent of the municipality hosting that roadway. Some municipalities designate truck routes and others do not (see TransLink 2009); in those municipalities with designated truck routes, all the MRN roads are included in the truck route.

Second, TransLink is an administrative site where regional transportation priorities are planned and legitimated. This is not to say it is the only such site; industry lobbying groups play an important role in promoting particular road-building agendas. Especially influential is the Greater Vancouver

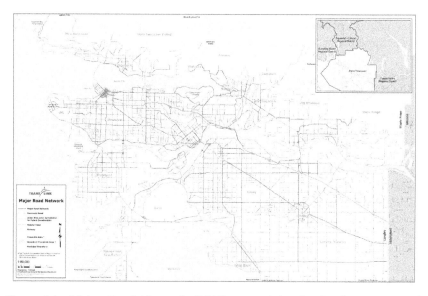

Figure 6.1 Major Road Network (2014) mapped by the South Coast British Columbia Transportation Authority.

Source: 2014 South Coast British Columbia Transportation Authority doing business as TransLink.

Gateway Council (GVGC), representing the goods movement industry, including the public port and airport authorities. Between the mid-1990s and mid-2000s, the GVGC successfully manufactured support for a package of infrastructure investments which the provincial and federal governments both came to claim as their own successes (Hall and Clark 2010). Federal and provincial governments also reach into the region to pursue specific infrastructure projects; the provincial government through its role in highway provision, the federal government through agencies responsible for ports and airports. However, TransLink cannot currently be entirely bypassed. In particular, it acts as the rule-making forum when municipalities seek to designate or de-designate truck routes. The clause in TransLink's governing legislation which gives it this power states that "despite the Community Charter, the Vancouver Charter or any other enactment... a municipality must not, without the approval of the authority, take, authorize or permit any action that would prohibit the movement of trucks on all or any part of a highway in the transportation service region" (British Columbia 1998, Part 2 Section 21(2)).

Vancouver ports have long handled exports of Canada's raw material wealth, but in the past 20 years the growth in throughput of containers has been remarkable. In the 1990s and 2000s, the number of container units imported and exported through the port grew at about twice the rate of all ports on the West Coast of North America. The number and pattern of port-related truck trips within the region has also expanded and evolved. Imported containers that are not loaded onto rail at the docks for transport to continental markets are taken by truck to import transloading facilities within the region, where their manufactured contents are unpacked and repacked for further distribution. The empty containers are then stored for a period, before being moved to export transloading facilities, where they are stuffed with raw materials (mostly grains, wood, and paper products) before being taken to the port terminals for export. In a small number of cases, import and export transloading and container storage are all done at the same site. However, a substantial proportion of imported containers makes at least four truck trips within the region (port to transload to storage to transload to port) before they are reexported.

Planners, industry advocates, and researchers observe that the majority of truck movements within the region are not related to port activity; data for 2008 indicates that just 37% of truck trips are associated with import and export activities of the "gateway" (TransLink 2014). However, these are the flows that attract most public and policymaker attention. They are, after all, the least locally serving movements. They are also the ones that attract provincial and federal funding, because unlike other segments of the goods movement industry, the port authority, federal and provincial governments, and industry lobbies have the kind of coherence that allows them to shape infrastructure-spending decisions in a way that parcel and grocery deliveries simply do not.

124   *Peter V. Hall*

I turn now to a three-part discussion of these intraregional, Gateway-related truck movements. First, I consider the processes by which roads are designed and built explicitly to handle truck traffic. Second, while regulation typically reinforces these construction choices, a consideration of the actual designation of truck routes shows that this may not always be the case. And third, while both construction and designation are influential, the routes that are actually driven result from the socially situated choices of truck drivers.

THE TRUCK ROUTES THAT ARE BUILT

The trucking implications of the Vancouver port's growth can be summarized as more circulation in general, as well as the desire to connect port terminals with inland suburban locations. Since many of the new transloading facilities that are integral to the import-export container business were located on lower-value lands south and east of the traditional metropolitan core, more circulation routes around the entire region were desired. In the metropolitan core, logistics land uses simply could not compete with the real estate development industry. The south- and eastward shift in logistics activity was also related to the opening in 1997 of the new Deltaport container terminal at Roberts Bank, located near the mouth of the Fraser River, some 40km south of the port terminals in Burrard Inlet. In turn, this entailed a reorientation of the previously dominant flows from Burrard Inlet directly eastwards to the continental rail yards in Port Coquitlam and Surrey, to a new southwest to northeast axis along the Fraser River (Hall 2012).

From the early 1990s, the Greater Vancouver Gateway Council and others advocated for a series of road and rail infrastructure projects to address the growing congestion generated by these new patterns of urban development, and to address preexisting congestion that had spurred such spatial reorganization in the first place. Several important but relatively minor roadway projects were funded; for example, a new overpass providing improved highway access for a transloading, warehousing, and container storage district in southeast Richmond was developed by the region's port authorities. There were also three major roadways included in the Gateway program, two of which have been fully built.

The first was a set of expansions to the provincially owned Highway 1, including the replacement of the Port Mann Bridge over the Fraser River. This project was undertaken despite opposition from Burnaby, the municipality that receives most of the east-west traffic on this expanded corridor. Also constructed was the South Fraser Perimeter Road (SFPR), a new road providing a direct link between the Deltaport container terminal and Canadian National's intercontinental railyard in Surrey. Long included in regional transportation plans, the SFPR tracks the Fraser River in some parts, and traverses agricultural lands in others. Environmentalist opposition to this roadway failed; seeing no way to halt this provincial highway, the municipality of Delta accepted a series of mitigations, including one which saw trucks rerouted away from a river-fronting residential area.

However, the third roadway expansion in the industry wishlist has stalled. Known as the North Fraser Perimeter Road (NFPR), this is a series of proposed road projects between the Canadian Pacific's intercontinental railyard in the northeastern suburb of Port Coquitlam and bridges which connect to the SFPR and, from there, with Deltaport. The NFPR is stalled because the municipality of New Westminster balked at the prospect of a segregated highway that would enlarge a truck route that is already a barrier between this suburban city and its waterfront. New Westminster has been able to hold out where Burnaby and Delta failed, in part by refusing to allow the expansion of a roadway that links New Westminster with the adjacent municipality of Coquitlam. Quite how this impasse will end is anybody's guess.

The larger point is that the truck routes that get built are the result of a social process of imagination, lobbying and resistance, planning, financing, and construction. In general, municipalities resist the construction of truck routes because they are unpopular with residents. In 2010, a research assistant and I used an inductive methodology to identify 42 separate instances of conflict between municipalities and port-logistics/Gateway-related infrastructure, land use, and related issues (see Hall 2014). Conflicts were found in relation to port terminal expansion, conversion of mills and agricultural land, parks and habitat, highway expansions, and rail corridor developments. Across these conflicts, we asked municipal representatives to rate 12 categories of impact on a simple scale. The highest-rated category of impact was 'traffic.' Furthermore, route-related impacts such as 'noise' and 'air quality' were ranked more highly than site-specific impacts such as 'views,' 'lighting,' and even 'wildlife, habitat, and ecosystem' concerns.

In many ways, this is not surprising, since route-related concerns by definition impact more jurisdictions than site-specific ones. But the findings also speak to the increasingly regional and route-oriented nature of the port's footprint. Truck routes are a tangible manifestation of the emerging, polycentric, and regional spatiality of contemporary urban economies.

THE TRUCK ROUTES THAT ARE DESIGNATED

Municipal governments use truck route designations to channel the flows that they understand to be inevitable. These choices typically give preference to those roads that were designed and built to handle trucks, but unlike construction, truck route designations can and may change after the fact. One such effort is the southern approach to the port terminals in the Burrard Inlet, where the City of Vancouver has invested in the Knight-Clark Street truck route through a series of incremental safety improvements and the introduction of turning lanes to reduce blockages. Such investment decisions have material consequences for the future form of the city. And they are controversial: The following examples compare the contests over truck route designation in a core urban and a suburban location. How communities respond to unwanted truck movements, and whether or not residents prevail, reveals much about the political and social processes behind the actual designation of truck routes.

The Clark Street Closure

In 2010 the Port Authority chose to close the Clark Street entrance to the Burrard Inlet (Vancouver) port terminals in order to achieve greater efficiencies for itself. Despite the rapid growth of Deltaport, there are two major container terminals in the Burrard Inlet. By closing the Clark Street entrance (to the south), the Port forced trucks to enter the Burrard terminal area from the eastern end, where there is plenty of space for truck staging. This ensures that the terminals would not have to wait for trucks to pick up and deliver containers. Faced with the closure of the usual entrance, truckers then altered their routes to the port, using a road adjacent to an area of detached, single-family houses. The road, Nanaimo Street, though designated as a truck route, had not been recently used as such. Residents were furious.

In 2011, a community group in the Grandview-Woodlands neighborhood of east Vancouver secured what can only be regarded as a victory in controlling truck movements on residential streets. The group, ACTORS (Advocating for Container Traffic Off Residential Streets) convinced the City of Vancouver to negotiate a 'pilot project' with the port, followed by a more permanent regulatory agreement, to restrict the movement of port-destined container trucks to only those roadways that form part of the MRN.[2] Support for the containment of truck traffic within the urban core reflects the deeper set of pro-livability practices, norms, and expectations in the City of Vancouver. Still, the agreement was unusual; it targeted only port-destined trucks, and it stands in sharp contrast with the second example presented below.

Here is the leader of ACTORS speaking when the agreement on the pilot project to restrict trucks was announced:

> It took several months after the closure of the Clark Drive entrance in August 2010 for the truckers to figure out that Nanaimo Street south of Broadway was the new Clark Drive and the best route to the Port entrance off of McGill. Individual residents noticed the unusual increase in container trucks, but it wasn't until early 2011 that people found out that the Clark gate had been closed. They began writing letters, some of them to the Grandview-Woodland Area Council where I serve as a director. In early Spring, two residents came out to GWAC and outlined the problem, the next month more residents came out to express their

concerns and ask what could be done. I met with several neighbors after that meeting and a group with the acronym ACTORS (advocating for container traffic off residential streets) was formed to inform the community and work toward getting these trucks off Nanaimo Street. We held strategy sessions, one member set up a Google group, one posted videos of trucks running red lights, and we all discussed the possibilities of petitions and picketing. By July, when representatives from Engineering and Port Metro Vancouver came out to speak at the monthly GWAC meeting, they were met with about 140 very angry and vocal residents.

(Statement by E. Mosca, ACTORS)

Quite why this neighborhood group prevailed is difficult to say. The political acumen of the Grandview-Woodlands residents, as well as the knowledge of port operations of some, should not be underestimated. It also helped that 2011 was a local government election year, and clearly ACTORS found at city hall a sympathetic political and administrative leadership. However, reducing automobile dependence in all its forms is a policy stance and part of the political culture of the city, which long predates the administration of the center-green Vision Vancouver party. So while it is likely that the timing helped, this alone does not explain why the City was willing and able to negotiate a deal with the Port that would satisfy residents.

The story highlights again the importance of multiple, overlapping jurisdictions and scalar politics in shaping the way decisions about transportation routes are actually made. In this instance we have the Port, an agent of the federal government, acting unilaterally in a way that imposed negative impacts on a neighborhood. Then, in response to community pressure, it came to an agreement with the local government to restrict truck movements in a way that avoided formal review of the truck route designation by the provincial authority TransLink. In this regard, the way the port authority chose to announce the pilot program is instructive: "[I]n an effort to mitigate the impact of container trucks accessing the Port via city streets, Port Metro Vancouver will implement a Truck Traffic Pilot Program requiring container trucks to use only Major Road Network (MRN) authorized routes in Vancouver. Nanaimo Street is not a Major Road Network route" (PMV 2011). The deal affirmed that the City of Vancouver could exercise some control over truck movements, but also that it could recruit the regulatory powers of the Port to ensure that port-destined trucks followed the TransLink-designated MRN. However, as the next vignette shows, designation as a truck route can be an unwanted and unchangeable social fact for a local community.

32nd Ave., Surrey

In 2013, relatively affluent residents along a two-lane portion of 32nd Ave. between 152 and 176 Streets in suburban Surrey requested the road be removed from the municipally designated truck route, but were denied by TransLink. The Surrey City Council had designated 32nd Ave. a truck route in 1998; prior to that time, all arterial roads in Surrey were truck routes. This is not a trivial difference compared with Vancouver's urban core, since it points to Surrey's evolution as a suburban municipality comprised of several rural village nodes connected by arterial roads and increasingly filled in by residential subdivisions. The practice of designating all arterial roads in such suburban places as corridors of interplace movement—and equally, of protecting all other residential streets from through movement (for example, through a Radburn street layout)—is consistent with this evolution of the suburban truck route network.

Still, opposition to the designation also had a long history. In 1999, the year after 32nd Ave. was confirmed as a truck route, residents requested it be redesignated a residential road. They were granted a temporary ban that extended until 2004. In 2010, the City identified 32nd Ave. for possible inclusion in the MRN and began a series of construction projects to widen the avenue to the east of the contested portion. It was this expansion of the corridor that spurred residents to again request that a portion of the avenue be removed from the truck route. In 2011, the city council supported the residents and forwarded their request to TransLink for technical review.

The TransLink (2013) final review report considered public health and air quality issues, agency responsibilities, empirical and academic evidence, noise and vibrations, traffic safety, connectivity, land use, truck and traffic volumes, road engineering considerations, and potential alternatives. But powerful industry voices also informed the decision. TransLink received input from the British Columbia Trucking Association and GVGC, as well as an adjacent municipality, the Township of Langley, all against the designation change. The authors of the report also argued that changes to the routing would "reduce local emissions but overall increase GHG emissions in the region due to longer travel times. Goods movement efficiency will also be compromised, as trucks will be forced to divert to less efficient travel routes" (TransLink 2103, p. 17).

Although 32nd Ave. is not part of the MRN, the request to de-designate the road as a truck route was denied. While it is true that the city council had requested the change, in many ways TransLink was reinforcing the prior practices of the City of Surrey (in contrast to those informing the City of Vancouver's response to ACTORS). In suburban Surrey, a presumption of a separation of land uses was confirmed; in Vancouver's urban core, a messier and more contingent construction of urban space is apparent. In this way, divergent decisions about truck route designation play a central role in reflecting and reinforcing the differences between urban core and suburban places in the same region. Or in other words, "by reformulating character of place as the mode of connection among unlike elements, and tradition as the mode of perpetuating these links, we gain a way to explain how place differences develop and persist" (Molotoch et al. 2000, p. 816).

THE TRUCK ROUTES THAT ARE DRIVEN

The statement of ACTORS greeting the pilot program starts with another small but important point, namely that "(i)t took several months after the closure of the Clark Drive entrance in August 2010 for the truckers to figure out that Nanaimo Street south of Broadway was the new Clark Drive and the best route to the Port entrance off of McGill." Truck routes are not only constructed through infrastructure investments and planned routes, but also through the behavior of truckers. Their routing choices are, of course, informed by technologies such as TransLink's "Truck Route Reference Guide" (2009) and a variety of commercial, GPS-enabled routing softwares. Ultimately, however, route choice rests with individual drivers who have to learn about the routes and who also have a rationality of route choice that may not neatly fit into the models of transport analysts.

In 2011 and 2012, my research assistants and I conducted in-depth work history interviews with immigrants working in the port-logistics industry in Vancouver (see Hall et al. 2013). The nine in-depth interviews we conducted with port truckers revealed that they went through a period of tutelage as codrivers before they could be hired as employee drivers or before anyone would be willing to lend them the money to buy their own truck rigs. While this tutelage typically took place on long-haul routes (since it is here that codrivers are most valuable to the trainer), learning about ways to access delivery and pickup locations was an important part of the training process. Furthermore, the mostly Punjabi truckers share real-time information about routing choices on their CB radios in their home language, described by one driver:

> You get all the information. Every five minutes you have update of highway. Where's accident, where's traffic. Oh, it's a good part of it. You have all the update. Every five minute you have update coming, oh, where's accident, where's the traffic, where's the cop. Oh, yeah, it's good. It's very good.

Truck drivers learn about routes from each other in both formal and durable ways, and also in a more up-to-the-minute way. This learning takes place in socially and spatially situated communities of practice. Hence, in greater Vancouver it makes a difference that the port trucking industry is a niche for immigrants and nonimmigrants of South Asian origin. According to 2006 population census data, three-fifths of immigrants in the trucking sector were of South Asian origin (Hall et al. 2013), and in a survey over half (55%) of all port drayage drivers reported Punjabi as their primary language (see Davies 2013).

Not only are the routing choices of truck drivers learned through selective social interactions, they also need to be understood in the context of the social relations of their work. As noted above, new (South Fraser Perimeter

Road) and expanded (Highway 1) routes have been built under the Gateway program. These two routes connect to the east of the expanded Port Mann Bridge, and thus, in theory, trucks driving between Deltaport and destinations in the northeast (such as the Canadian Pacific intercontinental railyards) should be able to follow this route. Instead, truckers have been observed to follow the more congested routes through New Westminster, further undermining support in this municipality for the proposed NFPR. One major reason for this route choice is the fact that the Port Mann Bridge is tolled. However, most port truckers are paid by the load, and the time savings of the expanded Bridge are not enough to allow most port truckers to carry an additional load per day. Hence, the routes that have been constructed and designated may not be the ones that are driven, because they are not optimal in the situated rationality of those who do the actual driving.

TRUCK ROUTES AND THE REGULARIZATION OF URBAN UNEVENNESS

The broader context for this exploration of truck routes is a "constellation of mobility" (Cresswell 2010) that is highly favorable to the movement of goods between and through urban space. Yes, the designation of certain roadways as truck routes restricts where the majority of truck movements occur, but at the same time designation clearly signals that truck movements are highly valued. The architecture of the particular mobility constellation that supports container truck movements in and through the Vancouver metropolitan region reflects a particular spatiality of political decision-making. This starts with the founding of Canada and the British North America Act (of 1867), in which the infrastructures of international and interprovincial trade (i.e., the port) were taken out of the hands of local and provincial governments. This constitutional preference for movement over place is reinforced through a set of provincial legislation, enforced by a regional agency (TransLink) to ensure that trucks are not restricted from using roadways, and a discursive claim (consensus is too strong) that the Gateway is good for the economy and residents of the region, as well as the national economy (see Hesse, this volume).

We have also seen that the actual pattern of truck route designation and use is both contingent on social/political acceptance and somewhat resistant to change. Indeed, there are important regularities in which roads actually get to carry trucks. These regularities both exploit and reinforce existing imaginaries of the proper use of particular places, pathways, and other socially referenced spaces in the city. A dialogue between mobilities and transportation geography perspectives is productive in understanding these regularities. Some of it has to do, as transportation geographers would assert, with the structure of the space economy, which informs the structural contours of what are the most direct, reliable, and quickest routes. However, existing transportation approaches typically do not explore how truck routes are implicated in the creation of urban space, nor do they address underlying questions about how existing truck routes were created.

Drawing on mobilities perspectives, this chapter has advanced an understanding of how truck routes come to be recognized as social facts. At one level, this has to do with the landscape of investment and regulation, where previous (or sunk) infrastructural spending decisions shape subsequent decisions to permit or not permit their use by trucks. These public decisions might be viewed as rational in the sense that they are derived from understandable (if somewhat opaque) technocratic and/or political decision-making processes. However, not only are these technocratic and political decision-making processes more social than they appear at first sight, they may also be reinforced or undermined depending on how landowners respond to the land value contours that are created around routes with more or less trucks; how existing neighborhood and community organizations respond to the negative impacts of trucking, and whether or not they prevail in public discourse; how truckers learn to use or not use these routes; and indeed, how the rationality of truckers and their decision-making is structured through broader employment and other social relationships.

So we need to understand the ways that the sense of the proper use of particular roadways is repeatedly transmitted to successive road and land users, regulators, and developers: "Both within the context of big events and mundane happenings, in regard to geographic units, but also, we suspect, a broader range of settings, interactional routines ratify differentiation and carry it forward" (Molotch et al. 2000, p. 819). This is not to say that rupture is impossible, but rather to recognize what is interesting and surprising about truck routes; namely, that despite a constantly shifting and restructuring metropolitan space economy, the conduits onto which one of the most flexible elements of that economy are channeled can and do display some resilience.

NOTES

1 This research is made possible by generous funding from the Social Sciences and Humanities Research Council of Canada. Thanks to Laura Benson, Ali Farahani, Kaleigh Johnston, and Choo-Ming Yeak for superb research assistance, and to Terri Evans, Pamela Stern, and the editors for valuable comments on earlier drafts.
2 In the interests of full disclosure, ACTORS contacted me in 2011 for advice. I met with the group, explained why I thought the gate closure was important to the Port, and by implication advised them on what kind of alternative arrangement might be acceptable to the Port.

WORKS CITED

(1998) *South Coast British Columbia Transportation Authority Act, 1998*, Victoria, BC: Queen's Printer.

Allen, J., Browne, M., and Cherrett, T. (2012) "Investigating relationships between road freight transport, facility location, logistics management and urban form," *Journal of Transport Geography*, 24:45–57.
Blomley, N. (2010) *Rights of Passage: Making Sidewalks and Regulating Public Flow*, Abingdon, UK: Routledge-Glasshouse.
Boerkamps, J. H., van Binsbergen, A. J., and Bovy, P. H. (2000) "Modeling behavioral aspects of urban freight movement in supply chains," *Transportation Research Record: Journal of the Transportation Research Board*, 1725(1):17–25.
Boyce, D. (2012) "Predicting road traffic route flows uniquely for urban transportation planning," *Studies in Regional Science*, 42(1):77–91.
Cresswell, T. (2010) "Towards a politics of mobility," *Environment and Planning D*, 28(1):17–31.
Dablanc, L., and Fremont, A. (2013) "The Paris region: Operating and planning freight at multiple scales in a European city," in P. V. Hall and M. Hesse (eds) *Cities, Regions and Flows*, Abingdon, UK: Routledge, pp. 95–113.
Davies, P. (2013) *Labour Force Profile of Port Drayage Drivers in Metro Vancouver: Final Report*. Prepared by Asia Pacific Gateway Skills Table.
Deleuze, G., and Guattari, F. (1987) *A Thousand Plateaus: Capitalism and Schizophrenia*, Minneapolis: University of Minnesota Press.
Fischer, M. J., Ahanotu, D. N., and Waliszewski, J. M. (2003) "Planning truck-only lanes: Emerging lessons from the Southern California experience," *Transportation Research Record: Journal of the Transportation Research Board*, 1833(1): 73–78.
Frank, W. C., Thill, J. C., and Batta, R. (2000) "Spatial decision support system for hazardous material truck routing," *Transportation Research Part C: Emerging Technologies*, 8(1):337–359.
Hall, P. G., and Pain, K. (2006) *The Polycentric Metropolis. Learning from Mega-Regions in Europe*, London: Earthscan.
Hall, P. V. (2012) "Connecting, disconnecting and reconnecting: Port-logistics and Vancouver's Fraser River," *L'Espace géographique*, 41(3):223–235.
Hall, P. V. (2014) "Port-city governance: Vancouver case study," in Y. Alix, B. Delsalle, and C. Comtois (eds) *Port-City Governance*, Le Havre: Fondation Sefacil, pp. 209–223.
Hall, P. V., and Clark, A. (2010) "Maritime ports and the politics of reconnection," in G. Desfor, J. Laidley, Q. Stevens, and D. Schubert (eds) *Transforming Urban Waterfronts: Fixity and Flow*, Abingdon, UK: Routledge, pp. 17–53.
Hall, P. V., with Farahani, A., Johnston, K., and Yeak, C.-M. (2013) *Pathways to Immigrant Employment in the Port-Logistics Sector*, Metropolis BC, Working Paper No. 13-02.
Hall, P.V., and Hesse, M. (eds) (2013) *Cities, Regions and Flows*, Abingdon, UK: Routledge.
Houston, D., Krudysz, M., and Winer, A. (2008) "Diesel truck traffic in low-income and minority communities adjacent to ports environmental justice implications of near-roadway land use conflicts," *Transportation Research Record*, 2067: 38–46.
Kozawaa, K.H., Fruin, S.A., and Winer, A.M. (2009) "Near-road air pollution impacts of goods movement in communities adjacent to the ports of Los Angeles and Long Beach," *Atmospheric Environment*, 43(18):2960–2970.
Lena, T. S., Ochieng, V., Carter, M., Holguín-Veras, J., and Kinney, P. L. (2002) "Elemental carbon and PM (2.5) levels in an urban community heavily impacted by truck traffic," *Environmental Health Perspectives*, 110(10):1009–1015.
Marshall, A. (1890) *Principles of Economics*, London: Macmillan.
McFarlane, C. (2011) "Assemblage and critical urban praxis: Part one, assemblage and critical urbanism," *City*, 15(2):204–224.

Metro Vancouver (2011) *Metro Vancouver 2040: Shaping Our Future*, BYLAW NO. 1136, 2010 of the Greater Vancouver Regional District.

Molotch, H., Freudenburg, W., and Paulsen, K.E. (2000) "History repeats itself, but how? City character, urban tradition and the accomplishment of place," *American Sociological Review*, 65(6):791–823.

O'Connor, K. (2010) "Global city regions and the location of logistics activity," *Journal of Transport Geography*, 18(3):354–362.

Perez, L., Künzli, N., Avol, E., Hricko, A.M., Lurmann, F., Nicholas, E., . . . and McConnell, R. (2009) "Global goods movement and the local burden of childhood asthma in Southern California," *American Journal of Public Health*, 99(S3):S622–S628.

PMV (2011, July 25) *Port Announces Truck Traffic Pilot Program*, News Release of Port Metro Vancouver. Retrieved May 5, 2014 from http://www.portmetrovancouver.com/en/about/news/2011/07/25/Port_Announces_Truck_Traffic_Pilot_Program.aspx.

Rodrigue, J.-P., Comtois, C., and Slack, B. (2009) *The Geography of Transport Systems*, Abingdon, UK: Routledge.

Ronen, D. (1988) "Perspectives on practical aspects of truck routing and scheduling," *European Journal of Operational Research*, 35(2):137–145.

Saccomanno, F. F., and Chan, A. W. (1985) *Economic Evaluation of Routing Strategies for Hazardous Road Shipments*, Transportation Research Board, No. 1020.

TransLink (2009) *Keeping Metro Vancouver Moving: Truck Route Reference Guide November 2009*, Vancouver: TransLink.

TransLink (2013) *Technical Review of 32 Avenue Truck Route in Surrey B.C. Final Report*, Vancouver: TransLink.

TransLink (2014) *Applied Freight Research Initiative Metro Vancouver Region: Freight Market Sectors Summary Report, January 2014*, Vancouver: TransLink.

Truck Routing, New York City Department of Transportation. (2014) Retrieved April 30, 2014 from http://www.nyc.gov/html/dot/html/motorist/truckrouting.shtml.

Urry, J. (2007) *Mobilities*, Cambridge: Polity.

Van de Riet, O., de Jong, G., and Walker, W. (2008) "Drivers of freight transport demand and their policy implications," in A. Perrels, V. Himanen, and M. Lee-Gosselin (eds) *Building Blocks for Sustainable Transport: Obstacles, Trends, Solutions*, Bingley, UK: Emerald, pp. 73–102.

Vernon, R. (1960) *Metropolis 1985*, New York: Doubleday.

Woudsma, C., Jensen, J. F., Kanaroglou, P., and Maoh, H. (2008) "Logistics land use and the city: A spatial-temporal modeling approach," *Transportation Research Part E: Logistics and Transportation Review*, 44(2):277–297.

7 Uncanny Trains
Cities, Suburbs, and the Appropriate Place and Use of Transportation Infrastructure
Julie Cidell

INTRODUCTION

> The dwelling places of modernity embody the material connections that make the social construction of bodies possible, by first materially constructing "others," in the form of natural or social processes, and then keeping them outside.
>
> (Kaïka 2004, p. 272)

Within the last three decades, North American railroads have gone from being on the verge of obsolescence to being at the center of controversy. After regulatory reform in 1976 and 1980 reduced government controls on pricing and allowed railroads to both merge and abandon rail lines more easily, the railroads slowly regained their competitiveness against trucking. They increased revenue while lowering prices and abandoning over 100,000 miles of track, leaving approximately 140,000 miles in place (Slack 2013; compare to 47,000 miles of the Interstate Highway System). Passenger rail continues to limp along in the form of Amtrak and metropolitan commuter rail (Minn 2013), but long distance freight rail has become an integral and growing component of logistics networks across the continent. Bulk goods such as wheat and chemicals, as well as auto carriers and intermodal containers, regularly travel on networks that were built in the 1800s, not only through rural areas but also alongside Main Street and through the heart of the largest cities. This juxtaposition of century-and-a-half-old infrastructure and modern urban and suburban environments, concentrated along many fewer miles of track than before, is increasingly leading to conflict. The increasing volume of train traffic through urban areas has meant longer and more frequent delays at grade crossings, more noise and vibration for nearby residents, and increasing opposition to expanded freight rail operations. Perhaps nowhere is this truer than Chicago, whose centrality within North American rail and global distribution networks has made it one of the most congested in the country. Efforts to restructure train traffic through Chicago can thus entail fierce politics of mobility and place.

Such opposition is not necessarily limited to concerns about the trains themselves or the goods they carry, even if their mobile nature already sets them apart from traditional understandings of risk (Cidell 2012a, 2012b). Rather, this opposition can include a fear of something more sinister than the inconvenience of waiting for a train to cross the road, a fear I interpret through the theoretical framework of the *uncanny*. Within urban geography, the concept of the uncanny has been most thoroughly explored by Maria Kaïka and Erik Swyngedouw, who draw on Freud to consider how the familiar or routine conceals the threatening, and how the threatening pops back up no matter how we try to suppress it or cover it over: "[w]hen the predictable nature of the familiar acts in unpredictable ways" (Kaïka 2004, p. 277). In an urban context, this means that the infrastructure that enables our daily lives and routines, and selectively lets some people and things flow while keeping others at bay, can become a threat when those flows (and the routines and places they help construct) are disrupted. Disruptions are undesirable not only in and of themselves, but because their surprise appearance points out inequities in access built into the system itself. The framework of the uncanny therefore becomes a way to understand how everyday, routine mobilities are interwoven with deeper meanings—including fear—in their construction and maintenance of urban and suburban places through inclusion and exclusion.

This fear of disruption and the uncanny can be seen through the recent suburban opposition to a railroad acquisition in the Chicago suburbs. In 2007, Canadian National (CN) began the approval process for the purchase of the Elgin, Joliet, & Eastern (EJ&E), a beltline railroad about 40 miles out from the center of Chicago (Figure 7.1), in order to reroute intracontinental train traffic from lines through the center of Chicago to this less congested bypass route. Although railroads are privately owned in the United States and not subject to environmental review regarding changes in their day-to-day operations, mergers must be approved by the Surface Transportation Board (STB). As part of the approval process, the STB required a full Environmental Impact Statement (EIS) to examine the implications of increasing traffic on the EJ&E line (from about five trains a day to about 30trains a day). As a public process, the EIS provided a window into the politics of mobility and place surrounding urban rail traffic as well as the fears that can surround such flows. The draft EIS met with fierce resistance from many of the communities located along the EJ&E, citing predictable concerns over increased traffic, blocked road crossings, and greater risk of hazardous material spills. But many speakers at the public meetings on the draft EIS also made somewhat different arguments, suggesting that the current rail network should remain unchanged because urban neighborhoods have 'adapted' to the presence of train traffic. These were curious statements to hear, particularly since many of these speakers were from fast-growing suburbs which are currently 'adapting' to other major changes like an influx of residents from closer-in suburbs or the central city, an influx accepted

136 *Julie Cidell*

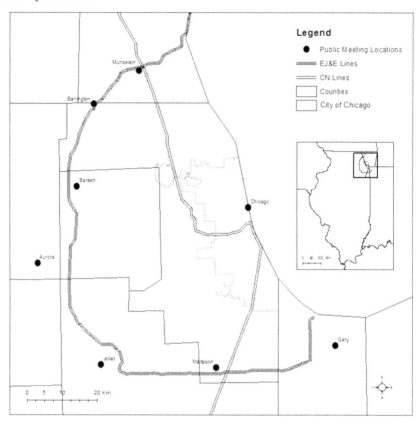

Figure 7.1 Map of the Chicago region, including CN and EJ & E lines.
Source: Author.

more unproblematically. Apparently, there was something different about trains and the disruptions they would cause, as compared to additional cars on the road, that led residents to particularly vociferous objections.

In this chapter, I argue that, for many suburban residents, transferring train traffic from urban to suburban lines represented something more than concern over additional traffic, something more deeply connected to the social and environmental inequalities embedded in the landscape, inequalities that they perhaps had moved to the suburbs in order to escape. In short, the politics of opposition became a politics of the uncanny, driven by fears of the familiar becoming unfamiliar, the predictable unpredictable, and that inequality would continue to haunt them through everyday activities like waiting for a train to cross the tracks. Through their opposition to the CN/EJ&E acquisition, suburban residents revealed particular constructions of the urban and the suburban through the flows and disruptions, or mobilities and immobilities, found in each type of place. To better understand those constructions—and the politics of fear they represent—the following section addresses four elements of the uncanny in more detail: the naturalization of the urban, the exclusion of the other, selective porosity, and disrupted rhythms. This chapter then sets out the details of the CN/EJ&E acquisition as one case study of how flows in the form of people and goods both structure and threaten urban and suburban environments. The conclusion considers how such conflicts about rail restructuring are distinct from other controversies over changing suburban land uses, and, as Paul Robbins reminds us, why looking at people's worries and anxieties is a useful starting point for understanding the politics of mobility in urban places of flow.

THE UNCANNY IN URBAN POLITICAL ECOLOGY

Infrastructure such as utilities and transportation corridors are fundamental to urban life, and yet they largely remain hidden, by design as much as by desire. In their discussions of urban infrastructure, Maria Kaïka and Erik Swyngedouw consider the extent to which incorporation of utility networks into the landscape hides not only the physical manifestation of the flows along those networks, but also the production process behind those flows:

> [A]lthough the urban is part and parcel of our everyday experience, the human labour and social power relations involved in the process of its production are forgotten. The production of the urban remains, therefore, unquestioned and the urban becomes "naturalized," as if it had always been there on the one hand, and as distinct and separate from nature on the other.
> (Kaïka and Swyngedouw 2000, p. 123)

They trace the example of the incorporation of water networks into the home, a history in which public spectacles meant to celebrate modernity and progress became taken-for-granted pieces of infrastructure meant to be hidden away (in part because the vision of modernity for all was impossible to achieve). They also argue that this veiling of such processes that go into commodifying nature and bringing it to our doorsteps is necessary to construct the home as a place of safety and security: "Thus, excluding socio-natural processes as 'the other' becomes a *prerequisite* for the construction of the familiar space of the home. . . . The inside becomes safe, familiar and independent not only by excluding rain, cold and pollution, but also through keeping fear, anxiety, social upheaval and inequality outside" (Kaïka 2004, p. 272, italics in the original). The *naturalized urban* that is excluded from the home through utility networks includes both the negative effects of the urban environment and the parallel social inequities.

Of course, complete separation of home and nature or home and city is not desirable: Commuters need to get to work, water needs to get into and out of the house, suburbs need to be leafy and green, etc. "By keeping outside the undesired . . . natural and social 'things' and processes, and by welcoming inside the desirable ones (filtered, produced and commodified), the modern home has acquired a *selective porosity* which is enabled by a set of invisible social and material connections" (Kaïka 2004, p. 275, italics in the original). Sociotechnical networks have to simultaneously keep in the good and exclude the bad, which means they also have to be able to parse out what is good and what is bad. Water within the home becomes desirable and undesirable, depending on where it is, how much of it there is, and what it is carrying. In transportation, *selective porosity* can be seen in the ways access is necessary to get to work, school, shopping mall, etc., but too much access for too many people, or the wrong kind of people, can be a problem.

It takes considerable work to maintain both these networks and the security they provide, work that cannot always be achieved. Inevitably, networks fail or are disrupted due to natural disaster, worn-out materials, human error, or deliberate actions. If the security of the home depends on the hidden yet functional state of utility networks, the disruption of those networks not only means a break in service but a crack in the façade of security. This results in an eruption of what Freud called the 'uncanny,' or "when the predictable nature of the familiar acts in unpredictable ways" (Kaïka 2004, p. 277). The anxiety brought about by the uncanny comes from not only the disruption itself—the failure of electricity, the leak in the roof—but also from the reminder of how fragile such networks are, and how difficult it is to keep them functioning. If security comes from not having to worry about how and where our networks are constructed, and what they connect and carry, the failure of those networks is a reminder of all of the people and places that we might choose to forget. Being able to *exclude the other* is a feature of sociotechnical networks, either deliberately or as a byproduct, and disruption threatens that exclusion.

The domestic is not the only sphere where the disruption of networks can bring forth anxiety, however. Gibas (2012) considers this same question at a broader scale in his study of the Prague metro, arguing that since the experience of the metro is defined as motion in a regular rhythm, any delay or disruption to mobility becomes a source of anxiety: "As for the metro, the rhythm—or better, the polyrhythmia of the everyday—gives birth to a preoccupation with the futile effort to perfect the rhythm, to prevent disruptions. . . . It gives birth to the fear of disturbance symbolically embodied in the struggle to purify the metro space" (Gibas 2012, p. 496). Gibas argues that that security and normality on the metro, in contrast to the home, consist of being in motion, enabling people to keep their regular routines of travel intact. *Disrupted rhythms* are therefore a disruption of security and space, even outside the home.

These four elements of the uncanny in urban sociotechnical networks—disrupted rhythms, excluding the other, selective porosity, and naturalizing the urban—enable us to consider how various spaces of routine and safety can be threatened through uncanny disruptions, not only of the house or metro, but also of the suburb. There is a long history of writing about the suburbs as non–city space, which, like Kaïka and Swyngedouw's commodified nature, require the city as the *other* in order to exist. However, most of this work neglects the suburbs and the city as spaces of flow, whether the daily travel of commuters or schoolchildren, the provision of goods to retail establishments, or longer distance flows that happen to cross suburban borders. When looking at the scale of the metropolitan area, we are reminded that "spatial claims made by the private sphere (domestic or other) are always translated into the deprivation of the public sphere from these same spaces and the reduction of spaces of the margin" (Kaïka 2004, p. 273). In order to claim space for private yards, public parkland was reduced. In order to enable swift commutes into the city, urban neighborhoods were bulldozed for interstate highways. In order to enable the easy travel of automobiles, pedestrian spaces were reduced or eliminated. Any action that might threaten the rhythm of automobile-based life in the suburbs must therefore be fought, not only on its own terms, but also as a reminder of the uncanny underside of the suburbs: People can only enjoy the lifestyle they have because others are excluded from it.

Past iterations of this process have come in the form of street widths and speed limits meant to discourage pedestrians, or zoning and lot sizes meant to exclude affordable housing. However, the shift of freight traffic from a central city railroad to a bypass line can also be seen as an infiltration of the suburbs by the urban. This new traffic therefore becomes something to be fought against, not only for its own sake, but also because of the threat it poses to daily routines and the safe spaces of the suburbs, as the following case study shows.

THE RAILROAD ACQUISITION PROCESS

As noted in the introduction, the rapid growth in freight rail traffic within North America is occurring on a greatly reduced network (compared to the mid-20th century) closely interwoven with urban spaces. The main reason CN wanted to purchase the EJ&E as a bypass was the rapid growth in global networks of container traffic and the resulting congestion at key urban chokepoints like Chicago. The shift to shipping goods in a standardized metal container from origin to destination helped lower labor costs and theft, dramatically lowering transport costs and allowing companies to change production locations to where costs of labor or environmental regulations were less (Levinson 2006). This prompted railroads to refine their network for global cargo transport, and CN has perhaps the most experience among the seven Class I railroads in North America (those with revenues of over $250 million per year). A series of mergers in the 1980s and 1990s, including purchase of the Illinois Central in 1999, gave CN not only a line stretching from the Pacific to the Atlantic, but also from the Great Lakes to the Gulf of Mexico (Madar 2002), connecting through the very center of Chicago.

Chicago's centrality within both the historical North American rail network and the new global distribution networks has made it one of the most congested in the country. The city is one of only two to host six of the seven Class I railroads (Kansas City is the other). As a result, almost all traffic from West Coast ports to Midwest distribution centers must pass through Chicago at some point. With land costs and environmental impacts limiting the construction of new lines, railroads must make better use of existing infrastructure to reduce congestion. The Chicago region already had a complete beltline railroad about 40 miles out from the city center (Figure 7.1). Known as the Elgin, Joliet, and Eastern, the EJ&E, or the 'J,' this line has connected up to 30 different railroads over its history, providing an important bypass function from its inception (Jaenicke and Eisenbrandt 2007). Although it did carry passengers in the early 20th century, it has always been, above all, a freight railroad, with about five trains per day in the early 2000s.

In the fall of 2007, CN applied to purchase the EJ&E in order to transfer its transcontinental, intermodal container traffic from the congested center of Chicago to this bypass. Under U.S. regulations, mergers and acquisitions have to be approved by the Surface Transportation Board in order to ensure competitiveness. The STB does not usually require an environmental review, as they consider economic issues above all. In fact, were CN simply shifting rail traffic from one line that it already owned to another, no approvals would be required, as it would be a private company's internal decision. However, in this case, the STB took the unusual step of requiring an Environmental Impact Statement as part of their acquisition review process, given the potential impacts to suburban communities. This case therefore offers a rare opportunity to examine public response to proposed changes in rail traffic through a major metropolitan area via comments submitted during the environmental review process.

The main impact of the acquisition would be to shift traffic from CN's radial lines into and out of downtown Chicago to the suburban beltline, increasing traffic on the beltline from about five trains a day to 25–40 trains a day. The draft environmental impact statement (DEIS) noted that urban neighborhoods and municipalities would see environmental benefits as traffic on the urban lines decreased. That traffic would be shifted, however, to the 36 suburbs along the 'J' that are more affluent and whiter than the inner suburbs and city neighborhoods traversed by CN's urban lines. The DEIS therefore also noted that existing environmental injustice would be alleviated by the shifting of traffic to suburban areas. But such shifts also played a role in how opponents of the acquisition framed their arguments. As part of the DEIS progress, public comments were solicited in response to the draft, through public meetings—in written or oral form—as well as by phone, email, or regular mail. The response was overwhelming. STB received approximately 9,500 comments on the DEIS in written or oral form, identifying over 55,000 individual issues (STB 2008). The analysis in this paper is taken from both the comments submitted to the STB and the transcripts of the public meetings, all of which are available on the STB's website, as well as the record of final decision issued by the STB. In the following section, I analyze these comments in relation to the four elements of the uncanny identified above—disrupted rhythms, selective porosity, excluding the other, and naturalizing the urban—to explore how fears over the proposed transaction can reflect deeper concerns about how flows of people and freight construct urban and suburban places.

UNCANNY TRAINS IN THE BEAUTIFUL SUBURBS

Disrupted Rhythms

One way in which the uncanny can be manifested in the urban environment is through disruption in the regular rhythms of life (Gibas 2012). Disruption was indeed of major concern to CN opponents. According to the STB's summary of the comments received on the DEIS, the most frequently raised issue was traffic delays and congestion, mentioned by nearly 5,500 of the 9,500 comments (STB 2008, p. 3.2-4). In particular, the increase in both frequency and length of trains would lead to longer periods of time with the gates down at at-grade crossings (Figure 7.2). People described how their daily routines involved crossing the tracks multiple times per day, expressing concern over the uncertainty of whether or not that crossing would be

Figure 7.2 At-grade crossing of EJ & E tracks at Old McHenry Road, Hawthorn Woods, IL.

Source: Author.

blocked by a CN train; in other words, "the predictable nature of the familiar act[ing] in unpredictable ways" (Kaïka 2004, p. 277). Representatives from different school districts spoke of the delays that buses would face, thus potentially disrupting the school day:

> Our bus transportation complex is located in the center of this village to provide the most efficient and economical transportation routes. Since the EJ&E lines cut through the heart of the district, we are also right on those train lines. Our busses cross EJ&E lines over 840 times per day. An additional 5.3 minute delay every time a train comes through this area will have a ripple effect, resulting in delays to the start of school and added travel times.
>
> (Superintendent, Barrington School District)

For many suburban residents, daily scheduling of themselves and their children is tightly choreographed, involving multiple car trips and carefully planned schedules with little room to spare. Having to wait for a train or drive to the nearest separated crossing would disrupt that schedule and throw the day into chaos. For school districts, scheduling buses can be difficult enough without having to factor in trains blocking critical crossing points, particularly as state funding for school bus transport has been reduced over the past decade.

The second most common issue for opponents was increased emergency response times, found in roughly 5,000 comments. Multiple police and fire representatives spoke at each public meeting about their concerns over having crossings blocked, some sharing personal experiences about having to wait for trains while trying to get to a fire or accident. Here, the disruption was emphasized to be not merely an inconvenience, but a matter of life and death. Relatively small suburbs were likely to have all of their emergency services on one side of the train tracks, putting certain neighborhoods at particular risk.

> The police department fears that with the additional trains, we'll have difficulty getting to where we are needed. When we're responding to nonemergency calls, waiting on trains is frustrating. However, when we have to get to an emergency call, it's much more than an inconvenience. It's life threatening.
>
> (Chief of Police, Aurora, IL)

> Like thousands of Bartlett residents, I live west of the EJ&E tracks. Unfortunately, all of the service that my neighbors and I potentially need during an emergency situation are located east of the tracks, including a new fire station, the two existing fire stations, the police station, and the nearest hospitals.
>
> (Resident, Bartlett, IL)

Here, fear of disruption tapped into deeper concerns about not having access to emergency services. Commenters spoke of senior centers, schools, and neighborhoods that would be temporarily inaccessible were a train to go by at the wrong time. The village of Barrington even made a video that counted down the seconds that the average train spent blocking a crossing, relating it to the urgency of getting help to a heart attack or stroke victim as soon as possible. The video was posted on an opposition website and shown at public rallies held by acquisition opponents. In response, in the final EIS the STB required that certain key intersections have cameras installed so that emergency dispatchers would know if a train was blocking the crossing and could reroute first responders accordingly, thus reducing the effects of this type of disruption.

Selective Porosity

Selective porosity is the capability of sociotechnical networks to allow some kinds of flows into the home but not others (Kaïka 2004). For many members of the public, this was what the STB's decision-making process should produce: allowing certain flows of people and goods to continue through their towns (namely, themselves on their way to work, school, or shopping) while keeping other flows out (freight traffic, residents of nearby towns taking shortcuts). For example, automobiles were considered a necessary part of life when they were in motion, taking children to school and commuters to work, but when stopped at crossings waiting for trains, they were demonized as contributing to both exhaust and carbon emissions. Even more pointedly, the contrast between commuter and freight trains was sharply drawn based on their function—not their physical presence:

> We appreciate the role that a railroad plays in the development of Barrington. Without the Northwestern, now Metra, we would not be among the most desired communities in which to live. However, the CN acquisition would overwhelm our community with increased freight traffic, a condition that if preexisting would have prevented the community from attaining the quality of life we now enjoy, one that is reflected in our home values and the tax base to support our excellent school system. That is the reality we must give priority to.
> (Director, Barrington Area Development Council)

The 'reality' being referred to here is that, for decades, the porosity of the transportation networks has allowed commuters to travel to downtown Chicago without having to drive, while keeping noisier and more hazardous freight traffic near the people who cannot afford to live in Barrington and its equally well-to-do neighbors (three municipalities neighboring Barrington are in the top 10 in the state in terms of per capita income, and Barrington itself is 39th of 1300). This quotation makes explicit the connection between the quality of life in the suburbs and the absence of freight trains, framed not through quiet and safety but through high property values. Barrington (and similar cities along the EJ&E line) would not be as large and prosperous as they are if they did not have rail access that enabled their residents to get to high-paying jobs in the central city. Yet, if residents had any complaints about having to wait for commuter trains at grade crossings, they did not voice them here. However, freight traffic had to be kept out of their backyards because, as quoted above, it would be coming in "overwhelm[ing]" volumes, different in character as much as in quantity, potentially harming that quality of life.

Excluding the Other

Beyond selective porosity, urban infrastructure also enables some people to completely keep out that which doesn't 'belong' as part of a safe home. There is a long history in the U.S. of suburban exclusion by race and by class, using tools such as redlining, steering, large lot sizes, and even outright violence to keep the 'wrong people' from moving out of the city. Transportation infrastructure itself can be part of the process of exclusion; the 'wrong side of the tracks' is a well-known euphemism for a disadvantaged neighborhood. Kaïka argued that this exclusion of the other is *necessary* in order to define the home (and by extension the suburb) as safe, in material terms of keeping particular bodies or objects out of sight as well as repressing the fear and anxiety associated with the 'other.' In the case of the EJ&E, that fear was on display among those who saw increased train traffic as posing a threat to home and community as place of safety and security. Many commenters, not only in the official record but also at public rallies held to increase awareness, spoke of how they had chosen to move to a particular place because of its peace, quiet, calmness, or other expressions of non-urbanness. For these people, the train would be a crack in the suburban façade, bringing with it the city and related fears:

> The people that choose to make Wayne their home do so because of the uniqueness and the character and the peace and tranquility; an oasis, if you will, in the center of urbanity.... [This transaction] will destroy the quality of life the families that have made Wayne home came out here looking for.
>
> (Mayor, Wayne, IL)

> I, along with many others, moved to this community and made time to shepherd its development for the same reason, that it is unique in the Chicago area.... Many consider it a kind of sanctuary to which residents can retreat after a day of heavy traffic, noise, and hassle experienced elsewhere.... [With transaction approval] increases in noise,

traffic, pollution, danger, and frustration would be our new normal. That's not what we came here for.

(Planning Commission Chair, Hawthorn Woods, IL)

The security and desirability of these communities is described in explicit contrast to *other* places outside the municipal boundaries: 'urbanity' that is 'elsewhere,' and which these people presumably moved from in order to escape. However, as with the commuters who need the train to get to work but reject the train carrying freight, these people are not denying their reliance on the 'other.' They need the city for employment, shopping, recreation, and other activities that are not possible in their 'sanctuary.' However, Kaïka would remind us that this is not the only way in which the 'other' is being excluded. Anxiety over social inequality is present as well, since both speakers are well aware that the only reason their municipality's residents can have a more rural lifestyle is by excluding others—placing them on the 'wrong' side of the tracks or, in this case, on the wrong set of tracks. The noise and emissions of CN trains would be a reminder that they are still part of an urban area, with all the economic and social inequality that entails, as much as they have tried to avoid it.[1] Moreover, a decline in property values might make it possible for previously excluded people to afford the same sanctuary, reducing its exclusivity and desirability.

Naturalizing the Urban

As utility networks become part of the landscape, they are naturalized or taken for granted (Kaïka and Swyngedouw 2000). Here, the connection between 'urbanity' and transportation infrastructure was explicitly developed and used by acquisition opponents to delineate the places where freight trains were and weren't appropriate. According to this argument, cities are where noisy, disruptive, hazardous trains have always been and should always be. In contrast, the suburbs are for peace and quiet and are not equipped to handle international flows of freight. This argument was made through reference to the material presence of infrastructure and the consequences of having it within the landscape:

> [If] you take a good look at the satellite photos for those tracks in Chicago, the vast majority of those tracks are covered by infrastructure that makes it able for the people to live in those areas to be able to handle it, infrastructure that we don't have, infrastructure in the form of overpasses and underpasses.
>
> (Alderman, Aurora, IL)

Remember that the communities Canadian National currently runs trains through have had many decades in which to adjust themselves to their current level of rail traffic. Houses, schools, and businesses have been located with an eye to the railroad tracks. Underpasses and overpasses have been planned and gradually constructed.

(Resident, Mundelein, IL)

People who currently live in the area today where the trains are running have dealt with this issue for many years. Their property values reflect already the high volume of train traffic. Who is going to pay our homeowners because of their home values being reduced because of these trains coming through?

(Village Trustee, Frankfort, IL)

[CN], I understand, has argued that the EJ&E has been there for a hundred years, so people get over it. You've had a railroad there, you knew that when you moved in. Well, we knew the character of the railroad and the way we look at it, the character has been sustained and has set a precedent for the last hundred years. Because of that precedent, because of the character of the railroad, communities have grown up around the railroad that are consistent with that character. It's too late to change the character of the railroad now.

(Resident, Frankfort, IL)

There are two components to this naturalization of the urban. First, 'the urban' supposedly has infrastructure that 'the suburban' does not, namely separated grade crossings that allow flows of automobile traffic to continue unhindered no matter how many trains go by. While Chicago did require all railroads to separate grade crossings in the 1800s, that practice did not continue in the adjacent suburbs.[2] The second quote at least acknowledges that the process of separating grades in the inner suburbs took time, but the first quote is representative of dozens who said the areas where CN trains currently run *are* physically adapted to train traffic via overpasses and underpasses. However, the draft EIS notes that there are 99 grade crossings listed along the EJ&E tracks, but 98 along the current CN line. In other words, there is no difference in terms of how many railroad intersections would be affected were train traffic to be shifted from one line to the other. Since the existing lines are within denser urban areas, it is likely that there are more cars waiting at those intersections than at the corresponding grade crossings in the outer suburbs, further casting doubt on opponents' claims.

The second point has to do with the supposed naturalization of other urban processes around transportation infrastructure. Opponents such as the Frankfort resident quoted above claim that outer suburban communities have grown up around the EJ&E tracks with their five trains a day and therefore should not be subject to more. However, this ignores the fact that residents of these same communities have themselves changed the character of the 'quiet' and 'rural' locations by moving there. The village of Frankfort, for example, has more than doubled its population in only 20 years, from less than 8,000 in 1991 to just over 18,000 in 2011. The increase in population has surely resulted in a similar increase in automobile traffic, and yet there was little awareness expressed by speakers that they might themselves have contributed to traffic congestion. Instead, they argued that inner suburbs have 'adjusted' to 30 or more trains a day by building appropriately. This 'adjustment' includes lower property values because of the undesirability of train traffic, something that acquisition opponents fear would be a consequence of CN acquiring the EJ&E. In other words, they argued for maintaining the status quo with regards to the presence of rail traffic and the uncanny reminder it poses of the urban landscapes they are trying to escape, while neglecting their own contributions to the increasing density of the suburban fringe. In fact, the STB's final report noted that the majority of delays at suburban railroad crossings were due to inadequate provision of road capacity in the face of rapid population growth, *not* increases in freight traffic, and therefore CN should only be financially responsible for a portion of a few new grade separations.[3]

CONCLUSIONS

It is not surprising that an individual or family who moved to the outer suburbs to avoid the congestion and hassles of 'the city' might be opposed to a rail transaction that would increase freight train traffic through their neighborhood. The increased noise, delay at road crossings, and risk of derailment associated with freight trains can be undesirable aspects in and of themselves. But their presence also serves as an uneasy reminder of the urban in what is supposed to be a tranquil, suburban environment. Disruptions in daily activities break down the barrier between the safety of the suburbs and the danger of the city, as the uncanny manifests itself through infrastructure that does not remain hidden the way it properly should. Assumptions about the separation between city and suburb are thus fundamental to such an understanding of the appropriate place and use of transportation infrastructure:

> Chicago loses and we lose if people cannot get to work in Chicago. And yes, we live in a global economy, but we do not have to destroy ourselves and get down to this potential level of pollution and probable hazardous material spills in our beautiful suburbs to even out the pain of the beautiful big cities that are train transportation hubs.
> (Resident, Barrington, IL)

The conflict over the CN/EJ&E acquisition therefore demonstrates the fundamental message of this book: the extent to which urban and suburban places are constructed through flows. In fact, the very distinction between city and suburb in this case is based on the presence and use of transportation infrastructure. Acquisition opponents portrayed 'the city' or 'the urban' as the appropriate place for freight trains, in part because of the supposed presence of overpasses and underpasses to enable automobile-based flows to continue, but also because the suburbs were supposed to be quiet, exclusive, free flowing, and otherwise non-urban. In order to make this claim, they had to ignore their own contributions to increasing suburban congestion as a result of moving to those locations and driving on a daily basis. They also had to ignore the fact that many suburbs would benefit from the transaction, instead categorizing everything inside the EJ&E beltline as *urban*. Accepting the presence of freight trains in their home community would mean that the suburban-urban line would have blurred, and with it their security and safety. For acquisition opponents, the daily routine of commuting, going to school, and enjoying a high quality of life was at stake. Importantly, this routine is based on the presence of rail and road infrastructure and suburbanites' ability to use that infrastructure for their own unrestricted mobility. Many of these residents had moved to this part of the region to reduce their exposure to congested roadways, so the potential for longer delays at grade crossings might be a reminder that they hadn't actually managed to escape that congestion. Allowing a significant increase in intersecting freight traffic would do more than cause a few minutes of delay: It would threaten the foundation of the suburban lifestyle, namely unfettered travel by automobile.

This chapter has avoided the language of NIMBY (not in my backyard), not because it is irrelevant—many opponents to the acquisition of the EJ&E by CN literally spoke about their backyards adjoining the rail line—but because the mobile nature of the threat complicates the picture. In contrast to the siting of a nuclear power plant, landfill, factory, or other undesirable land use that stays in one place, the proposed increase in train traffic poses a more complex sort of risk: Sometimes it is there, and sometimes it isn't (Cidell 2012b). This is what makes the concept of the uncanny valuable for understanding this transportation controversy and others like it. The 'sometimes there/sometimes not' nature of the risk posed by trains is embodied in the uncertainty over disruption of daily routines that might occur through waiting for a train to clear the crossing—disruption that can bring to mind other undesirable features of 'the urban' that may encroach on previously safe suburbs. The uncanny also offers us a way to understand how the daily routines of driving to school, work, and shopping are connected to global flows of freight, through threatened disruption that blurs the carefully drawn and reinforced distinction between city and suburb.

Nevertheless, a just city would confront the uncanny, not push it aside:

> Demonstrating the ideological construction of private spaces as autonomous and disconnected and insisting on their material and social connections calls for an end to individualization, fragmentation and

disconnectedness that are looked for within the bliss of one's home. It calls for engaging in political and social action, which is, almost invariably, decidedly public.

(Kaïka 2004, p. 283)

Although we have far to go in meeting that call for a more just city, this chapter contributes by exploring the charged connections between public and private city and suburb, as seen through a conflict over the suburbanization of freight rail traffic. I do not mean to brush aside opponents' concerns as completely unjustified: Access to emergency services and the threat of hazardous material spills in areas that rely on local aquifers for drinking water are very real concerns (albeit ones that inner suburban and city residents also face). In fact, as geographer Paul Robbins reminds us, anxiety and concern is often the place where we should start when we think about human-environment relationships and their consequences (Robbins 2007; Robbins and Moore 2013). This is why understanding the ways in which transportation networks and flows of materials along those networks construct and threaten the places they travel through is vital. Rail freight traffic continues to grow in volume across the continent, especially with regard to hazardous materials such as crude oil. Being able to parse out Kaïka's 'ideological construction of private spaces' and the anxiety it engenders from the real threats to lives and livelihoods that inadequately regulated transportation networks might pose is necessary to be able to justly resolve future conflicts over uncanny infrastructure.

NOTES

1 As a nine-year-old from Barrington wrote in an official comment submitted at a public meeting, "my mom tells me [Chicago Mayor] Richard Daley is going to move the getto [sic] to the suburbs," a reminder that terms like 'hassle,' 'danger,' and 'urbanity' can serve as code words used by adults to avoid sounding openly racist.
2 Ironically, the Chicago Region Environmental and Transportation Efficiency Program (CREATE), a public-private partnership to reduce rail and road congestion in the Chicago area that was cited by many opponents as the answer to the problem of CN's congestion instead of shifting traffic to the suburbs, includes the conversion of 27 existing grade crossings within the inner suburbs to overpasses or underpasses.
3 This critique of the suburban opposition to the EJ&E acquisition should be balanced against some questionable decisions and practices of both CN and the regulatory authority. The STB approved the acquisition of the EJ&E by CN on December 24, 2008—arguably the day of the year on which opponents would least be able to mobilize a response. All the STB required of CN was that it pay the majority of the costs of two grade crossings and the installation of monitoring equipment at a few more intersections to enable dispatchers to reroute first responders, and that it report on the number of minutes their trains blocked grade crossings along the EJ&E for five years. Approximately

a dozen of the municipalities located along the EJ&E made their own private mitigation agreements with CN. After only a year, however, automatic monitoring of the crossings demonstrated that CN was underreporting the number of incidents where a crossing was blocked for more than 10 minutes by a staggering factor of 100, leading to the first ever fine issued by the STB (Eldeib 2010). CN claimed they had reported a mere 14 incidents because they thought they were only required to report when *stopped* trains blocked crossings, whereas the automatic monitors' observations of 1,457 incidents were based on *slow-moving* trains.

WORKS CITED

Cidell, J. (2012a) "Fear of a foreign railroad: Transnationalism, trainspace, and (im)mobility in the Chicago suburbs," *Transactions of the Institute of British Geographers*, 37:593–608.
Cidell, J. (2012b) "Just passing through: The risky mobilities of hazardous materials transport," *Social Geography*, 712:13–22.
Eldeib, D. (2010, December 21) "Canadian National fined $250,000: Regulators say railway underreported Chicago-area crossing delays," *Chicago Tribune*.
Gibas, P. (2012) "Uncanny underground: Absences, ghosts and the rhythmed everyday of the Prague metro," *Cultural Geographies*, 20:485–500.
Jaenicke, P., and Eisenbrandt, R. (2007) *Elgin, Joliet and Eastern Railway*, Charleston, SC: Arcadia Publishing.
Kaïka, M. (2004) "Interrogating the geographies of the familiar: Domesticating nature and constructing the autonomy of the modern home," *International Journal of Urban and Regional Research*, 28:265–286.
Kaïka, M., and Swyngedouw, E. (2000) "Fetishizing the modern city: The phantasmagoria of urban technological networks," *International Journal of Urban and Regional Research*, 24:120–138.
Levinson, M. (2006) *The Box: How the Shipping Container Made the World Smaller and the World Economy Bigger*, Princeton, NJ: Princeton University Press.
Madar, D. (2002) "Rail mergers, trade, and federal regulation in the United States and Canada," *Publius*, 32:143–159.
Minn, M. (2013) "The political economy of high speed rail in the United States," *Mobilities*, 8:185–200.
Robbins, P. (2007) *Lawn People: How Grasses, Weeds, and Chemicals Make Us Who We Are*, Philadelphia, PA: Temple University Press.
Robbins, P., and Moore, S. (2013) "Ecological anxiety disorder: Diagnosing the politics of the Anthropocene," *Cultural Geographies*, 20:3–19.
Slack, B. (2013) "Rail deregulation in the United States," in J.-P. Rodrigue *The Geography of Transport Systems* (3rd ed.). Retrieved May 8, 2014 from https://people.hofstra.edu/geotrans/eng/ch9en/appl9en/ch9a1en.html.
Surface Transportation Board (STB) (2008, December 24) Canadian National Railway Company and Grand Truck Corporation—Control—EJ&E West Company. STB Finance Docket No. 35087. Decision No. 16.

Part III
Networks
Cities and Regions in Wider Context

8 Place-Making, Mobility, and Identity

The Politics and Poetics of Urban Mass Transit Systems in Taiwan

Anru Lee

This chapter argues that a mass rapid transit system—as a component of urban infrastructure traversing a particular locality—and the meanings and interpretations that it helps to engender are mutually interdependent. I will address this issue through an ethnographic inquiry into the construction of the Kaohsiung Mass Rapid Transit System (hereafter, Kaohsiung MRT).[1] Kaohsiung is located in Southern Taiwan and is the country's second largest city, its hub of heavy industry, and a world-class port. Not long after the groundbreaking ceremony of the Kaohsiung MRT in 2001, a telling story circulated in Kaohsiung City. According to the story, an old man stood by a Kaohsiung MRT construction site, looking over the fences with a solemn look on his face. People wondered why he bore such sadness. "Is it because the construction is blocking the city's traffic?," they asked. The old man answered, "oh no, how can I grieve over such a wonderful event? I came to pay homage to the huge construction machines. It has been many years since I last saw them in our city." The fact that this story was part of the Kaohsiung City government's campaign to muster popular support for the Kaohsiung MRT did not make it less powerful.

Although apocryphal, this story effectively captured the deep sentiment among Kaohsiung City residents surrounding the project, underlain by a sense of injustice that their city had not seen major public investment since the Ten Major Construction Projects in the early 1970s.[2] This general sentiment was best captured in the words of Mr. Hsu, the person who brought my attention to this story. A civil servant in his 40s and a native of Kaohsiung, Mr. Hsu laughed at the contrived nature of the story. Yet, he was also zealous about the importance of an MRT to his city. He did not care how and why the MRT project came to be in Kaohsiung. "It might very well be the result of political calculation," he said,

> But so what? Kaohsiung has always been ready for big tasks. We have a well-educated population who are highly motivated, but who in the past had to seek [professional, high-skilled] jobs elsewhere because they couldn't find suitable jobs in Kaohsiung. The question is how we can create an environment—an infrastructure—to embrace their talents.

What we need are opportunities and adequate resources. Now we've finally got the [government's] recognition. This will be our best chance.[3]

But others are more skeptical. Around the same time, I had a conversation in Taipei (the capital and financial center of Taiwan, located in the north) with a transportation engineer who was involved in the planning of several MRT projects, including the Kaohsiung MRT in the late 1980s. Having just returned from a Kaohsiung MRT panel discussion in Kaohsiung, she commented:

Nobody took it seriously when we were commissioned to do the planning. None of us thought this was a feasible project. We knew there wouldn't be enough passengers to make the system financially viable. A light rail would satisfy the need of Kaohsiung City—or they could simply put two thousand more buses on the streets of Kaohsiung if they really care about public transportation. Even today, when the [Kaohsiung MRT] construction is well underway, I still can't believe that they are doing it. They know it's not going to work. This cannot be real.

She is not alone in her skepticism. The potential for Kaohsiung MRT ridership has been called into question before, during, and after its construction by Taiwan's planning circle, the media, and the general public (including Kaohsiung residents). Indeed, the number of Kaohsiung MRT passengers upon its grand opening fell far short of previous government estimates: Original estimates projected a daily ridership of 450,000 by 2010 (KRTC 2007), but by the end of May 2014, only 165,000 passengers rode the Kaohsiung MRT each day (KRTC 2014). Perhaps more revealing than the number of daily passengers, however, is the phenomenon that more people travel on the Kaohsiung MRT on weekends and holidays than on weekdays. A survey conducted by the Kaohsiung Rapid Transit Corporation (KRTC, the private company in charge of the building and day-to-day operation of the Kaohsiung MRT) in 2009 indicated that only 7% of respondents utilized public transportation (including buses and the Kaohsiung MRT), while roughly 65% used a motorcycle, and close to 20% a private automobile, as their primary means of transportation (see Table 8.1 for acronyms). In addition, 25% of the Kaohsiung MRT users surveyed happened to be visitors from out of town, indicating the actual percentage of local residents to use the Kaohsiung MRT regularly is likely lower than 7%. In short, Kaohsiung MRT is less a means of daily commuting than of leisure and entertainment (Chibin and Hsaio 1996; Yong-Hsiang 1996).

This chapter takes its departure from the discrepancy between the enthusiasm for the coming of the Kaohsiung MRT and its underutilization, and examines the circumstances under which this discrepancy was produced. Specifically, I ask how mobility was discursively constructed and represented (cf. Cresswell 2010) leading to the completion of the Kaohsiung MRT, and

Table 8.1 List of Acronyms

KAOHSIUNG MRT	Kaohsiung Mass Rapid Transit System
KRTC	Kaohsiung Rapid Transit Corporation
MTBU	Kaohsiung City Mass Rapid Transit Bureau
TAIPEI MRT	Taipei Mass Rapid Transit System
TRTC	Taipei Rapid Transit Corporation
ROC	Republic of China
KMT	Nationalist Party [Kuomintang]
DPP	Democratic Progressive Party

what we can learn about the materiality and spatiality of mobility, using the Kaohsiung MRT as a case study. I draw my inspiration from Sheller and Urry (2006), who assert that there is no increase in mobility without extensive systems of immobility, because "all mobilities entail specific and often highly embedded immobile infrastructure" (p. 210). As such, "[m]obility is always located and materialized, and occurs through mobilizations of locality and rearrangements of the materiality of places" (ibid.); it is "a resource to which not everyone has an equal relationship" (ibid., p. 211). I seek here to extend Sheller and Urry's point of the concurrent existence of mobility and immobile infrastructure through an emphasis on the notion of locality. What matters metaphysically is not simply that mobility is conditioned by the presence of located infrastructure, but also where the infrastructure is located. For example, to build a public transit system in a small rural area or major urban area could mean different things; or to build a mass transit system in two cities comparable in size and economic activity might entail very different connotations. That the Kaohsiung MRT is an urban infrastructure traversing a particular metropolitan area is reflexive to the meanings and implications of mobility that it has facilitated to shape and generate.

Thus, the purpose of this chapter is twofold. First, it addresses the question of what kind of mobility has been envisioned as enabled by urban mass transit systems in the case of the Kaohsiung MRT, arguing that its significance derives from the fact it was conceived as more than a public transit system. From the beginning, the Kaohsiung City government has been capitalizing on the novel image of urban mass transit system to craft a vision of prosperity, centered on the Kaohsiung MRT. Since the early 2000s local government has implemented a series of urban renewal projects to create a better living environment and transform the city into an attractive tourist and investment destination, reviving Kaohsiung's deindustrialized city economy (Lin 2006), reflecting place-making trends observed elsewhere (Chang 2000; Harvey 1989; Smart and Smart 2003; Steven and Paddison 2005; Yeoh 2005). In a metaphorical sense, therefore, the Kaohsiung MRT has been imagined as a vehicle of not only physical movement but of change, breaking away, and becoming. In other words, the Kaohsiung MRT enabled not only the possibility of flow for the city population, but also the flow of the city into a different and brighter future.

Because the built environment of any given city could always be altered without the construction of a mass transit system, one might ask: What is distinctive about mass transit systems in this context? Thus the second purpose of this chapter is to highlight the significance of spatiality in the understanding of mobility, approaching the 'why mass transit' question in two different but interrelated ways. To build a mass transit system involves a massive scale of creative destruction of urban space such as city streets, presenting an obvious and convenient opportunity to carry out other renovations towards global competitiveness. Given the substantial amount of money needed for such projects, however, the question of acquiring the financial resources for construction is a significant one. Many major cities in the world (especially those in the Asian Pacific region) have made major investments in urban infrastructure like public transportation (Lo and Marcotullio 2000; Olds 2002), yet many of these megaprojects have also failed to fully deliver anticipated benefits. To explain the disparity between the continued popularity of urban megaprojects and the unfulfilled promises, Siemiatycki (2005, 2006) postulates that we should take into account not only the tangible gains (such as global economic competitiveness) but also intangible benefits of these projects. Beyond their functionality, it is perhaps spatially, temporally, and culturally rooted symbols, meanings, mythologies, and imageries that generate the widespread political and public support needed for urban megaproject investment (Richmond 2005).

Accordingly, public discourses about the Kaohsiung MRT touch upon a set of emotionally provocative, yet politically potent, questions about citizens' rights writ large in the language of national identity. The quotidian presence of the Kaohsiung MRT embodies Taiwan's historical-spatial inequality, which is both a product of and conducive to the national geopolitics. In this sense, the final product of an urban megaproject is no longer simply urban mega-infrastructure, but it represents something more or something else altogether.

To understand the values and meanings associated with the Kaohsiung MRT, this chapter explores four related issues. I start with "From North to South: The Moral Geography of Taiwan," presenting Kaohsiung City's geopolitical-economic and spatial-regional context, essential for comprehending popular support for the Kaohsiung MRT and political pressure for its funding. It highlights the salience of locality—and locatedness—in understanding mobility, through a case in which huge funding allocated for its construction was perceived as a belated step towards regional equality in Taiwan, and as spearhead for the transformation of Kaohsiung to excel on the world stage. The next two sections, "Taipei MRT as a Reference Point" and "Envisioning the Kaohsiung MRT," further address the extended role of mass rapid transit systems beyond the function of public transportation. While the "art of being global" (Roy and Ong 2011) is always a process of modeling and/or interreferencing, the Taipei MRT itself provided a major (and hugely expensive) reference and prime target against which the popular discourse in support of the Kaohsiung MRT was formulated. Looking at how the Taipei MRT changed the people and city is a conduit to understanding different meanings and popular imageries explored in "Envisioning the Kaohsiung MRT," which discusses the various urban renewal projects carried out by city government in conjunction with MRT construction, representing an official vision of—and practice about—the potential of mass transit systems and, by extension, the future of Kaohsiung City itself. That the Kaohsiung MRT is an urban transportation infrastructure embedded in a particular locale and, correspondingly, a specific web of sociocultural and political-economic dynamics is integral to the meaning of mobility that it has helped to engender. The last and concluding section ties together this chapter and the larger theme of this collected volume, in which Prytherch and Cidell (this volume) call for cross-fertilization of transportation geography and mobilities studies by proposing a new urban geography of networked flow with a new focus on transportation. The chapter reflects on the politics and poetics of the Kaohsiung MRT as a locale "whose form, function, and meaning are not self-contained but woven with the networked social organization of flow" (this volume, p. 45).

This chapter is derived from nearly a decade of ethnographic fieldwork coinciding with the construction of the Kaohsiung MRT (2001–2008), including archival research, formal interviews, and participant observation/ informal conversations, from 2001 to 2009. Transportation geography has traditionally relied on spatial analysis and behavioral science methods, especially when studying the impact and effect of transit systems. To approach the meanings and imaginaries associated with transit and urban (re)development, I offer ethnographic research methods as a more appropriate alternative for the realm of imagination, which is itself a realm of indeterminacy (Larkin 2013; Sneath et al. 2009). Ethnographic research, with its attention to specificities of time and space and its intention to give thick description to social and cultural life, is therefore crucial for exploring the processes of imagination and the possibilities of imagining enabled by these processes.

FROM NORTH TO SOUTH: THE MORAL GEOGRAPHY OF TAIWAN

To analyze urban transportation as 'a place of flow,' one must understand the interdependence of an urban system with the meanings and implications of mobility it helps to generate and shape. For the Kaohsiung MRT, that means accounting for the political-economic context—including (factual and perceived) South-North regional disparities—from which the project emerged. Interviews can manifest such contexts, in which words exchanged embody not only affective expression but also certain political potency.

One year after the grand opening of the Kaohsiung MRT, in 2009, I interviewed two young coworkers in their 20s about their experiences with the Kaohsiung MRT. One had lived most of his life in Taipei and only recently relocated to Kaohsiung for a job; the other was a Kaohsiung native who had lived in the city for all her life, except for the years away in college. The Kaohsiung native said she was in high school when the construction of the Kaohsiung MRT was announced. "How did you feel then? Were you excited?," I asked. She tilted her head slightly, thought for a few seconds, and said, "not really. We didn't pay much attention. [It's not like] the subways were leading to anywhere." There was no stop planned near where she lived or where she went to school at the time, but since then, she added, she and her family had consciously moved to a new residence close to the Kaohsiung MRT. However, even though her current company was located at the end of one of the lines, she did not take the Kaohsiung MRT but continued to ride her motorcycle to work. As a matter of fact, she hardly ever used the system except for the few weekends when she went shopping at Kaohsiung's largest department store.

Her words seemed to trigger some strong feelings that her coworker from Taipei had held for a while, who quickly said,

> I have never quite forgiven . . . ["You do mean 'forgiven?' Not just 'understood?,'" I teased him half jokingly.] Yeah, I have never quite forgiven [laugh], given the size of its population, Kaohsiung had to build a heavy-capacity mass transit system but not [for example] a medium-capacity light-rail system, even though it costs much less to build a light-rail. Besides, one could enjoy the scenery along the way while riding a light-rail.

"Maybe because there is nothing [for the passengers] to look at on the ground . . . I know a light-rail loop system has been planned to connect with the existing Kaohsiung MRT lines," his Kaohsiung-native coworker felt obliged to respond. She continued: "Mass transits are important. It's like . . . I don't have a car. Now we have moved to close to the MRT, I can take it to go around. It's convenient." "CONVENIENT? When do you ever take the MRT?," her coworker could not help but cut in, "I don't understand. I have heard so many times [from Kaohsiung residents] that they like the MRT because it's convenient. But only once in a blue moon they will ride it. What's the point [of having an MRT] then?" "S-E-C-U-R-I-T-Y! Just in case you need to use it," said his Kaohsiung-native coworker.

This young woman described the Kaohsiung MRT project "like the coming of a dream" that people in Kaohsiung had been waiting to happen, although she never clearly articulated the detail of the dream. If the exchange of words between her and her coworker bears any larger significance, it is the seeming callousness and persistent questioning from people in Taipei—in this case, her coworker—under which people in Kaohsiung feel they are unduly subjected. The indignant and defiant sentiment voiced in the young woman's words, though expressed jokingly, resonates with that of Mr. Hsu, who was introduced in the beginning of this chapter.

At the heart of such comments is a discourse about the Taiwan government's developmental policy, rooted in the country's history and political-economic context. Critiques of perceived unequal resource distribution and regional disparity between Southern Taiwan/Kaohsiung and Northern Taiwan/Taipei surfaced after the political democratization since the late 1980s and were widely circulated among people in Southern Taiwan, but have a deeper history. The development of Kaohsiung as a modern city began in the Japanese colonial period (1895–1945), during which Taiwan was perceived as a colony of high economic and military value in the expanding Japanese empire. In the empire's blueprint, Kaohsiung was not only to be a fishing and commercial port but also an important military base for the Japanese Imperial Army's southward advancement, especially after the onset of the Pacific War in 1941. To accomplish this, the Japanese worked out detailed urban plans to transform Kaohsiung into a modern city. They built the Kaohsiung harbor, constructed roads and railways to connect Kaohsiung with the surrounding regions and the rest of the Taiwan Island, set up modern amenities of electricity and running water, and established the gridiron of streets. By the early 1930s, the Japanese also built up industrial infrastructure, including steel plants and oil refineries.

These construction efforts greatly influenced Kaohsiung's development, state policy overall, and local perceptions. On the one hand, they indicated the importance of Kaohsiung in Taiwan's modern economic history; on the other hand, they also underlined the colonial state's power in shaping natural environments and livelihoods in Kaohsiung. This trend continued after World War II, when the defeated Japanese turned over Taiwan to the victorious China, at the time represented by Chiang Kai-shek's Nationalist Party (Kuomintang or KMT). Equipped with the port facilities and other infrastructure from the Japanese period, Kaohsiung quickly developed after World War II into a manufacturing center important to Taiwan's rapid industrialization. Under Taiwan's industrial policy at the time, Kaohsiung became a base of container logistic centers, steel plants, shipyards, and shipbreaking, scrap metal, cement (both mining and processing), and petrochemical industries (Hsu and Cheng 2002), most of which were highly polluting and have had great impact on the health of metropolitan Kaohsiung residents (Lü 2009). Furthermore, due to Taiwan's tax structure, the profits earned by these companies were taxed by the local government relative to where the companies' headquarters were located (namely, Taipei), but not by the local government of where the production actually took place (i.e., Kaohsiung). Consequently, it was the Taipei City government and the residents of Taipei who enjoyed the fruits of these industrial endeavors, while Kaohsiung City government and residents bore the consequences of environmental and health degradation, and lack of city funds for public investment and social development. Over time, deeply seared in the mind of Kaohsiung City residents was the sentiment that they were treated as secondary citizens and their welfare was overlooked by the central government (Lee 2007a).

Designated with different functions in Taiwan's post–WWII economic development (Hsu and Cheng 2002), Taipei and Kaohsiung were thus affected differently by Taiwan's recent economic restructuring (Lee 2004). On the one hand, the role of Taipei as Taiwan's command and coordination center has been reinforced. Dubbed by Hsu (2005) as an "interface" city, after the 1980s Taipei emerged to be a node in the global flow of capital, knowledge, and technology connecting high-tech industries in the Taipei-Hsinchu corridor, the technology hub in Silicon Valley, and high-tech production in Shanghai (aided by foreign direct investment from Taiwan). Taipei is also the site of the corporate headquarters and the market center for the majority of Taiwan's leading companies. On the other hand, the manufacturing-based economy of Kaohsiung has been hard hit due to capital outflow and the consequent deindustrialization. The unemployment rate of Kaohsiung is among the highest in the country. What's more, as capital city, Taipei was said to have been the jewel of the central government and given a lion's share of resources. The new 'global' status of Taipei has only reinforced the sense of unfairness in resource distribution, which has been appropriated by opposition party politicians to marshal electoral support.

These discourses have framed the planning of Kaohsiung MRT over the past 20 years, as it has been transformed from transportation plan on paper to a transit infrastructure under construction. First proposed in the late 1970s, Kaohsiung MRT remained an idea until 1989, when the Taiwan government finally made the decision to build it, and only in the early 1990s were routes determined. Up to this time, the KMT was still the ruling party, and the Kaohsiung City mayor was an appointee of the central government (the first direct mayoral election took place in 1994 after the city was granted status of special municipality in 1979). The actual construction of the Kaohsiung MRT finally materialized in 2001, after Chen Shui-bian, the then opposition Democratic Progressive Party (DPP) candidate, won the presidential election, running on a Taiwan-centered, South-based (as opposed to a China-friendly, Taipei-focused) campaign, and when Kaohsiung City mayor Hsieh Chang-ting was a fellow DPP member. The announcement of the Kaohsiung MRT construction also came at a time when the Taipei MRT was just completed and had proven to be hugely popular among metropolitan Taipei residents. The Kaohsiung MRT, therefore, served as a potent political symbol that indicated the determination of the DPP-led government to balance the (perceived and factual) South-North disparity.

TAIPEI MRT AS A REFERENCE POINT

To comprehend why the Kaohsiung MRT, among other possible public works projects, played such a significant role in the DPP's political discourse, one has to look at the Taipei MRT as a reference. The Taipei MRT's construction began in 1988, and its primary network was completed in 2000. The vast funding provided by Taiwan's central government to bankroll the Taipei MRT construction (NT$441.7 billion [US$13.4 billion] in total) was only one of the factors, albeit a crucial one.[4] Equally—if not more—important was the observation of how the Taipei MRT functioned as not only a highly effective transportation system but also a catalyst in transforming the urban culture and civic identity of Taipei City (Lee and Tung 2010).

Originally intended by government officials and planning professionals as a solution to Taipei's worsening traffic problems, the Taipei MRT quickly took on other meanings. Upon the grand opening of the Taipei MRT, the Taipei Rapid Transit Corporation (TRTC), the city agency in charge of the daily operations, waged a zealous public campaign to educate passengers about 'proper' rider behavior. Signs were erected at the top and foot of nearly every escalator inside Taipei MRT stations to remind passengers they should stand on the right-hand side to let others in a rush pass by. Lines were drawn on the platform so that people could stand in line while waiting, ensuring passengers would not push or scramble, but get on the train in an orderly manner. To keep the environment clean, passengers were forbidden to eat, drink, or chew gum inside the stations or in the carriages; anyone who violates these regulations is fined. The Taipei MRT also hired an army of middle-aged female workers to sweep the floor, wipe the walls, and dust every surface—high and low—inside Taipei MRT stations. These women's hard work has kept the stations dirt free and spotless since the very beginning. As a result, the stations are exceptionally clean; there is simply no littering. On the whole, passengers observe a behavioral code of order that one does not normally see anywhere in Taiwan outside mass transit systems, although we have begun to see these influences extend beyond the space of the Taipei MRT. While not the first civility campaign attempted by the Taipei City government, it was the first supported and followed by most citizens (Lee 2007b).

Step by step, a collective identity began to take shape among Taipei residents based on their shared experience as MRT riders. The Taipei MRT is decidedly punctual, so morning commuters are no longer compelled to leave home half an hour earlier. Soon after opening, it was also observed that an increasing number of passengers began to dress up for the ride, wearing designer clothes and high heels or coordinated outfits purchased in department or brand-named stores, because they no longer needed to race frantically to catch a bus. Passengers nowadays are also more willing to yield their seats to the needy or the elderly, for their trips are made short and pleasant by the efficiency of the Taipei MRT, even if they have to stand. This is a great departure from riding Taiwan commuter railway trains in previous decades, when people would not hesitate to climb through the windows of a train so that they could be quick enough to get a seat. Ultimately, the Taipei MRT not only changed the habits of its passengers but also helped to initiate a new model for—and image of—metropolitan Taipei residents. People in Taipei seemed to pay greater attention to their own city and appraise it (and themselves) in a larger, global framework.

The novelty of the Taipei MRT also quickly turned the mass transit system into a tourist attraction. In the first few years after its opening, on weekends even the most casual observers could easily spot big tour buses unloading travelers on the roadside outside Tamsui Station, a charming harbor town and the final destination of one of the Taipei MRT's most scenic lines. These out-of-towners then joined the stream of crowds who traveled via the Taipei MRT from the surrounding metropolitan region for a day's excursion. The Muzha Line, the shortest among the routes operating between downtown and the zoo, also became a part of the standard tour package for schools in central and Southern Taiwan. As a matter of fact, the son of Mr. Hsu, who was introduced in the opening story of this chapter, was the first in his family to ride the Taipei MRT on his elementary school graduation trip. Mrs. Hsu, who had been listening to my conversation with her husband, chided in and told me: "The kids liked it. It was a new experience. My son enjoyed it, though he said the car quivered at times." The young boy's joy was obviously communicable, and the parents shared the excitement created by the novel technology. The changes in the urban life of Taipei observed upon the grand opening of the Taipei MRT had also served as a reference for their high anticipation for the impact of the Kaohsiung MRT.

In addition to such (extratransportation) implications, the Taipei MRT is first and foremost a public transit system for Greater Taipei residents, serving close to 2 million passengers (in a metropolitan area of 7 million people) on a daily basis (TRTC 2014). The convenience and punctuality of the Taipei MRT, the ease it has made of one's daily commute, the extensive distance one can travel with it, its effect on Taipei's streets and air quality, the gradual change in etiquette and behavior among metropolitan Taipei residents, and the overall transformation in Taipei's civic culture are all part and parcel of the structure of feeling engendered by the Taipei MRT as a transportation technology (cf. Thrift 1994, in Cresswell 2006, p. 46).

Yet, however contradictory this might sound, it is also true that the transportation function did not dominate the public discussion about MRTs in Kaohsiung or elsewhere in Taiwan. Most out of town visitors as tourists in Taipei did not take the Taipei MRT with commuters during rush hours on a weekday; neither did they come to ride the Taipei MRT to experience its efficiency and advantage as a means of public transportation. They did not become interested because they wanted to learn how a mass transit system could help to solve the traffic problems of a congested city. Rather, novelty aside, it was the amalgam of the sleek and orderly image of the Taipei MRT, the busy commercial activities, the bustling urban life, Taipei as Taiwan's primary and globalizing city, and the seeming sense of self-confidence—or arrogance—among Taipei citizens that was leaving a lasting impression in the minds of out of town visitors. In sum, MRTs in general—and the

Taipei MRT in particular—became the personification of progress and were marveled at in themselves as objects of admiration, fascination, and desire. Their attraction lay in the promise that they were carrying a better future (cf. Kaika and Swyngedouw 2000, p. 129).

ENVISIONING THE KAOHSIUNG MRT

However, can this structure of feeling be replicated—or another kind of structure of feeling generated—with the MRT technology but without the habitual practice of using the MRT? The Taipei MRT has become a reference for urban development in Taiwan, but can its lessons be extended? To answer my own questions, the current section addresses how MRT was envisaged in the context of Kaohsiung City. Specifically, I focus on the official (re)presentations by the city government and transit agencies (KRTC) behind the Kaohsiung MRT and urban policies.

The government's decision to build the Kaohsiung MRT engendered deep skepticism among Taiwanese planning professionals, especially those based in Taipei, which they suspected was based on political calculation to attract the support of southern voters. The feeling of lack of urgency was also echoed by both the media and residents in Kaohsiung, who saw their city as having broader streets, less traffic, and more parking space than Taipei. Many of them, like the planning experts in Taipei, also questioned whether there would be enough passengers. According to my interview with Chou Li-liang, a former MTBU director-general whose term (1998–2004) covered the inceptive and defining period of the Kaohsiung MRT construction in the early 2000s, there was a discrepancy of 2 million people between the population of Taipei City during the day and at night. That is, 2 million people commuted from neighboring towns and cities to work or school in Taipei on a daily basis, and, as such, the need for public transportation was pressing. Prior to the construction of the Taipei MRT, a well-developed bus system was already in place to serve Taipei's commuters. In comparison, there were approximately 1.5 million people in metropolitan Kaohsiung both during the day and at night. The number of people who commuted to Kaohsiung City from the surrounding areas was small. Moreover, only a tiny fraction of commuters in Kaohsiung—mainly high school students who possessed no better means—used the city's underdeveloped, inadequate bus system. The majority of Kaohsiung City residents relied on personal motorcycles as their primary means of transportation. Similarly, time and again I was told during the course of my research that there were only two routes planned for the Kaohsiung MRT, which, together, would cover the downtown area already served by most of the existing bus lines. As such, what good could the Kaohsiung MRT do for the commuters in the city? Even the executive secretary at the Kaohsiung City government spokesperson's office confessed with some embarrassment that he would continue to ride a motorcycle to work, after he earnestly informed me of the wonderful things that the Kaohsiung MRT would bring to his city.

Former MTBU Director-General Chou emphasized to me that the Kaohsiung MRT "is not about the present but about the future." Using a metaphor from traditional Chinese medicine, he referred to Conception and Governor, two of the vessels in the human body. It is said that a martial arts master will become invincible if s/he has these two vessels open to allow unimpeded movement of *chi* in the body.

> Transportation is basic infrastructure; it is like the Conception and Governor vessels [ren du er mai]. If you have these two vessels open, you will have all kinds of possibilities. [The Kaohsiung MRT] is to give Kaohsiung such a chance.

By evoking this metaphor, Director-General Chou conjured up a popular cultural imaginary parallel to the MRT as a medium of flow. Similarly, Mr. Cheng, the chief secretary under Chou at the MTBU, explained how much of a struggle it was to reconcile with himself that the Kaohsiung MRT was a necessary project, in spite of the concern over its future ridership. "No city could have a successful mass rapid transit system without a well-established, widely-utilized bus system," he said. At the end, he reckoned:

> This is an age of intercity competition. Have you ever seen a service-based city that doesn't have a mass transit system? Kaohsiung might still not have a chance [in the current stage of global competition] with the MRT system. But we will definitely not make it without an MRT.

Opportunity for change or chance for transcendence was very much a part of the official discourse regarding the Kaohsiung MRT. If the indignation derived from regional disparity provided the Kaohsiung City government the moral justification to obtain funding for the construction of a mass transit system (NT$181.3 billion [US$5.46 billion] in total),[5] the transformative effect that an MRT could have on city culture and image (as observed in Taipei) has served as an inspiration for practice. One difference between Taipei and Kaohsiung, however, was the level of urgency of the need for alternative urban transportation plans. Accordingly, as opposed to the Taipei MRT that was 'demand driven,' the Kaohsiung MRT was promoted as a 'supply-oriented' system in the Kaohsiung City government's public campaign. Furthermore, the Kaohsiung MRT was taken by both the city government and urban planners and transportation experts to be a key to the urban renewal and economic revival of the city. The Kaohsiung MRT was expected to not only facilitate the flow of people but also of goods and capital, and art and culture.

To put this expectation into practice, local transit agencies commissioned several world-renowned architects and artists to design, or incorporate their

Place-Making, Mobility, and Identity   165

works into the structure of, a handful of "Special Stations" (MTBU 2009). Among these Special Stations, the most prominent is likely the Formosa Boulevard Station, which is located at the traffic circle of the busy intersection of Chung-shan Road and Chung-cheng Road (the two main boulevards in Kaohsiung), wherein the two MRT lines meet. The intersection is also where the Formosa Incident, a watershed event in Taiwan's struggle for political democratization, happened in 1978. Installed in the ceiling of the grand concourse of the Formosa Boulevard Station is *The Dome of Light*, the largest single piece of glasswork in the world designed by Italian artist Narcissus Quagliata (Figure 8.1). The station structure above the ground, named *Praying*, is the work of Japanese architect Shin Takamatsu, comprised of four station exits of identical shape made of glass panels standing at the four corners of the intersection. Jointly, they take the shape of four hands coming together to pray. Both of these works address the historical event of Formosa Incident with contemporary artistic interpretations. Simultaneously, the MTBU spearheaded a "Formosa Boulevard" project, which involved transforming a section of Chung-shan Road adjacent to the Kaohsiung MRT into a tree-lined boulevard with expanded pedestrian sidewalks wide enough to accommodate open-air cafes and street art performances—that is, "like Champs-Élysées" (MTBU 2006; Nan Zhu-jiao 2003). An earlier proposal for the Formosa Boulevard project also called for mobile bookstalls on the sidewalks, "so that there will be not only fragrance of flowers [from the trees] and coffee but also fragrance of books," Mr. Cheng, the aforementioned MTBU chief secretary, explained to me by invoking a popular expression in the Chinese language. Unfortunately, the

Figure 8.1   "*The Dome of Light,*" Kaohsiung MRT Formosa Boulevard Station.
Source: Perng-juh and Peter Shyong (Dimension Endowment of Art in Taipei), with permission.

idea of mobile bookstalls was dismissed due to the lack of regulatory laws. "Besides, who's going to enjoy browsing books while standing under [Kaohsiung's tropical] scorching sun?" Cheng continued, "but just imagine! What if it weren't open-air bookstalls but enclosed book kiosks with air conditioning? And what about encasing the metal frameworks of these kiosks with glass, like this restaurant in Rome?" Cheng showed me some photos of this restaurant that he took on one of his official trips to Europe and said:

> We could have five of these book kiosks on each side [of Formosa Boulevard]. Just imagine! When the light comes up at night inside these book kiosks, and when the light shines through the glass, these kiosks will be transformed into luminous pearls. With a string of gleaming gems along Formosa Boulevard, what a gorgeous picture that would be!

The idea of mobile bookstalls or book kiosks was never realized. However, in conjunction with the MTBU endeavor in station architectural design, the Kaohsiung City government has been pursuing a series of urban renovation projects in recent years. These include renovating the harbor into a pedestrian-friendly waterfront and converting vacant port facilities into exhibition spaces and art studios, in addition to greening and widening the sidewalks, as well as creating parks near the Kaohsiung MRT (Lin 2006). More specifically, Ai River (literally 'Love River'), which flows through the heart of Kaohsiung and was once considered by many as the soul of the city but turned into a huge open sewer over the past few decades as a result of the wastewater dumped by factories and private households, was cleaned up. Concomitant to the effort to clean the river was the endeavor to improve the landscape along the riverbanks. Today, sightseeing cruises sail along much of Ai River. One can also take a stroll along the riverside promenade, spend time at the Museum of History, listen to music at the Concert Hall and its plaza, watch a movie at the Municipal Film Archives, or simply sit down, have a cup of coffee, and "enjoy the exotic, romantic ambiance of Love River, day and night, as if sitting on the Seine riverbank in Paris" (Yeh 2004). Through these efforts the Kaohsiung City government is attempting to create a renewed civic identity, as well as announce to the country—and to the world—that Kaohsiung is Taiwan's southern capital, a better alternative to Taipei.

Over time, *The Dome of Light* has garnered huge popularity and become one of the most trendy photo locations for both local residents and out of town visitors; it has gradually grown into a landmark of Kaohsiung City. The KRTC has been providing guided as well as audio public art tours at the Formosa Boulevard Station. The company also sponsored photography competitions and musical concerts under *The Dome of Light*. The station concourse as an open space has also been utilized by different groups in the city for assorted purposes, such as a romantic wedding ceremony for 78 couples (Hsieh 2013). In 2012, the Formosa Boulevard Station was chosen by the BootsnAll travel website as the second most beautiful subway stop in the world.[6] Both the KRTC and the MTBU publicly advertised this on their websites. In spite of the Kaohsiung MRT's less than ideal volume of ridership, for both of these agencies, as well as city residents, this seemed to indicate a moment of vindication that their effort had not gone to waste but was slowly winning world recognition.

CONCLUSION: URBAN TRANSPORTATION AS A PLACE OF FLOW

What is distinctive about mass rapid transit systems? How do they inform us about transportation as a place of flow? This chapter takes on these queries by addressing a seemingly self-contradictory question: how and why an urban mass transit system can be considered as urgently needed while, in reality, infrequently used. An immediate answer to my own question is that technologies are unstable things (Larkin 2013). A mass rapid transit system is never just a piece of transportation technology. Rather, it is loaded with meanings, and these meanings are embedded in the political-economic context and cultural-ideological dynamics wherein the mobility is taking place. This brings to the fore that all mobility stories are essentially local stories, however much they are conceptually understood as technical issues or discursively constructed as global and/or universal phenomena. The material aspect involved in an act of movement, be it manifested in a corporal body or a piece of infrastructure—or both, such as in the case of urban mass transit systems—made the act inevitably a spatially grounded practice and experience. That the Kaohsiung MRT is a piece of urban infrastructure traversing a particular metropolitan area is reflexive to the meanings and implications of mobility that it has facilitated to generate and shape.

This chapter offers new insights to advance our understanding of urban transportation as a nexus of space and flow. First of all, let me go back to the question: What can a mass rapid transit system be other than a means of public transportation? In the current world of networked societies, urban mass transit as networked infrastructure takes on the function of linking not only local communities but also increasingly the local and the global, both physically and metaphorically. Although MRT systems and interurban competition have, up until now, been considered as largely separate phenomena in academic literature (McLellan and Collins 2014), the case of Kaohsiung shows that they are obviously closely connected in the minds of those in charge of MRT planning and construction. Often counted in the worldwide circle of transportation practitioners are the economic factors (such as increasing land values near MRT stations, and lower costs for fuel, shipping, and vehicle maintenance due to reduced road congestion) and extra-economic factors (such as quality of life, healthy lifestyle, and environmental quality) thought beneficial for a city to move up the ladder of global economic hierarchy, even though a direct causal relationship is yet to be openly established (McLellan and Collins 2014). What is highlighted in the

Kaohsiung MRT is the growing significance of the notion of place-making as manifested in transportation infrastructure. That is, it is no longer just the function of mass transit systems as facilitators of movement but, rather, the physical presence of mass transit systems as sites of cultural and artistic exhibitions that is increasingly emphasized. But, of course, a 'place' of flow such as an MRT can become exhibitional precisely because it is a place of 'flow' wherein the large number of passengers and/or passersby could be turned into potential spectators and consumers.

However, why should a mass rapid transit system be the focus of place-making among other possible subjects? As there is no palpable exchange value in such an effort, the conviction in the transformative effect of mass transit systems has to be—at least partially—based on something symbolic that is not readily exchangeable but locally identifiable. In Taiwan, MRTs came to symbolize renewed civic cultural formation and (subsequently) economic revitalization against the background of Taiwan's unequal regional development and the resulting differential positioning of these regions in the Asian Pacific economic ordering. Yet, these symbolisms were not merely products of active imagination. What we learned from the experience of Taiwan is that they were made real—or they could be realized—exactly because an urban mass transit system as a transportation venue is a space in which not only flows of goods and capital occur but, through structured interactions and routinized behavior on a daily basis, where shared sociality and a collective identity are cultivated (such as in the case of Taipei). The meaning of mobility is thus embodied and spatialized.

This takes us to the third and last question: How might a municipality acquire the financial resources to pay for its infrastructural construction? While infrastructural development including mass public transportation is widely deemed necessary for economic vitality, not every city in the world that shares this inspiration can garner the resources to make it a reality. The way that the Kaohsiung MRT project was funded underscores for us that urban transit is a medium of spatial-political relations, the material condition of which plays a role in constituting and contesting the effectiveness of government. Through MRT a new means of anticipating the future emerged in Taiwan, not randomly, but as a result of its governance structure and the consequent regional disparity. The heightened expectation brought about by the MRT—and the concurrent indignation of being deprived of an MRT in one's own city—thus became the basis for political redress that provided the moral justification for the construction of the Kaohsiung MRT.

NOTES

1 For a listing of acronyms that appear in this chapter, see Table 8.1.
2 The Ten Major Construction Projects were national infrastructure projects embarked on by the Taiwan government in 1973. They involved massive government investment and had been considered as a major factor in bringing

Taiwan out of an international recession and accelerating its economic and social development. Three of the Ten Projects were located in Kaohsiung: China Shipbuilding Corporation Shipyard, China Steel Corporation, and an oil refinery of Chinese Petroleum Corporation (all of these three corporations were state-owned enterprises then).

3   All the interviews referred to in this article were conducted in Mandarin Chinese. The English translations of the interview quotations used in this article are my own.
4   This covered the Taipei MRT's primary network of six lines, with 79 stations and 76.8 kilometers in length.
5   This cost covered the construction of the current system, which comprises two lines with 36 stations covering a distance of 42.7 kilometers. Different from the Taipei MRT, whose construction was funded in entirety with public money, the Kaohsiung MRT was a BOT (Build-Operate-Transfer) project with a public-private partnership. However, the government provided most (83.2%) of the funding, despite the fact that this was a BOT project.
6   The BootsnAlltravel guide website. Retrieved February 19, 2015 from http://www.bootsnall.com/articles/11-11/15-of-the-coolest-subway-stops-in-the-world.html. The Central Park Station, another Kaohsiung MRT special station, was ranked number four in this contest.

WORKS CITED

Chang, T. C. (2000) "Renaissance revisited: Singapore as a 'global city for the art,'" *International Journal of Urban and Regional Research*, 24(4):818–831.

Chibin, W. (王啟彬), and Hsiao, Y. (蕭永秀) (1996) "Xinjiapo jieyun neng, women weisemo buneng? (新加坡能, 我們為什麼不能? [If the Singapore MRT can, why can't we?])," *Yingjianzhixun* (營建知訊), 166:38–47.

Cresswell, T. (2006) *On the Move: Mobility in the Modern Western World*, New York: Routledge.

Cresswell, T. (2010) "Towards a politics of mobility," *Environment and Planning D: Society and Space*, 28:17–31.

Harvey, D. (1989) "From managerialism to entrepreneurialism: The transformation of urban governance in late capitalism," *Geografiska Annaler*, 71B:3–17.

Hsieh, L.-T. (2013, September 16) "Zhonggang 78 dui xinren wen zai guangzhiqiongding xia" (中鋼78對新人吻在光之穹頂下 [78 couples of the China Steel kissed under *The Dome of Light*]), *United Daily*. Retrieved September 29, 2013 from http://udn.com/NEWS/DOMESTIC/DOM6/8165669.shtml.

Hsu, J. (2005) "A site of transnationalism in the 'ungrounded empire': Taipei as an interface city in the cross-border business networks," *Geoforum*, 36:654–666.

Hsu, J., and Cheng, L. (2002) "Revisiting economic development in post-war Taiwan: The dynamic process of geographical industrialization," *Regional Studies*, 36(8):897–908.

Kaïka, M., and Swyngedouw, E. (2000) "Fetishizing the modern city: The phantasmagoria of urban technological networks," *International Journal of Urban and Regional Research*, 24:120–138.

Kaohsiung Rapid Transit Corporation (高雄捷運公司 [KRTC]). Retrieved September 11, 2007 from http://www.krtco,com.tw/answer/ans_3_1.htm.

Kaohsiung Rapid Transit Corporation (高雄捷運公司 [KRTC]) "Statistics." Retrieved July 11, 2014 from https://www.krtco.com.tw/about_us/about_us-6.aspx.

Larkin, B. (2013) "The politics and poetics of infrastructure," *Annual Review of Anthropology*, 42:327–343.

Lee, A. (2004) *In the Name of Harmony and Prosperity: Labor and Gender Politics in Taiwan's Economic Restructuring*, Albany: SUNY Press.

Lee, A. (2007a) "Southern green revolution: Urban environmental activism in Kaohsiung, Taiwan," *City and Society*, 19(1):114–138.

Lee, A. (2007b) "Subways as a space of cultural intimacy: The mass rapid transit system in Taipei, Taiwan," *The China Journal*, 58:31–55.

Lee, A., and Tung, C. (2010) "How subways and high speed railways have changed Taiwan: Transportation technology, urban culture, and social life," in M. Moskowitz (ed) *Popular Culture in Taiwan: Charismatic Modernity*, London: Routledge, pp. 107–130.

Lin Chin-rong, C. (林欽榮) (2006) *Cheng shi kong jian zhi li de chuang xin ce lüe: san ge Taiwan shou du cheng shi an li ping xi, Taibei, Xinzhu, Gaoxiong* (城市空間治理的創新策略:三個台灣首都城市案例評析, 台北, 新竹, 高雄 [*Innovative Strategies in the Governance of Urban Space: Three Taiwanese Cases, Taipei, Hsinchu and Kaohsiung*]), Taichung: Xi zi ran zhu yi (新自然主義).

Lo, F., and Marcotullio, P. (2000) "Globalization and urban transformation in the Asia-Pacific region: A review," *Urban Studies*, 37(1):77–111.

Lü, H. (2009) "Place and environmental movement in Houjin, Kaohsiung," *Kaogu renlei xuekan* (考古人類學刊 [*Journal of Archaeology and Anthropology*]), 70:47–78.

Mass Rapid Transit Bureau, Kaohsiung City (高雄市捷運工程局 [MTBU]) (2006) *Guanjian banian* (關鍵八年: 高雄捷夢想的實現1998–2006 [The Crucial Period to Fulfill the Kaohsiung MRT Project]), Kaohsiung: Kaohsiung City Mass Rapid Transit Bureau.

Mass Rapid Transit Bureau, Kaohsiung City (高雄市捷運工程局 [MTBU]) (2009) *Kaohsiung Jie yu: che zhan zhuan ji* (高雄捷運: 車站專輯 [Kaohsiung Mass Rapid Transit: The Stations]), Kaohsiung: Kaohsiung City Mass Rapid Transit Bureau.

McLellan, A., and Collins, D. (2014) "'If you're just a bus community... You're second tier': Motivations for Rapid Mass Transit (RMT) development in two mid-sized cities," *Urban Policy and Research*, 32(2):203–217.

Nan Zhu-jiao magazine. (2003) "Jieyun xin fengmao, manbu Meilidao dadao—Zhuanfang Gaoshi jieyun juzhang Chou Li-ling (捷運新風貌, 漫步美麗島大道: 專訪高市捷運局長周禮良 [MRT new look, sauntering through the Formosa Boulevard: Interview with MTBU Director-General Chou Li-liang])," *Nan Zhu-jiao* Magazine (南主角), 27:29–60.

Olds, K. (2002) *Globalization and Urban Change: Capital, Culture, and Pacific Rim Mega-Projects*, Oxford: Oxford University Press.

Richmond, J. (2005) *Transport of Delight: The Mythical Conception of Rail Transit in Los Angeles*, Akron, OH: University of Akron Press.

Roy, A., and Ong, A. (eds) (2011) *Worlding Cities: Asian Experiments and the Art of Being Global*, New York: Routledge.

Sheller, M., and Urry, J. (2006) "The new mobilities paradigm," *Environment and Planning A*, 38:207–226.

Siemiatycki, M. (2005) "Beyond moving people: Excavating the motivations for investing in urban public transit infrastructure in Bilbao, Spain," *European Planning Studies*, 13(1):23–44.

Siemiatycki, M. (2006) "Message in a metro: Building urban rail infrastructure and image in Delhi, India," *International Journal of Urban and Regional Research*, 30(2):277–292.

Smart, A., and Smart, J. (2003) "Urbanization and the global perspective," *Annual Review of Anthropology*, 32:263–285.

Sneath, D., Holbraad, M., and Pedersen, A. (2009) "Technologies of the imagination: An introduction," *Ethnos*, 74(1):5–30.

Steven, M., and Paddison, R. (2005) "Introduction: The rise of culture-led urban regeneration," *Urban Studies*, 42(5/6):833–839.

Thrift, N. (1994) "Inhuman geographies: Landscapes of speed, light, and power," in P. Cloke (ed) *Writing the Rural: Five Cultural Geographies*, London: Paul Chapman, pp. 256–311.
Taipei Rapid Transit Corporation [TRTC]. Retrieved October 28, 2014 from http://www.trtc.com.tw/ct.asp?xItem=1366864&CtNode=27643&mp=122031.
Yeh, L. (2004, April 23) "Kaohsiung aspires to be cultural icon," *Taipei Times*. Retrieved May 4, 2004 from http://www.taipeitimes.com/News/taiwan/archives/2004/04/23/2003137736.
Yeoh, B. (2005) "The global cultural city? Spatial imagineering and politics in the (multi)cultural marketplaces of South-east Asia," *Urban Studies*, 42(5/6): 945–958.
Yong-hsiang, C. (陳永祥) (1996) "Gaoxiong jieyun qiwen qida (高雄捷運七問七答 [The Kaohsiung MRT Q&A])," *Yingjianzhix un* (營建知訊), 167:7–14.

9 Contesting the Networked Metropolis
The Grand Paris Regime of Metromobility

Theresa Enright

> Transportation, again transportation, always transportation. . .
> (Paul Delouvrier, quoted in Merlin 1967, p. vii)

Grand Paris has a dynamic and contested history. Referring most generally to the Greater Paris region, the term has been periodically invoked by planners and politicians for over a century to refer to the relationship between Paris and its environs. From Baron Haussmann's massive Second Empire annexations, through the regional development plan of Henri Prost in 1934 and the post–World War II suburbanization programme of Paul Delouvrier, Grand Paris is an ever-changing project of urban assemblage. Notably, each of these pivotal moments of rethinking and reconfiguring the city has involved spectacular infrastructure projects of mobility and circulation. Grand Paris is not just the physical territory *outré-muros* (outside of the walls of Paris proper) but it is a *horizon of movement and connection*, a set of multiscalar social and spatial relationships constituting the dynamic flux of metropolitan life.

In its most recent iteration, Grand Paris has become an umbrella term for a number of regional agglomeration and redevelopment initiatives launched in 2007 by the central state and piloted by then President Nicolas Sarkozy. Grand Paris aims to improve regional cohesiveness and bolster Paris's position as a globally competitive city (RépubliqueFrançaise 2010). While Grand Paris invokes a variety of institutions and policies—a high-profile architectural competition in 2009 on the future of the region, a proposed regional administrative body, new targets for housing and plans for city greening, as well as thousands of site-specific works—above all, it is shorthand for an ambitious extension of the Métropolitain (métro) transit network into the suburbs (*banlieue*). With territorial and economic development oriented toward greater capacities for circulation, both the architectural exhibitions and the laws directing the initiative stress the need to improve regional mobility above all other concerns. "It is surely transportation," asserts Sarkozy (2009),"that will play the most decisive part."

The placement of mobility and flow at the center of urban-regional development responds to an urgent and long-standing crisis of transit in metropolitan Paris. It also attests more generally to the fact that the form of the urban and its networks of movement are inseparable (Bridge and Watson 2011; Dupuy 1991; Offner 1993; Virilio 2006). Urban mobility systems fundamentally shape political-economic patterns of production, consumption, and distribution at multiple scales (Sheller and Urry 2006; Straw and Boutros 2010). Transit conditions—individual and collective movements throughout the city—are essential elements in how inhabitants experience, imagine, and identify with each other and their environment. The Paris métro is exemplary in its phenomenological and symbolic import (Augé 2002; Dallas 2008; Delaney and Smith 2006; Ovenden 2009). Networked infrastructures are intertwined with social services and facilities (housing, schools, parks, healthcare, etc.) and thus comprise an important site of regional politics and planning. They also, as Hesse (this volume) points out, constitute an aspirational "spatial imaginary" that frames particular notions of modernity and prosperity. Changes in transit infrastructure not only affect who moves, where, how, and why, but also define the uneven contours of regional growth, quality of life, and governance.

Conflicts over transit projects are thus also conflicts over the essence of the metropolis. The official and unofficial conversations about Grand Paris must thus be understood as highly contested discourses where new spatial politics are emerging (Offner 2007). Debates over transportation are crucial sites where the meaning, form, and function of the 21st-century city are being worked through (see also Orfeuil and Wiel 2012; Orfeuil 2012).

There are a number of issues the Grand Paris emphasis on mobility brings to the fore. This chapter follows three main lines of inquiry. First, what are the main functions of rapid collective transit that make it a privileged vector of contemporary urban restructuring? Second, how are mobility megaprojects and the desired 21st-century city coproduced? Third, what are the political and ethical stakes of the emerging mobility regime of Grand Paris? This chapter addresses these questions by analyzing contemporary discourses of transportation and development in Paris, focusing on a series of public debates held from September 2010 through January 2011.

I argue that while the debates reveal the links between transportation and myriad social, environmental, political, economic, and ethical objectives, this diversity of meaning still affirms an overall planning priority on infrastructure as an economic asset. The most audible voices in the Grand Paris transit debates assume that mobile infrastructures must be pursued in order to attract increasingly mobile capital into the built environment, to catalyze urban rent production, and to increase territorial competition and economic growth. I define this form of neoliberal accumulation based on contemporary networked connectivity as a *"regime of metromobility."*[1] This regime is clearly expressed through Grand Paris, but its scope extends well beyond the region. Indeed, the Paris case is useful for thinking more generally about the ways in which infrastructures of transport are linked to processes of urbanization in the contemporary era, and it highlights some of the constitutive tensions in the coproduction of mobility and the metropolis in the 21st century.

This perspective is a departure from much of the literature on Grand Paris that has tended to emphasize interinstitutional struggles and the challenges of metropolitan governance in the Île-de-France (Béhar and Estèbe 2010; Behar 2013; Mongin 2010; Subra 2012; Wiel 2009). While governance and infrastructural provision are undeniably connected, often overlooked in this perspective are the particular meanings of mobility that characterize transit infrastructures (see Lee, this volume). In looking at how rapid urban rail is theorized, as well as its real and imagined relationships to other services and sectors, much is revealed about the multiple material and symbolic *uses* of mobility in urban-regional restructuring.

In what follows, I first contextualize the framework used to understand the restructuring of Paris's transport system. I then closely analyze a series of public debates over proposed Grand Paris transit infrastructures. I end the chapter by outlining the regime of metromobility that is the backbone of the networked global city and the political and ethical implications of such a regime.

MOBILE URBAN INFRASTRUCTURES IN THE NEOLIBERAL ERA

In order to assess the multiple linkages between transit networks and metropolitan processes, I engage three main literatures: the new mobilities paradigm, a politics of infrastructure, and critical accounts of neoliberal urbanization. Taken together, these enable an approach to mobility in broad social context, escaping both the technical myopia of infrastructural planning and the equally narrow, subjective perspective of much cultural analysis of movement (Prytherch and Cidell, this volume).

The new mobilities paradigm (see, for example, Cresswell 2006, 2010; Sheller and Urry 2006) is a starting point for thinking about the functions of mass urban transportation in the contemporary conjuncture. Authors such as John Urry and Mimi Sheller have done much to reorient social science away from the stationary categories of subjects, spaces, and order toward the dynamics of relationality, flows, and processes. Providing rich, qualitative accounts of broadly conceived 'movements,' this scholarship shows how mobility can best be understood in relation to contingent political, cultural, economic, and social forces (Urry 2007).

Transportation is never merely a technology to move people or goods from one point to another, but is a system that is mediated through existing social worlds and that gives rise in turn to particular types of society. Urry (2007) refers to these transportation assemblages as "mobility systems." The systems framework makes visible constitutive connections between multiple aspects of movement, and it allows for historicizing and contextualizing a given mode of mobility in relation to its material conditions of possibility and to other mobility systems. In particular, this chapter considers the complex assemblage of rapid mass urban transit, or metromobility, against and alongside the system of automobility that has been hegemonic across much of the world for the past 50 years (Featherstone 2004; Sheller and Urry 2000; Thrift 2004; on automobility in France, see Offner 1993; Orfeuil 2001; Ross 1996).

Whereas the new mobilities scholarship offers rich accounts of the experiences of movement across a variety of contexts, it often deemphasizes the economic frameworks shaping these movements, as well as political conflicts embedded in networks. The language of mobility "regimes" suggested by Steffen Böhm et al. (2006), however, captures the structured patterns of control of interlinked social movements, as well as the relations of force and domination that accrue in and through transportation technologies. The focus on conflictual regimes of urban transportation rather than autopoetic systems dovetails with a related analytical endeavor to study what Colin McFarlane and Jonathan Rutherford (2008) call "political infrastructures." Infrastructures (of mobility or otherwise) are always highly contested material, symbolic, cultural, and political artifacts (Young and Keil 2010). In order to appreciate the political character of networked infrastructures, they need to be understood not merely as organic ecologies or the neutral results of rational planning, but as expressions of "congealed social interests" (Graham and Marvin 2001; see also Amin and Thrift 2002; Gandy 2005; Lorrain 2002). In a fundamental sense, mobility assemblages are products of local struggles, ideological plans, and uneven cementations of capital investment and disinvestment.

Attention to flows of capital through urban space, often conspicuously absent from mobilities literature, is particularly important to understanding regimes of metromobility in an era of neoliberalization.[2] Transit infrastructures are absolutely central to this transformation, providing amenities for elite residents, enterprises, and tourists (Harvey 1989; Graham and Marvin 2001), enabling interactions in a polycentric regional economy and the growth of clusters of competitive advantage (Sassen 2001; Amin and Thrift 2002), and serving as a symbol for the emancipatory possibilities of networked urban life (Castells 1996). Infrastructures of transportation in this era, however, also fracture or "splinter" urban space, segregating territories, differentiating mobilities by class, and serving as an instrument of racialized systems of surveillance and penality (Graham and Marvin 2001; on Paris in particular, see Wacquant 2008; Silverstein 2004).

The regime of metromobility is thus an essential political terrain where the logics and contradictions of neoliberalism unfold (see also Grengs 2005). Grand Paris brings into focus competing infrastructures, plans, development agendas, forms of regional growth, and modes of urban life. I turn to the Grand Paris transportation plan as an expression of the situated politics of mobility infrastructure in contemporary Île-de-France.

A CRISIS OF TRANSIT IN THE ÎLE-DE-FRANCE

The landscape of mass transportation in metropolitan Paris consists of the iconic metro and regional rail lines (*Réseau Express Régional*, or RER), tramways, and buses. The mass transit network is operated jointly through the state-owned Autonomous Operator of Parisian Transports (RATP), the National Society of French Railways (SNCF), and the regional organizing authority, the Île-de-France Transport Union (STIF).[3] While some 10 million users partake of the network on a daily basis, making over 3 billion trips each year (RATP 2013), the existing transportation infrastructure is ill equipped to meet the needs of the region and is the subject of much criticism.

Grand Paris identifies a generalized 'crisis of transit' and claims that an overhaul of regional transportation is needed in order to improve daily life and boost regional growth. Existing offerings are saturated and prone to disruption, suburb-suburb links are almost nonexistent (as a result, 80% of these commutes are done in cars), commutes are long and arduous (often upwards of 90 minutes), and vast areas of the region remain unconnected to one another in any viable fashion (INSEE 2007, 2010). Furthermore, westward shifts in employment away from Paris and toward the suburbs present a problem for static infrastructures and less mobile residential markets (INSEE 2006). A number of recent studies have concluded that Paris has one of the most unequal and centralized urban transit networks in the world (Panerai 2008).

Over the past 40 years, there has been a series of failed attempts to renovate the transportation network, but most have been set aside due to concerns over cost or abandoned due to interinstitutional disputes between the region of Île-de-France—which took primary control of mass transit through the decentralization laws of the 1980s—the central state, and local collectivities. The 1994 regional master plan of the Île-de-France (SDRIF) was especially influential for focusing development on suburban regions, based on an extensive transportation network outside of the Paris proper. What were originally local plans to increase regional coherence and equality of opportunity through improved daily service in and between *banlieues* have been challenged in more recent years by large-scale national initiatives to drastically reconfigure the metropolitan region as a prime space of economic accumulation, thus reestablishing Paris's rank among global cities.

Indeed, a number of highly developed local transit projects following the perspective of the regional master plan, including the Arc Express (see below), were in negotiations when Grand Paris was announced in 2007. In creating the Society of Grand Paris (SGP), an appointed public establishment to oversee the new transportation project, Sarkozy strategically removed power from the regional and municipal levels and partially reestablished transit and land use planning as capacities of the central state (Subra 2012;

Wiel 2009).[4] Christian Blanc, the first head of the SGP, criticized the master plan for its "localism" and lack of economic vision, and swiftly rejected the popular Arc Express. Instead, Blanc offered more a business-friendly alternative, the 'Great Eight' (also known as the Double Ring, for its two main loops), which would create rapid conduits of travel through the main airports of the city and which would connect hubs of business and enterprise on the peripheries with the core of Paris.

The Arc Express and the Great Eight not only represent regional versus national development priorities, but they propose divergent solutions to the transit crisis and two quite different visions of the future transportation network. Between 2007 and 2009 these competing visions—one local, with an emphasis on social inclusion, and one national, with an emphasis on economic competition—became locked in a stalemate. In order to break the impasse between the two alternatives, in 2009 the National Commission on Public Debate (CNDP), a nonpartisan entity, announced a series of official 'public debates' on the draft documents of Arc Express and the Great Eight, now officially renamed the Public Transport Network of Grand Paris (RTPGP). These double debates were meant to provide a venue for stakeholders to review the plans and a forum to more generally discuss the crisis of regional transportation, as well as potential solutions.

The debates comprised one of the largest public planning exercises ever implemented. In addition to attending public meetings, stakeholders (*intervenants*) could contribute brief written opinions, more argumentative contributions, or comprehensive actor workbooks. Interested parties could also review technical plans and post questions on the respective websites (http://www.debatpublic-arcexpress.org/ and http://www.debatpublic-reseau-grandparis.org). Over 20,000 people attended in person, while hundreds of organizations contributed written submissions and almost 300,000 visitors were logged at the respective websites (CNDP 2011a, 2011b).[5]

The debates are self-described as "an exceptional mine of information" (CNDP 2011c, p. 3), and, indeed, they are a remarkable archive of various meanings of urban mobility, problems of commuting, and perspectives on the nexus of transportation infrastructure and metropolitan development. Yet the limits of the forum are patent. Many mayors complained that no meetings were held in their communes, pointing to the exclusionary nature of public sphere participation. One participant called them a "democratic illusion," as the SGP retained full discretionary power over collective transit planning in the last instance (CNDP 2011b, p. 24). Perhaps the strongest criticism rests on the fact that many of the most influential participants were not individual citizens, but lobbyist organizations such as business improvement associations, corporations, and property developers. Despite, or indeed because of, these limits, the debates reveal much about the contested system of metromobility and the struggles over urban meaning in metropolitan Paris.

DEBATING REGIONAL FUTURES: FROM COMPETITION TO COMPROMISE

The plans for both Arc Express and Great Eight/RTPGP are rooted in a shared belief that it is in the interests of Paris and of France to strengthen and expand the current transit network and that public authorities should direct these changes. The two plans differ, however, in terms of their goals and features and in terms of the values of urban life that they espouse.

Arc Express is a moderate plan designed to bring large parts of the city into the same finely knit transit web and to encourage communication and interaction between disparate parts of the regional territory. The authors of Arc Express stress that the residents of the near *banlieue* are their top priority, and they promise a 'fine service' of interwoven transit options that will '*desenclaver*' (liberate) marginal territories and communities. The project is oriented primarily toward territorial and social development—favoring social equality and cohesion, a re-equilibration of services, a reduction of territorial inequalities, and an increase in living standards—but supporters highlight that the improvement of transit services is also a matter of increasing competitiveness, opening up territory to investment and stimulating the economy. In conjunction with its social aims, the Arc Express will, for example, "develop a dynamic Île-de-France and maintain its global influence" (STIF 2010, p. 9).

Great Eight/RTPGP, however, is intended to serve both the center of Paris and the near and far peripheries, with a priority placed on connecting Roissy and Orly airports with the business district La Défense and existing and emerging scientific, informational, and cultural 'hubs.'[6] The Society of Grand Paris argues that the capital region must create transit amenities to welcome high-value added industries, to restore a lost regional attractiveness, and to realize the potential values of underutilized peripheral land. Christian Blanc sums up the RTPGP agenda of gentrifying and mobilizing space for advanced production, stating that "[i]n this knowledge world, territory is not the framework of the economy, it is its motor" (quoted in Société du Grand Paris 2010, p. 36). In contrast to the Arc Express, the emphasis here is undoubtedly on economic development, but economic objectives are seen to be coextensive with social ambitions of territorial balance and solidarity.

Materially, the two transit systems would consist of different structures and would move people in divergent fashions around the city. Arc Express is an underground automatic rail line consisting of a 60 km ring around the inner suburbs. The stations are located 1.5 km apart and include correspondences with 38 existing RER, metro, and tramlines. The vehicles of Arc Express travel 25–40 km/h, and the network has a capacity of 30,000 passengers in peak hours and 1 million passengers per day (CNDP 2011a, p. 4). The estimated cost for the loop, €7.1–€8.3 billion, is to be financed locally by the region and local tax initiatives.

The Great Eight/RTPGP consists of three high-speed underground rail lines completing two large loops that cut through Paris and extend toward targeted peripheral hubs at a total length of 130 km. The automated trains run at an average of 65 km/h, significantly reducing travel times between targeted poles. According to estimates, the network would be capable of providing 2–3 million trips per day (CNDP 2011c, p. 5). The gains in attractiveness are said to justify the large state expenses of the RTPGP, estimated between €21.3 and €23.5 billion, to be financed by stakeholders, new taxes, and partially by debt.

The Arc Express and Great Eight/RTPGP can thus be understood according to two imperatives of urban growth and two conceptions of what a transportation network should do. Whereas Arc Express sees the transportation system as being necessary to facilitate daily commutes, mainly for employees in the *banlieue* to their place of work, RTPGP sees the network as primarily easing travel between nodes of international connection. The region and the Île-de-France Transport Union define the city as a place of inhabitation and transit as a public good that should be in the service of all, while the Society of Grand Paris defines the city as an economic motor of the national economy and transit as a means of channeling investment and clustering competitive industries.

There are clear divisions between them, and yet, as the debates progressed, dozens of actors called for a "fusion," a "rapprochement," or a "convergence" between the two plans (CNDP 2011b, p. 19). In mid-November 2010, halfway through the debate proceedings, Jean Paul Huchon, the regional president of Île-de-France, announced a merger of the two projects and declared that the remaining scheduled debates should consider not the merits of one or the other plan, but the possibility of their consolidation. The CNDP (2011c, p. 4) followed suit and officially reframed the terms of the debate from comparison, competition, and argumentation to reconciliation, complementarity, and convergence.

This move would prove decisive in the struggles over the future of mass transit in Paris. In what has come to be called an 'unprecedented compromise' between the state and the region of Île-de-France, Maurice Leroy, Minister of Cities charged with Grand Paris, and Huchon announced in January 2011 the blueprint for the combined Grand Paris Express (GPE). On May 26, 2011, Huchon and Leroy, along with other key players in regional governance, officially unveiled the GPE network that would finally realize a 'Paris for all Parisians.' The main priority of the Grand Paris Express is to build a new, high-speed (60 km/h), automatic, and 24-hour ring metro. This ring will have a length of 150 km and consist of over 50 stations, while additional branches in the network will connect business poles and major national and international infrastructures. The GPE aims to connect existing productive districts to one another and to the center of Paris, and to catalyze the development of a number of new strategic, economic, cultural, and commercial sites in the suburbs.

Representing a vast public investment, the GPE network has a projected cost of €32 billion (€9 billion from the state and region, respectively, €7 billion from new taxes, and €7 billion from debt) and is set to be operable by 2025. Although costlier and more ambitious than either of the initial proposals, the GPE closely resembles the state-sponsored RTPGP. Despite this, there was remarkable approval of the GPE across the political spectrum, and most conservative, socialist, and green party members found it satisfactory.

Public debates through the CNDP were neither a technical appraisal of two blueprints nor a battle between purely partisan or institutional interests. The discussions passed well beyond "infrastructural technocratics" (Orfeuil and Wiel 2012, p. 110) into questions of housing and employment, the growing disconnect between employment and residence, and the social and spatial inequalities that cut through the region. Issues of governance, land use, and urban services were frequently invoked, but so too were those concerning quotidian experiences of commuting. Interventions also spoke to more esoteric issues: the meaning of the metropolitan scale, the collective identities of neighborhoods and communities, the link between urban and economic development, the value of 'natural space,' and the substance of democracy, the public, equality, and solidarity. They undoubtedly represent a rich dialogue on the heterogeneous ways that transit is understood and experienced in the city and the manifold stakes of new infrastructure initiatives.

The broad discourses of mobility utilized in Grand Paris, however, narrowed greatly over the course of the initiative. The eventual consensus in support of the Grand Paris Express establishes a regime of metromobility in which slippery social, economic, and ecological values of global city development are pursued together. With this obfuscation, possibilities for a critical intervention in the region are reduced.

THE REGIME OF METROMOBILITY

After decades of neglect, mass urban transit is today a crucial aspect of spatial and economic planning in the Île-de-France. The Grand Paris transit debates and the broad discussions they engendered point to an interest in non-car modalities of movement, and highlight a coherent approach treating infrastructures of mass transportation as an integrated element in the politics of urbanism (Merlin 2012). The métro may not be replacing the car as the driving force of urbanization, but metromobility is one of the key sociotechnical institutions through which the 21st-century urban region is being produced.

The Grand Paris regime of metromobility has several main aspects. The Grand Paris Express, like the Grand Eight/RTPGP, is poised to capitalize on space by improving land values around new stations and providing a substratum for comparative advantage clusters where the knowledge and information of the new global economy are produced. What the multipolar model at the base of both of these plans provides above all is a map for 'spatial fixes' to ensure that the surplus capital of the urban is reinvested in areas that are ripe to produce future profits. This invariably targets the high-income segments of the population and high-productivity economic activities. The main beneficiaries of this speculative process are thus not inhabitants themselves, but financial and real estate actors who profit from closing rent gaps and trading attractive debt on securities markets (JP Morgan 2013).

It is clear that infrastructures will continue to reproduce the conditions for accumulation and capitalist flow, but in a neoliberal climate these infrastructures are also necessarily bound up with social reproduction and inhabitation. Indeed, the traditional contradictions in transport between use value and exchange value collapse under the regime of metromobility, as transit networks invariably impact quality of life, the 'public good,' and democratic urbanity. The métro promotes a mobile *way of life* that, in an era of place marketing and competition, can itself be sold to the world. Mass transportation does much symbolic work today beyond moving people (Enright 2013; Siemiatycki 2005a,2005b; Lee, this volume).

Due to the fact that the Grand Paris Express is so thoroughly embedded in urban life, it is at the heart of the metropolitan question more generally. While the GPE presents itself as a panacea to the region's woes, it is important to keep in perspective that the emerging regime of metromobility is a site of struggle, inequality, and antagonism. Mass transit and networked infrastructures are not absolute 'goods,' but are wrought by political battles and constitutive tensions. These tensions were raised in the debates, but were glossed over, thus obscuring the contradictory goals to which the two prospective transit systems aimed. Paradoxically, what the debates reveal more than anything is the extremely limited scope of thinking around the meaning of transportation and the possibilities of infrastructural creation.

There is an overriding assumption within Grand Paris (exemplified in the compromise on the GPE) that the multiple objectives of transportation are complementary and mutually reinforcing. This "illusion" (Orfeuil and Wiel 2012) rests on extremely undialectical understandings of urbanization and of mobility as without conflict and without contest (on the need for a dialectical account of mobility, see also Addie, this volume). Perhaps most importantly, the ideology of mobility embedded in the transit network displays a willed ignorance of patterns of uneven development and especially of the way in which the production of ground rent—perhaps *the* defining feature of the contemporary metropolis—is itself tied to the inequality and sprawl that mass transportation purports to address.

The privatized intensification and diversification of transit-oriented development upon which GPE is thus based consists of a contradiction: In increasing ground rent around new transit stations, it risks further peripheralizing marginal populations, thus reducing access to the city and mobility and precluding, if not outright defying, goals of balance and equity (Orfeuil and Wiel 2012). Furthermore, there is what Young and Keil call a "technico-material bias" to such a linear system that "predestines the places located between designated destinations to lie in a fallow land of unsatisfactory access" (2007, p. 88). Merlin (2012, p. 194) echoes this concern. This model, writes Merlin, may increase certain forms of densification around stations, but will likely have the opposite effect elsewhere.

If the land-rent contradiction brings to the fore a clear tradeoff between equity and economic aims, similar tensions also exist in the transit plans between speed and service, regional and local interests, megaprojects and sustainability. The GPE dissimulates a unity of urban planning practice but fails to interrogate the precise mechanisms linking transit to urban development and the conflicts that underlie this relation.

The ideological force of metromobility is particularly obfuscating because 'mobility' and 'networks' continue to have overly positive connotations. Unquestioned support for mass transit is rooted in the notion of a *maillage*, which at once invokes the fine mesh of service and the corporate-speak of 'networking' and connecting. In this framework, all areas and peoples are bound to each other through various nonhierarchical means. The networked, connected metropolis is a powerful symbol that promises universal access, freedom, and prosperity, and benefits in a dense network are presumed to accrue to all equally. Yet it should be noted that the benefits of networks are not undifferentiated, nor are they guaranteed (Leitner and Sheppard 2002).

CONCLUSION

Beyond the ideological and stakeholder differences of distinct plans like Arc Express and Great Eight, advocates agree that transportation networks must be built to ensure development in the region on the premise that circulation is good for economic and social life. And both sides agree that high-value developments focused around the new stations will be good for the region. In one reading, this can be understood as the growing centrism of mainstream French political culture and the cross-party use of a shared populist rhetoric. This interpretation is not wrong, but it alone is insufficient. Another interpretation is also needed—one that acknowledges the sheer complexity of transit development in the contemporary conjuncture.

Continued attention to the material and ideological significance of mobility systems is necessary, yet this task is daunting, especially when the broad language of mobility systems is incorporated into policy. The discursive breadth of megaprojects of transportation, like new megaprojects more generally, makes them difficult to oppose and question (Lehrer and Laidley 2008). Promises of widespread social, cultural, and ecological benefits leave little room for criticism, yet the diverse invocations of mobility and networked urbanism do not weave seamlessly into a neoliberal rationality that reveals itself in the idea of the global city. Transit-oriented metropolitanization is a contradictory and contested terrain that largely exceeds conventional theorizations.

The concept of metromobility is useful for grounding analyses of mobility more firmly in critical soil and for drawing out the inherently political aspects of infrastructure. While this chapter has focused on the particular dynamics of metromobility in the context of Grand Paris, similar analytics can, and should, be applied elsewhere. Given the global interest in mass urban transit today, such a critical, comparative framework is urgently needed. Metromobility—at once about movement and about the metropolis—is an influential regime that demands further attention.

NOTES

1. I understand metromobility as primarily based on metro (subway) networks in urban regions. But this system is necessarily multimodal and is intimately linked, for example, with tramlines, buses, local light rail, and regional rail lines, and increasingly, international airports, bicycle routes, and pedestrian paths.
2. France has often been seen as an outlier in the wave of Anglo-American style policies that have swept the globe, but recent scholarship has demonstrated the pervasiveness of neoliberalism *à la francaise*, in which the *dirigiste* state leads the way in deregulation, privatization, and spatial reorganization (Baraud-Serfaty 2011; Dikeç 2007; Jobert and Théret 1994; Renard 2008).
3. The Professional Transport Organization of Île-de-France (Optile) also regulates select suburban bus routes under the authority of the STIF.
4. These intergovernmental disputes as a result of decentralization are also overlaid with partisan battles. During Sarkozy's presidency, the central state was led by the right-wing Union for a Popular Movement, while the region and the majority of its communes were headed by communist and socialist officials.
5. There was extensive media attention to the debates at both the national and regional level. The press coverage outside of the region underlines the fact that the debate was not just a local matter, but concerned metropolitan and mobility issues at multiple scales.
6. Depending on the iteration, there are between six and nine main poles to be serviced by the RTPGP.

WORKS CITED

Amin, A. (1995) *Post-Fordism: A Reader*, Malden: Wiley-Blackwell.
Amin, A., and Thrift, N. (2002) *Cities: Reimagining the Urban*, London: Polity Press.
Auge, M. (2002) *In The Metro*, Minneapolis: University Of Minnesota Press.
Baraud-Serfaty, I. (2011) "La nouvelle privatisation des villes," *Esprit*, 3:149–167.
Béhar, D. (2013, January 28) "Les paradoxes du rôle de l'Étatdans la gouvernance du Grand Paris," *Métropolitiques*. Retrieved January 24, 2014 from http://www.metropolitiques.eu/Les-paradoxes-du-role-de-l-Etat.html.

Béhar, D., and Estèbe, P. (2010) "Grand Paris: l'Impasse du grand soir institutionnel," *Esprit*, 5:177–180.
Bridge, G., and Watson, S. (eds) (2011) "Reflections on mobilities," in *The New Blackwell Companion to the City*, Malden: Blackwell, pp. 155–168.
Böhm, S., Jones, C., Land, C. and Paterson, M. (2006) "Impossibilities of automobility," in *Against Automobility*, Malden: Wiley Blackwell, pp. 3–17.
Castells, M. (1996) *The Rise of the Network Society: The Information Age: Economy, Society, and Culture*, Malden: Wiley Blackwell.
Commission Nationale du Débat Public (CNDP) (2011a) "Compte-rendu du débat public Arc Express." Retrieved January 24, 2014 from http://www.debatpublic-arcexpress.org/debat/compte-rendu-bilan.html.
Commission Nationale du Débat Public (CNDP) (2011b) "Réseau de transport public du Grand Paris: Compterendu du débat public." Retrieved January 24, 2014 from http://www.debatpublic-reseau-grandparis.org/site/DEBATPUBLIC_GRAND PARIS_ORG/DEBAT/COMPTE_RENDU_ET_BILAN_DU_DEBAT.HTM.
Commission Nationale du Débat Public (CNDP) (2011c) "Bilan du débat public sur le réseau de trasnport public du Grand Paris." Retrieved January 24, 2014 from http://www.debatpublic-reseau-grandparis.org/site/DEBATPUBLIC_GRAND PARIS_ORG/DEBAT/COMPTE_RENDU_ET_BILAN_DU_DEBAT.HTM.
Cresswell, T. (2006) *On the Move: Mobility in the Modern Western World*, New York: Taylor and Francis.
Cresswell, T. (2010) "Towards a politics of mobility," *Environment and Planning D*, 28(1):17–31.
Dallas, G. (2008) *Métro Stop Paris: An Underground History of the City of Light*, New York: Bloomsbury Publishing.
Delaney, A., and Smith, G. (2006) *Paris by Metro: An Underground History*, Northampton, MA: Interlink Books.
Dikeç, M. (2007) *Badlands of the Republic: Space, Politics and Urban Policy*, Malden, MA: Wiley-Blackwell.
Dupuy, G. (1991) *L'Urbanisme des Réseaux: Théories et Methods*, Paris: Armand Colin.
Enright, T.E. (2013) "Mass transportation in the neoliberal city: The mobilizing myths of the Grand Paris Express," *Environment and Planning A*, 45:797–813.
Featherstone, M. (2004) "Automobilities: An introduction," *Theory Culture & Society*, 21(4–5):1–24.
Gandy, M. (2005) "Cyborg urbanization: Complexity and monstrosity in the contemporary city," *International Journal of Urban and Regional Studies*, 29:26–49.
Graham, S., and Marvin, S. (2001) *Splintering Urbanism: Networked Infrastructures, Technological Mobilities and the Urban Condition*, New York: Routledge.
Grengs, J. (2005) "The abandoned social goals of public transit in the neoliberal city of the USA," *City*, 9(1):51–66.
Harvey, D. (1989) "From managerialism to entrepreneurialism: The transformation in urban governance in late capitalism," *Geografiska Annaler. Series B. Human Geography*, 71(1):3–17.
Institut National de la Statistique et des Études Économiques (INSEE) (2006) No. 265: Déplacements domicile-travail: un desserrment de l'emploi parisienvers la grand couronne.
Institut National de la Statistique et des Études Économiques(INSEE) (2007) No. 1129: Les déplacements domicile-travail amplifiés par périurbanisation.
Institut National de la Statistique et des Études Économiques(INSEE) (2010) No. 331: Les Franciliens consacrent 1 h 20 par jour á leur déplacements.
Jobert, B., and Théret, B. (1994) "France: La consécration républicaine du néolibéralisme," in B. Jobert (ed) *Le Tournant Néo-libéralen Europe*, Paris: L'Harmattan, pp. 21–86.
JP Morgan (2013, February 11) *All Aboard the Grand Paris Express*, Europe Equity Research.

Lehrer, U. and Laidley, J. (2008) "Old mega-projects newly packaged? Waterfront redevelopment in Toronto," *International Journal of Urban and Regional Research*, 32(4):786-803.
Leitner, H., and Sheppard, E. (2002) " 'The city is dead, long live the net': Harnessing European interurban networks for a neoliberal agenda," *Antipode*, 34:495–518.
Lorrain, D. (2002) "Capitalismes urbains: La montée des firmes d'infrastructures," *Entreprises et Histoire*, (3):7–31.
McFarlane, C., and Rutherford, J. (2008) "Political infrastructures: Governing and experiencing the fabric of the city," *International Journal of Urban and Regional Research*, 32(2):363–374.
Merlin, P. (1967) *Les Transports Parisiens*, Paris: Masson et Cie.
Merlin, P. (2012) *Transports et Urbanismeen Île-de-France*, Paris: Documentation française.
Mongin, O. (2010) "Le Grand Paris et la réforme des collectivités territoriales," *Esprit*, 3:120–131.
Offner, J-M. (1993) "Les 'effets structurants' du transport: Mythe politique, mystification scientifique," *Espace Géographique*, 22:233–242.
Offner, J-M. (2007) *Le Grand Paris, Problèmes Politiques et Sociaux*, 942, Paris: La Documentation française.
Orfeuil, J.P. (2001) "L'automobileen France: Comportements, perceptions, problèmes, perspectives, Colloque International de l'Institut Pour La Ville En Mouvement, Marne La Vallée." Retrieved August 13, 2014 from http://ville-en-mouvement.pagesperso-orange.fr/interventions/Jean_Pierre_Orfeuil.pdf.
Orfeuil, J-P. (2012) "Le Grand Paris, ambitions et intendance," *Urbanisme*, 385:4.
Orfeuil, J-P., and Wiel, M. (2012) *Grand Paris, Sortir des Illusions, Approfondir les Ambitions*, Paris: Scrineo.
Ovenden, M. (2009) *Paris Underground: The Maps, Stations, and Design of the Métro*, New York: Penguin Books.
Panerai, P. (2008) *Paris Métropole: Formes et Échelles du Grand-Paris*, Paris: Editions de La Villette.
RégieAutonome des Transports Parisiens (RATP) Accueil (2013) Retrieved March 19, 2015 from http://www.ratp.fr/.
Renard, V. (2008) "La villesaisie par la finance," *Le Débat*, 148:106–117.
RépubliqueFrançaise (2010) "LOI n° 2010–597 du 3 juin 2010 relative au Grand Paris." Retrieved November 3, 2013 from http://www.legifrance.gouv.fr/affich Texte.do?cidTexte=JORFTEXT000022308227&dateTexte=&categorieLien=id.
Ross, K. (1996) *Fast Cars, Clean Bodies: Decolonization and the Reordering of French Culture*, Cambridge: The MIT Press.
Sarkozy N. (2009) "Inauguration de l'exposition 'Grand Pari(s).'" Retrieved April 20, 2012 from http://www.legrandparis.culture.gouv.fr/videosdetail/153/Inauguration%20de%20l%27exposition.
Sassen S. (2001) *The Global City: New York, London, Tokyo*, Princeton: Princeton University Press.
Sheller, M., and Urry, J. (2000) "The city and the car," *International Journal of Urban and Regional Research*, 24(4):737–757.
Sheller, M., and Urry, J. (2006) "The new mobilities paradigm," *Environment and Planning A*, 38:207–226.
Siemiatycki, M. (2005a) "Beyond moving people: Excavating the motivations for investing in urban public transit infrastructure in Bilbao Spain," *European Planning Studies*, 13(1):23–44.
Siemiatycki, M. (2005b) "The making of a mega project in the neoliberal city: The case of mass rapid transit infrastructure investment in Vancouver, Canada," *City*, 9:67–83.
Silverstein P. A. (2004) *Algeria in France: Transpolitics, Race, and Nation*, Bloomington: Indiana University Press.

Société du Grand Paris (2010) "Le Réseau de transport public du Grand Paris; Le dossier du maître ouvrage." Retrieved January 24, 2014 from http://www.debatpublic-reseau-grandparis.org/site/DEBATPUBLIC_GRANDPARIS_ORG/_SCRIPT/NTSP_DOCUMENT_FILE_DOWNLOADF0A3.PDF.
STIF (2010) "Cahier Central: Arc Express: Débatsur le métro de rocade." Retrieved January 24, 2014 from http://www.debatpublic-arcexpress.org/_script/ntsp-document-file_download5c2b.pdf?document_id=92&document_file_id=106.
Straw, W., and Boutros, A. (2010) *Circulation and the City: Essays on Urban Culture*, Montreal: McGill-Queen's University Press.
Subra, P. (2012) *Le Grand Paris: Géopolitique d'une ville mondiale*, Paris: Armand Colin.
Thrift, N. (2004) "Driving in the city," *Theory Culture & Society*, 21:41–59.
Urry, J. (2007) *Mobilities*, London: Polity.
Virilio, P. (2006) *Speed and Politics*, Los Angeles: Semiotext(e).
Wacquant, L. (2008*) Urban Outcasts: A Comparative Sociology of Advanced Marginality*, New York: Polity Press.
Wiel, M. (2009) *Le Grand Paris, Premier Conflit Né de la Decentralization*, Paris: L'Harmattan.
Young, D., and Keil, R. (2010) "Reconnecting the disconnected: The politics of infrastructure in the in-between city," *Cities*, 27:87–95.

10 Towards a City-Regional Politics of Mobility
In Between Critical Mobilities and the Political Economy of Urban Transportation

Jean-Paul D. Addie

INTRODUCTION

Questions surrounding the spatial politics of urban areas have gained increased prominence as the chameleon-like 21st-century metropolis "shifts shape and size [as] margins become centers, centers become frontiers; regions become cities" (Roy 2009, p. 827). Flows of people, capital, and information now integrate extended urban agglomerations via dynamic networks of connectivity and propinquity, while intensive processes of neoliberal globalization animate new scales of urbanization and urban governance (Scott 2012). Urbanity is expressed and codified at the regional scale in a manner that defies simple narrative of functional or spatial change. Our established notions regarding the territorial logics of *metropolitan urbanization*—characterized by political, social, and morphological binaries between the urban core and 'traditional' suburbs—no longer contain the relational flows and processes of a polycentric, globally integrated, *city-regional urbanization*.

The global nature of regional growth may be a key element of contemporary urban development, but such "internalized globalization" often sits in tension with the territorially defined interests of many local actors (Keil 2011). Service provision and policy formation remain predominantly conducted in and through bounded political units, despite the networks, flows, and social relations that transcend their increasingly porous borders. Capital may flow through global circuits, but accumulation regimes are necessarily grounded in particular spatial arrangements. Territoriality therefore remains a vital consideration for politics of representation, social provision, and mobility (Cox 2013; Morgan 2007). If the epistemological challenge of the 21st century is theorizing the city as a territorial space and circulatory system (Prytherch and Cidell, this volume), the task at hand is to account for how notions of mobility, connectivity, and their spatial politics are embedded within the topological and scalar dimensions of what McCann and Ward (2010) term the "relationality/territoriality dialectic."

Transportation infrastructures—both *technical* systems of highways, rail lines, and airports and *social* institutions and informal practices—provide a provocative lens to uncover how city-regions are produced, rendered visible, and governed. A significant governmental and civic consensus, from local authorities to global development agencies, now frames transportation as vital to building regional resilience by connecting technology to territorial expansion and selective densification. Mobility holds a further position of prominence in populist urban literature, whether in the form of international migration flows (Saunders 2010), global connectivity generated through airport-integrated development (Kasarda and Lindsay 2011), or technologies to tackle sprawl and proclaim "the triumph of the city" (Glaeser 2011). Yet normative understandings abound across these policy programs. Assumptions regarding the economic and environmental resilience of urban areas, as well as the transformative capacity of transport systems, are multiple (Brenner and Schmid 2014).

In practice, transportation planning and politics in an era of city-regional urbanization capture the global metropolis in a mix of rhetorical, technological, and socio-spatial change (Addie and Keil forthcoming). The contested development of transportation infrastructure at the city-regional scale divulges an ongoing, multiscalar negotiation of diverse communities, interests, and space-times. To understand the complex spatiality and spatial politics of city-regions we must conceptually account for both political-economic territoriality *and* multiple, co-present mobilities—from global economic activity to local practices of everyday life—which sit at the core of contemporary urbanization. This chapter seeks to reconceptualize city-regional politics of mobility through the insights, tensions, and political projects at the intersection of dialectical materialist urban analysis and critical mobilities studies. Their different underlying spatial ontologies invoke divergent theories of how urban power relations are produced and structured. But I suggest the sparks generated between the sociology of mobility and the political economy of transportation—when operationalized through Cresswell's (2010) concept of "constellations of mobility"—constructively illuminate contradictory tensions in the ways mobility is utilized, codified, and experienced under city-regional urbanization.

The argument presented here draws from an analysis of the global city-region of Toronto, Canada. At the present juncture, local, regional, and provincial politics in Southern Ontario are fundamentally conditioned by debates over urban transport. Mettke (this volume) analyzes the contested politics of mobility swirling around attempts to integrate the urban fabric of the amalgamated *municipality* of Toronto. In this chapter, I shift scalar focus to stress the importance of (often overlooked) transformations in the metropolitan periphery in reconstructing and redefining how *city-regions* are experienced, governed, and connected. Beyond the global city center, new built forms, globalizing infrastructures, and rhythms of everyday life are radically redefining how the metropolis functions and is understood. Because most urban growth now takes place in dynamic, multilayered, and splintered landscapes that are not fully urban, suburban, or exurban (Keil 2013), I engage Toronto as a pertinent case that highlights broad relations and mobilities internalized within a global era of city-regional development. 'Smart growth,' for example, has placed polycentric densification at the heart of contemporary regional planning. Yet, as new urbanizing hubs rise across the city-region, ubiquitous suburbanized spaces of the *Zwischenstadt* (or in-between city) (Sieverts 2003) are rendered "mere empty vessels to be filled with connective tissue meant to produce centralities elsewhere" (Young and Keil 2014, p. 1599). Here, I argue that illuminating the divergent production and lived experience of Toronto's *Zwischenstadt* opens the potential for a progressive politics of mobility to materialize from the emergent polycentricism of the neoliberal city-region.

THEORIZING TRANSPORTATION IN AN ERA OF CITY-REGIONAL URBANIZATION

Transportation, Dialectical Materialism, and the Urban

The central argument guiding this chapter is that city-regions and their transportation infrastructures evolve in a dialectical relationship through which urban space is produced, differentially experienced, and transformed. The urban is neither a simple generic site over which social relations and restructuring processes unfurl, nor a pre-given universal condition or form. Rather, it is a theoretical category: a conceptual abstraction delineated by dynamic dialectical interconnections, conflicts, contradictions, and change within urbanization processes (Brenner 2014; Lefebvre 2003). Opposed to normative imaginaries of the city as a collection of physical objects (buildings, streets, people, infrastructure, etc.), urban regions appear "as the concrete, local articulation of processes of more general technological, economic and regulatory change" (Kloosterman and Lanbregts 2007, p. 54). Urbanization, by extension, is a "polymorphic, variable and dynamic" historical process *and* contested social product constituted via the urbanization of capital, consciousness, and social practice (Brenner and Schmid 2014, p. 750).

Within this dialectical materialist context, I seek to theorize *urban* transportation, not in terms of place-based notions of infrastructure 'in the city,' but through a focus on the sociotechnical systems that connect and mediate abstract, yet essential, social relations and the concrete spaces and practices of everyday life (following Lefebvre 2003, pp. 79–81). Conceptualizing urban transportation via this mode of inquiry foregrounds the key socio-spatial processes facilitated through the production, maintenance, and dissolution of transportation infrastructures.

As a key sector for strategic state intervention, urban transportation both enables/reflects the complexity of neoliberal territoriality *and* illuminates the relational mobilities of everyday spatial practice. Infrastructure restructuring discloses both deterritorialization associated with globalization (and the rise of the network society) and reterritorialization through which new social and scalar relations are produced. On one hand, as Graham and Marvin (2001) have influentially argued, publicly managed infrastructures have been increasingly 'splintered' from collective public systems via neoliberal processes of deregulation and privatization. The unbundling of existing infrastructure systems establishes 'premium network spaces' (e.g., toll roads, privatized express rail links) that integrate places into selective global frameworks through specialized development funds and public-private partnerships. On the other hand, restructuring urban transport planning, management, and governance reconfigures the internal structure and governance of city-regions themselves, via a "new territorial politics of collective provision" (Jonas 2013). Here, strategic investments in city-regional infrastructure offer a potential (albeit temporary) spatial fix to secure the conditions for social reproduction amidst neoliberal restructuring (Addie 2013; Enright, this volume; Jonas et al. 2014).

Urban transportation is therefore inextricable from evolving modalities of political and economic power. Transformations are realized through the interaction of new technical/spatial innovations (e.g., new technologies, or reconfigured networks and governance regimes) with particular developmental trajectories and their associated social, institutional, economic, and environmental obduracies (Maassen 2012; Mettke, this volume). New topological networks may tie together a privileged archipelago of elite global nodes, for example, but in doing so they also establish differential access and processes of infrastructure 'bypassing,' with uneven development, marginalization, and exclusion the result (Graham and Marvin 2001; Young and Keil 2010). This dynamic is well captured by the "dialectic of centrality" described by Lefebvre (1991, pp. 331–334). Rather than refer to a physical location ('the city') concentrating things, activities, and processes, Lefebvre understands centrality as a social product that internalizes moments of gathering-inclusion and dispersal-exclusion. Privileged products, people, and symbols are brought together at the same time that dissident or undesirable elements are peripheralized. The center and periphery exist in a complex relation that reflects the logic (i.e., political rationality) and contradictions of urban space more widely. Such tensions between materially embedded spatial structures and social dynamics lead city-regions to internalize an amalgam of dialectics: centrality/marginality, concentration/extension, fixity/fluidity (Harvey 1996, p. 419). The mobilities of everyday life—whether for social reproduction, work, or play—are thus always emergent and contested, over multiple pathways and scales.

Bringing in the Mobilities Turn

Recent commentaries, including the contributions to this volume, have asserted that the new mobilities paradigm offers a powerful lens to conceptualize how cities are experienced, reconfigured, and reimagined (Jaffe et al.

2012; Skelton and Gough 2013; Watt and Smeats 2014). Amidst this surge of literature, there appears to be the potential for constructive synergies between the projects of dialectical materialist urban analysis and the mobilities turn to illuminate the spatiality, politics, and possibilities of city-regional urbanization. Both theoretical frameworks reject normative interpretations of the city as constituted by a static, universal ensemble of material objects. A focus on territorial boundaries may assist in locating spatially defined institutional responsibilities, but MacLeod (2011, p. 2651) argues shifting our ontological focus to mobility, and networks can enable us to identify those processes of connection and fissure that shape city-regions. By presenting a world in which "mobility is an ontological absolute" (Adey 2006, p. 76), critical mobilities research challenges conceptions of fixity, stasis, and stability so historically central to the social sciences. Indeed, the ontological elevation of mobility is a potentially productive move that draws attention to the experience and regulation of spatial practice in a manner often overlooked in political economy accounts of regional restructuring.

Reconciling dialectical materialism and the logic of the mobilities turn, however, is a conceptually and politically problematic maneuver. Much critical mobilities scholarship attempts to leverage the idea of dialectics to draw out the tensions between mobility and immobility. Urry (2003), for instance, places significant emphasis on the "mobility/mooring dialectic" as a means to provide movement with context. The central contradiction pivots on the assertion that mobility is premised on immobility; infrastructure systems are moored in place, but facilitate numerous mobilities that provide the mechanisms and context through which modern life functions. Yet the notion of mobility and movement at the core of critical mobilities analysis does not simply equate to concepts of change, contradiction, and sublation central to dialectical materialist thinking, methods, and critique (see Ollman 2003). As a program concerned with asserting the conceptual primacy of mobility itself, the mobilities turn remains decidedly one-sided and reticent to operationalize strong dialectical reasoning. Mobilities analysis may be utilized to engage conceptual and political issues beyond a central concern with unveiling the social and symbolic meanings of peoples and objects' movements (Hannam, Sheller, and Urry 2006). Still, the paradigm's tendency to focus on issues of governmentality, representation, and experience and sociocultural analyses of personal mobility, identity, and affect limits its capacity to illuminate the political and economic challenges of contemporary urbanization. By contrast, dialectical materialism presents a strong explanatory framework to this end, grounded in the concepts of accumulation, class, property, rent, and uneven development (Brenner et al. 2012). Dialectical materialism and critical mobilities studies also function with divergent theories of relationality. Whereas poststructural strands of the 'relational turn' construct the relational (as network based) in opposition to the territorial (as place based) (e.g., Amin 2004; see Jacobs 2012), dialectical approaches frame relational space as an epistemological lens that is held in contradictory tension with the absolute and relative qualities of space (Harvey 2006).

Although significant tensions exist between the ontological foundations of the mobilities turn and dialectical materialism, I argue that critical mobilities' mandate for a politics of mobility does provide scope to generate some productive sparks when interrogated through strong dialectical reasoning. In particular, Cresswell is concerned with uncovering the geographical and historical specificity of distinct "constellations of mobility" that together help illuminate "the ways in which mobilities are both productive of . . . social relations and produced by them" (2010, p. 21). As spatially and temporally contingent formations of movement, practice, and discourse, constellations of mobility contextualize the meaning and politics of mobility at specific junctures; what Harvey (1996) would construct as historically specific socio-spatial "permanences." The concept and political provocation of constellations of mobility strongly resonate with debates on city-regions that: (1) foreground the contradictory flows and fixity underpinning the relationality/territoriality dialectic and (2) prompt questions of centrality, access, and the right to produce and use urban space. We can therefore place the theoretical and empirical challenge of *connectivity*—the material and symbolic networks that facilitate (and prohibit) practices of mobility—at the nexus of critical urban theory and the mobilities turn. This move, I suggest, holds the potential to yield innovative engagements with contemporary urbanization and social struggles within the complex, contested, and contradictory landscapes of globalizing city-regions.

Engaging Terrains of City-Regional Mobility Through the *Zwischenstadt*

Whereas the city-region may be the principal scale at which people experience everyday life (Storper 2013, p. 4), processes of functional specialization and segregation render much city-regional space highly fragmented and amorphous. In this context, the concept of the *Zwischenstadt* provides a provocative frame to engage the challenge of connectivity amidst city-regional urbanization. At once an indistinct environment produced to be transgressed at high speeds by privileged groups and a place of everyday spatial practice and inhabitation, the *Zwischenstadt* reframes the city-region through an "unbounded yet also newly re-hierarchized" architecture of urban spaces (Keil and Young 2011, p. 4). Rather than being reduced to spatially static forms that can be readily drawn on a map or identified through positivistic indicators, the in-between city is best conceived of as a relational space (in the dialectical sense) which is represented, perceived, and experienced in qualitatively differentiated ways by different users (*pace* the mobilities turn).

The emergent interactive patterns of the city-region beyond the urban core "are less like its blocky spatial layout and more like the entwined overlay of paths and nodes in a rainforest, where clearings and connections for different uses are mixed together, connected by twisting links, lacking any easy visible order" (Kolb 2008, p. 160). The ability to move through this landscape—that is, the ability to experience and actualize the connectivity of city-regional space—is habituated by highly unequal power relations that invoke specific forms of "kinetic elitism" (Cresswell 2006). As Sieverts puts it, the *Zwischenstadt* "permits the widest variety of action spaces and connections or as a 'menu' with the help of which inhabitants can put together for themselves *à la carte*, provided they can afford it" (2003, p. 71). In the following, I unpack the constellation of mobility emerging through the in-between spaces of the Toronto city-region to highlight the multiscalar and multifaceted nature of urban transportation and point towards how a critical dialectical reading of this analytical lens can foster a progressive city-regional politics of mobility.

OPENING "CONSTELLATIONS OF MOBILITY" IN THE TORONTO CITY-REGION

The challenge of providing, scaling, and governing urban transportation holds a prominent position in Toronto's regional politics. Postwar experiments in metropolitan government and infrastructure provision secured Toronto's reputation as 'the city that works.' Yet the nearly three decades of largely uncoordinated neoliberal growth ushered in after the breakdown of Toronto's spatial Keynesian fix in the 1970s fostered expansive and largely unsustainable auto-dependent development across Southern Ontario (Addie 2013).

Following their rise to power in 2003, the Ontario Liberal Party has promoted an integrated program of land use, environmental, and transportation policies for the Toronto city-region to address the pressing issues of sprawl and congestion. The mutually reinforcing Places to Grow and Greenbelt Acts, introduced over 2005–2006, established a legal planning framework to structure regional development in Southern Ontario. In addition to these classic governmental land use regulation tools, the provincial government also established Metrolinx as a sectoral instrument to oversee transportation planning and implementation in the Greater Toronto and Hamilton Area (GTHA). Metrolinx's regional transportation plan, The Big Move (2008), detailed a system of growth centers, transport corridors, and no-growth areas around which Toronto's city-regional urbanization is now directed. The Big Move responded to the limitations of the region's existing technical and social infrastructure, yet its top-down assertion of regionalization also internalized the imperatives of global economic competitiveness and resilience within the province's strategic transportation plans.

Producing Peripheral Centralities

Taking their lead from the province's mandate, many of Toronto's municipal neighbors actively embraced a reframed planning agenda intended to support polycentric, smart, and sustainable urbanization. Newly planned and competing suburban 'downtown' developments have rapidly risen up along the regional arteries of Highway 7 and Highway 427. As sites of emergent urbanity, provincial growth hubs centered in the cities of Mississauga, Brampton, Vaughan, and Markham challenge the primacy of the city of Toronto and radically reorient the center-periphery dynamics of the city-region. Polycentric development discourses now lock in the normative goal of transforming suburbs into cities in their own right; as the city of Vaughan's 2020 strategic plan puts it, "[transitioning] from a growing suburban municipality to a fully urban space" (2011, p. 1). Urbanization in this context encapsulates the desire to concentrate economic and social activity, and direct prioritized modes of regional mobility, into new centralities located in the previously peripheral spaces of the city-region.

Urban transportation and mobility are central to understanding city-regional urbanization and state intervention in Southern Ontario. Densification, mixed-use development, and multimodal transport planning attempt to restructure everyday suburbanism away from metropolitan lifestyles traditionally understood and experienced through automobility and the single-family home. As urban growth accelerates in areas lacking in established transit infrastructure, there is an emergent recognition across the GTHA that transit service cannot stop at municipal boundaries, nor concentrate on moving people downtown. This issue does not simply rest on service integration or the introduction of interjurisdictional routes, but on the establishment of common visions, practices, and political synergies.

Transit networks, coordinated with the goals of the Places to Grow framework, provide a sociotechnical infrastructure to guide the ongoing development on the GTHA's urban periphery. Most prominently, Viva, York Region's bus rapid transit (BRT) system, has introduced vital material and governance technologies supporting new centralities north of the city of Toronto. The public-private partnership underpinning Viva's establishment in 2005 not only brought together global capital and expertise to introduce BRT service but also supported a restructuring of the region's physical infrastructure. Express bus service initially operated in mixed traffic along the Yonge Street and Highway 7 corridors (with connections at rapid transit terminals in Northern Toronto), but in 2009 construction commenced on a network of dedicated, bus-only 'rapidways' and 'Vivastations' in the center of York Region's key arterial routes. The material reimagining of regional mobility has been matched by the concerted cultivation of a high-end transit experience to reach people who would traditionally view riding the bus as a step down (see York Region Transit Corporation 2014). Evident in this moment of infrastructure restructuring is the extension of both technical (unbundling express regional and local routes) *and* cultural splintering that belies the complex sociopolitical dynamics found within Toronto's city-regional constellation of mobility. As such, drawing from the central concerns of the mobilities turn and dialectical materialism highlights the importance of generating new representations and experiences of regional mobility for Viva while disclosing the resulting production of distinctly uneven geographies of connectivity within contemporary city-regional space.

Copresent Mobilities and Capital Across the Urban In Between

City-regional urbanization integrates metropolitan space, but in doing so produces fragmentation and differentiation, as captured in Lefebvre's 'dialectic of centrality.' Dramatic demographic and socioeconomic change in Southern Ontario has engendered the proliferation of numerous, qualitatively distinct practices of everyday life. Large-scale population growth driven by both endogenous, non-European communities and new immigrants, mostly from South and East Asia, has transformed suburban populations—and the lived realities of city-regional urbanism itself—towards greater heterogeneity. The emergence of social diversity in the GTHA's suburbs has also been accompanied by the suburbanization of poverty, as societal issues long thought of as belonging to the inner city—including concentrations of transit-dependent populations—are brought to the fore across the *Zwischenstadt* (Young and Keil 2014). At this socioeconomic intersection, transit riders—especially new immigrants who now call Toronto's suburbs home—face disproportionately lengthy commutes in terms of time and distance (Axisa et al. 2012; Turcotte 2011).

The diversity, densification, and social centrality emerging on the edges of the GTHA offer the potential to reimagine autocentric and fragmented spatial arrangements. Some novel programs have attempted to respond to evolving mobility requirements in Toronto's in-between landscapes. However, the city-region's morphology and established transport infrastructure have compelled the implementation of more individualized movement through polycentric city-regional space. For example, carpooling and workplace shuttle programs operated by Smart Commute (a collection of local transportation management authorities financed by Metrolinx and regional employers) provide an adaptive response to, and potentially sustainable fix for, the mobility challenges of city-regional urbanization. Yet, their limited capacity to move large numbers of commuters and strategic focus on supporting economic activity reflects the difficulty in realizing public transportation options that can accommodate complex mobilities over the vastness of the global city-region.

Although investments in specific transport technologies are not mutually exclusive, the province's understanding of regional space and territory—principally founded upon privileged network components which optimize regional competitive advantages—are prioritized among the multiple, overlapping mobilities of lived city-regionalism. Leveraging urban transportation to promote competitive regionalization is not unique to the GTHA. Enright (this volume) documents comparable struggles to scale a "regime of metromobility" in Grand Paris. City-regional transportation planning in both cases represents clear attempts to direct global flows of capital, territorialize them in place, and secure locational advantages. Urban densification and the infrastructure of regional mobility foster new social centralities that create new use and exchange values, which themselves are fundamentally tied to the production of urban land markets. This is evinced as private firms have been empowered to develop public sector air rights over high-order transit hubs as part of the public-private partnership directing BRT in York Region. Global capital underwrites the construction of condo and office towers around the province's targeted growth hubs as developers and local governments attempt to realize ever more profitable rents and tax bases. Consequently, the imperatives of capitalist urbanization, in its recent parasitic articulation, continue to underpin the territorial organization of the Toronto city-region (Harvey 2009).

The politics of local movement here is inexorably integrated into multiscalar political-economic circuits and power relations. Yet the globally focused regionalism emerging in the center of the Toronto city-region's constellation of mobility has become notably problematic for inner suburban communities that developed at the height of the postwar boom but now find themselves bypassed by regional integration and on the wrong side of intensified socioeconomic polarization. The economic limits to mobility faced by individuals prohibit access to premium network spaces; whether associated with the costs of purchasing and running a car, paying to access tollways, or covering the fares on regional express bus service.

Differential access is inscribed in the socio-spatial structure of Toronto's *Zwischenstadt*. As Mettke (this volume) argues, low-income and visible minority residents in the city of Toronto's (predominantly inner suburban) 'priority neighborhoods' lack rapid transit connections to downtown Toronto. But they are also problematically disconnected from the GTHA's emerging growth hubs. The Jane-Finch neighborhood, for example, is a mere five-minute drive from Vaughan's new downtown complex, yet for residents without access to a car, the trip would take at least half an hour on two buses and require the payment of two fares. The Toronto Transit Commission's Spadina Subway extension should bring rapid transit to Vaughan Metropolitan Centre by Fall 2016, yet although a planned station located at Keele Street and Finch Avenue is in close proximity to Jane-Finch, the line bypasses the neighborhood's existing low-income residential districts to target service at York University and areas amenable to new build development.

Intersecting Places and the Dialectic of Centrality

The relative distance between the downtowns emerging along Highway 7 and Toronto's inner suburbs starkly illuminates the need to integrate disconnected areas of the *Zwischenstadt* to the GTHA's new city-regional centers.

However, Toronto's in-between spaces often find themselves caught between local, regional, and global scales of mobility and the persistent territorial geographies of a previous era of metropolitan urbanization. Whereas the province has focused on building a network of regional mobility hubs, the city of Toronto's Transit City LRT plan looked to integrate marginalized, transit-deficient inner suburbs into the city's, rather than the city-region's, urban fabric (Addie 2013; Mettke, this volume). Dispersed growth beyond the compactness of the urban core remains conditioned by the logics of automobility. The cultural capital and availability of the car engrain it as an essential part of the *Zwischenstadt*, enabling the (selective) freedom to move but exerting disciplinary control over everyday life (Howe 2002), even as new traditions and innovations are overlaid upon existing political, social, and technological obduracies. This reflects both the challenges presented by the sheer scale of city-regional landscapes and the entrenchment of social, political and morphological divisions that continue to codify constituent parts of the GTHA.

Toronto's future transportation and, indeed, city-regional development are presently arrayed in a reductionist war of urbanisms. After the release of The Big Move and the province's approval of Transit City, right-leaning politicians and media outlets lamented that "Toronto's 'anti-car' council has focused almost exclusively on public transit . . . despite an additional 10 million cars bought by North Americans each year, Toronto has no plan to accommodate the extra vehicles" (Yuen 2009). Such arguments recast both the city of Toronto and Metrolinx's agendas as a 'war on cars.' Conservative Toronto city councilor Denzil Minnan-Wong editorialized:

> Our city doesn't need a transit policy—we need a mobility plan. A mobility plan recognizes that many people who drive in our city have to do so because of a host of life circumstances that transcend mere preference. The city's undeclared but very active war on cars is really a war on people who, for the most part, lack alternatives.
>
> (2009, p. A7)

Conversely, advocates of the central city's Jane Jacobs-style, middle-class progressivism insisted, "downtown density will prevail over the slums of suburbia" (Hume 2008, p. A10). Vast swaths of the wider city-region were discursively reduced to auto-dependent "flat, grey sprawl" (Macfarlane 2008, p. 15). The spatial politics codified through these discourses reinforce path dependencies that perpetuate normative understandings of city and suburb (see Cidell, this volume). In doing so, they obfuscate the interconnected nature of city-regional space and the emergent conceived and lived differentiation of movement within the *Zwischenstadt* (Fiedler 2011; Watt and Smeats 2014). The consequent framing of transportation politics overlooks the proliferation of regional spaces within the GTHA and the necessity of rescaling the politics of social reproduction, an issue especially vital for the nonprivileged global city workforce that needs to traverse fragmented, in-between city spaces to access affordable housing and employment opportunities (Goonewardena and Kipfer 2005; Jonas 2013).

Toronto's bourgeois central city transportation politics and spatial imaginaries obscure the urbane city center's dependence on a web of globalized infrastructure networks increasingly located in the urban periphery. The suburbanization of global freight operations lies at the heart of the production regimes, morphologies, and multiscalar flows of post-Fordist city-regions such as the GTHA (see Erie 2004). The extended complex of single-story distribution warehouses, industrial buildings, and factories—crisscrossed by a network of superhighways and rail lines and centered on Pearson International Airport—serves as the discursive and functional foil to the glamorous face of "global Toronto" (Keil and Young 2008). At once, this clearly discloses the dialectic of centrality as particular functions, social groups, and urban forms must be expelled to the periphery to produce Toronto's famed (image of) progressive urbanism. Yet concomitantly, it discloses the production of an alternate global centrality as a new suburbanized infrastructure fix supersedes the now antiquated facilities of previous accumulation regimes (see Hall, this volume). Accordingly, municipalities that are deeply integrated in global trade infrastructures "are at the leading edge of the new global logistics network *and* the leading edge of suburbanization" (Cidell 2011, p. 833).

The global is clearly territorialized through the diverse sociotechnical infrastructures found on the urban periphery, as much as the command and control functions concentrated in the core of the global city. A progressive city-regional politics of mobility therefore cannot simply be constructed in opposition to the automobilist, middle-class privatism traditionally associated with the suburbs. The amalgam of mobilities housed in and across Toronto's *Zwischenstadt* indicates such metropolitan dichotomies cannot hold. Rather, a socially just politics of collective provision must center on the multiplicity of urbanisms expressed within the lived city-region, contra to the tripartite power of the state, capital accumulation, and private authoritarianism.

CONCLUSION

This chapter has deployed a dialectical understanding of urban transportation to examine how city-regions are produced and governed as territorial and relational urban spaces. Developing an argument through the concrete spaces and grounded flows of the Toronto global city-region, I have asserted that the spatial politics of city-regional urbanization operate through a diverse constellation of social and spatial mobilities. Yet the urban cannot be understood without the territorial institutions, rules, and practices that arise from and regulate class and property relations (Lefebvre 1996,

p. 106). Urban land economies have emerged as vital markets through which overarching city-regional dynamics unfold. Contemporary urbanization processes disclose an accumulation regime in which "coercive laws of competition . . . force the continuous implementation of new technologies and organizational forms" and in turn structure the internal form and external relations of city-regional space (Harvey 2009, p. 316). Examining the transformations on the outskirts of the GTHA through the scalar lens of the global city-region demonstrates that strategic investments in urban transportation are reflective of, and in turn generate, new state strategies. These attempts to integrate and rationalize previously fragmented space are increasingly linked to the imperatives of neoliberal rent-seeking and global economic competitiveness.

The connectivities facilitating particular practices of mobility are therefore a central object of class struggle between abstract processes of 'urbanization as accumulation' and the production of the urban as a mode of everyday life. Although geographical distance between rich and poor may collapse across city-regional space, relative connectivity and the symbolic distance between center and periphery are greatly exacerbated. Everyday spatiality, as critical mobilities studies demonstrate, is perceived and lived in a fragmented and partial manner. Here, neoliberal logics of infrastructure provision have shifted towards the valorization of individual choice and atomized mobility in a manner that obscures the continued reliance on public infrastructures that enable such movement. The privileging of transport infrastructures that most benefit the productive capacity of the Toronto city-region forges distinct kinetic elitism by elevating the importance of one particular set of spatiotemporal rhythms within the relationality/territoriality dialectic. Here, the dialectic of centrality galvanizes the structural (rather than explicit) perpetuation of urban injustice for disconnected communities.

Whereas the city-region appears as the territorial spatial form *de jour* for global capitalism, the agora of the *Zwischenstadt* illuminates the simultaneous production of "space for forms of living which conflict with the globalized economy, for the slowness of an unmotorized existence and for withdrawal into self-sufficiency in times of crisis" (Sieverts 2003, p. 73). The emergence of such qualitatively distinct modes of city-regional urbanism could, however, open both conceptual and material spaces for forging more equitable pathways of social change, via a new politics of mobility. Lefebvre (1996) argued the introduction of centrality into peripheral zones offered the potential to transform marginalized spaces into actual *urban* space by extending the struggle against exclusions from space. Social centrality, though, operates simultaneously at different scales with tremendous repercussions for the spatial practices of urban inhabitants. Recognizing the structural complexity and multiple mobilities evident in global city-regions' particular expression of the *Zwischenstadt* is a necessary step in breaking past physical, mental, and social dichotomies reified through previous rounds of metropolitan urbanization (Kolb 2008). Pan-regional rapid transit, such as Viva-style BRT, presents a potential infrastructure fix to the challenges of city-regional urbanization, yet the introduction of new transportation routes and modes must negotiate an array of required uses and scales of mobility—local movement with frequent stops and fast, regional trips with limited access—if it is to avoid reproducing the marginality of many communities in the in-between city. Considering how people understand, live, and move through their daily lives within specific spatiotemporal contexts becomes a central concern for any spatial politics. This, of course, is a key component of critical mobilities studies and a challenge to which the lens of constellations of mobility is attuned.

Engaging dialectical materialist urban studies alongside critical mobilities analyses can help illuminate the constant material and experiential transformation that defines the urban process. Differential mobilities in the blurred, multiply topological space of city-regions establish new, urban "kinetic hierarchies" (Cresswell 2010, p. 29). When interrogated through a political-economic critique, such theoretical abstractions can inform an adaptive urban politics capable of incorporating and mobilizing new connectivities, centralities, and overlapping political relations. At the same time, it argues persuasively for democratizing their governance. The challenge here is twofold. First, it is necessary to recognize the diverse form, function, and structure of city-regional space, and second, expose the contradictory, crisis-prone tendencies evident in the commodified core and centralizing infrastructure. Constituting a politics of mobility based in uncovering the dialectical disturbances, displacements of centrality, and differential experiences of urban transportation can then offer the possibility for new forms of social justice, innovation, and creativity to develop among the dispersed and horizontal fissures of polycentric city-regional space.

WORKS CITED

Addie, J.-P.D. (2013) "Metropolitics in motion: The dynamics of transportation and state re-territorialization in the Chicago and Toronto city-regions," *Urban Geography*, 34(2):188–217.

Addie, J.-P.D., and Keil, R. (forthcoming) "Real existing regionalism: The region between talk, territory and technology," *International Journal of Urban and Regional Research*.

Adey, P. (2006) "If mobility is everything then it is nothing: Towards a relational politics of (im)mobilities," *Mobilities*, 1(1):75–94.

Amin, A. (2004) "Regions unbound: Towards a new politics of place," *Geografiska Annaler B*, 86(1):33–44.

Axisa, J.J., Newbold, K.B., and Scott, D.M. (2012) "Migration, urban growth and commuting distance in Toronto's commuter shed," *Area*, 44(3):344–355.

Brenner, N. (ed) (2014) *Implosions/Explosions: Towards a Study of Planetary Urbanization*, Berlin: Jovis Verlag.

Brenner, N., Madden, D.J., and Wachsmuth, D. (2012) "Assemblages, actor-networks, and the challenges of critical urban theory," in N. Brenner, P. Marcuse, and

M. Mayer (eds) *Cities for People, Not for Profit: Critical Urban Theory and the Right to the City*, New York: Routledge, pp. 117–137.
Brenner, N., and Schmid, C. (2014) "The 'urban age' in question," *International Journal of Urban and Regional Research*, 38(3):731–755.
Cidell, J. (2011) "Distribution centers among the rooftops: The global logistics network meets the suburban spatial imaginary," *International Journal of Urban and Regional Research*, 35(4):832–851.
City of Vaughan (2011) *Vaughan Vision 2020: The City of Vaughan Strategic Plan*, Vaughan, ON: City of Vaughan.
Cox, K.R. (2013) "Territory, scale, and why capitalism matters," *Territory, Politics, Governance*, 1(1):46–61.
Cresswell, T. (2006) *On the Move: Mobility in the Modern Western World*, New York: Routledge.
Cresswell, T. (2010) "Towards a politics of mobility," *Environment and Planning D: Society and Space*, 28(1):17–31.
Erie, S.P. (2004) *Globalizing L.A.: Trade, Infrastructure, and Regional Development*, Stanford, CA: Stanford University Press.
Fiedler, R.S. (2011) "The representational challenge of the in-between," in D. Young, P. Wood, and R. Keil (eds) *Inbetween Infrastructure: Urban Connectivity in an Age of Vulnerability*, Keolowna, BC: Praxis (e)Press, pp. 67–85.
Glaeser, E. (2011) *Triumph of the City: How Our Greatest Invention Makes Us Richer, Smarter, Greener, Healthier and Happier*, New York: Penguin.
Goonewardena, K., and Kipfer, S. (2005) "Spaces of difference: Reflections from Toronto on multiculturalism, bourgeois urbanism and the possibility of radical politics," *International Journal of Urban and Regional Research*, 29(3):670–678.
Graham, S., and Marvin, S. (2001) *Splintering Urbanism: Networked Infrastructures, Technological Mobilities and the Urban Condition*, New York: Routledge.
Hannam, K., Sheller, M., and Urry, J. (2006) "Mobilities, immobilities, and moorings," *Mobilities*, 1(1):1–22.
Harvey, D. (1996) *Justice, Nature and the Geography of Difference*, Malden, MA: Blackwell.
Harvey, D. (2006) "Space as a keyword," in N. Castree and D. Gregory (eds) *David Harvey: A Critical Reader*, Malden, MA: Blackwell, pp. 270–294.
Harvey, D. (2009) "The right to the city," in *Social Justice and the City* (revised ed.) Athens, GA: University of Georgia Press, pp. 315–332.
Howe, J. (2002) "Vehicle of desire," *New Left Review*, 15(May–June):105–117.
Hume, C. (2008, March 3) "Downtown density will prevail over slums of suburbia," *Toronto Star*, p. A10.
Jacobs, J.M. (2012) "Urban geographies I: Still thinking cities relationally," *Progress in Human Geography*, 36(3):412–422.
Jaffe, R., Klaufus, C., and Colombjin, F. (2012) "Mobilities and mobilizations of the urban poor," *International Journal of Urban and Regional Research*, 36(4): 643–654.
Jonas, A.E.G. (2013) "City-regionalism as a contingent 'geopolitics of capitalism,'" *Geopolitics*, 18(2):284–298.
Jonas, A.E.G., Goetz, A.R., and Bhattacharjee, S. (2014) "City-regionalism as a politics of collective provision: Regional transport infrastructure in Denver, USA," *Urban Studies*, 51(11):2444–2465.
Kasarda, J.D., and Lindsay, G. (2011) *Aerotropolis: The Way We'll Live Next*, New York: Farrar, Straus and Giroux.
Keil, R. (2011) "The global city comes home: Internalized globalization in Frankfurt Rhine-Main," *Urban Studies*, 48(12):2495–2517.
Keil, R. (ed) (2013) *Suburban Constellations: Governance, Land and Infrastructure in the 21st century*, Berlin: Jovis Verlag.

Keil, R., and Young, D. (2008) "Transportation: The bottleneck of regional competitiveness in Toronto," *Environment and Planning C: Government and Policy*, 26(4):728–751.
Keil, R., and Young, D. (2011) "Introduction: In-between Canada—The emergence of the new urban middle," in D. Young, P. Wood, and R. Keil (eds) *Inbetween Infrastructure: Urban Connectivity in an Age of Vulnerability*, Keolowna, BC: Praxis (e)Press, pp. 1–18.
Kloosterman, R. C., and Lanbregts, B. (2007) "Between accumulation and concentration of capital: Toward a framework for comparing long-term trajectories of urban systems," *Urban Geography*, 28(1):54–73.
Kolb, D. (2008) *Sprawling Places*, Athens, GA: University of Georgia Press.
Lefebvre, H. (1991) *The Production of Space*, Oxford: Blackwell.
Lefebvre, H. (1996) *Writings on Cities* (E. Kofman and E. Lebas, Trans.), Oxford: Blackwell.
Lefebvre, H. (2003) *The Urban Revolution*, Minneapolis, MN: University of Minnesota Press.
Maassen, A. (2012) "Heterogeneity of lock-in and the role of strategic technological interventions in urban infrastructure transformations," *European Planning Studies*, 20(3):441–460.
Macfarlane, D. (ed) (2008) *Toronto: A City Becoming*, Toronto, ON: Key Porter Books.
MacLeod, G. (2011) "Urban politics reconsidered: Growth machine to post-democratic city?," *Urban Studies*, 48(12):2629–2660.
McCann, E., and Ward, K. G. (2010) "Relationality/territoriality: Toward a conceptualization of cities in the world," *Geoforum*, 41(2):175–184.
Metrolinx (2008) *The Big Move: Transforming Transportation in the Greater Toronto and Hamilton Area*, Toronto, ON: Metrolinx.
Minnan-Wong, D. (2009, May 22) "Fight congestion with mobility," *Toronto Star*, p. A27.
Morgan, K. (2007) "The polycentric state: New spaces of empowerment and engagement?," *Regional Studies*, 41(9):1237–1251.
Ollman, B. (2003) *Dance of the Dialectic: Steps in Marx's Method*, Urbana, IL: University of Illinois Press.
Roy, A. (2009) "The 21st century metropolis: New geographies of theory," *Regional Studies*, 43(6):819–830.
Saunders, D. (2010) *Arrival City: How the Largest Migration in History is Reshaping Our World*, New York: Vintage Books.
Scott, A. J. (2012) *A World in Emergence: Cities and Regions in the 21st Century*, New York: Edward Elgar Publishing.
Sieverts, T. (2003) *Cities Without Cities: An Interpretation of the Zwischenstadt*, London: Spon Press.
Skelton, T., and Gough, K. V. (2013) "Young people's im/mobile urban geographies," *Urban Studies*, 50(3):455–466.
Storper, M. (2013) *Keys to the City: How Economics, Institutions, Social Interaction, and Politics Shape Development*, Princeton, NJ: Princeton University Press.
Turcotte, M. (2011) *Commuting to Work: Results of the 2010 General Social Survey*, Ottawa, ON: Statistics Canada.
Urry, J. (2003) *Global Complexity*, Cambridge: Polity.
Watt, P., and Smeats, P. (eds) (2014) *Mobilities and Neighborhood Belonging in Cities and Suburbs*, Basingstoke: Palgrave Macmillan.
York Region Transit Corporation (2014) "VivaNext." Retrieved May 31, 2014 from http://www.vivanext.com.
Young, D., and Keil, R. (2010) "Reconnecting the disconnected: The politics of infrastructure in the in-between city," *Cities*, 27(2):87–95.

Young, D., and Keil, R. (2014) "Locating the urban in-between: Tracking the urban politics of infrastructure in Toronto," *International Journal of Urban and Regional Research*, 38(5):1589–1608.

Yuen, J. (2009) "Toronto's war on cars," *Toronto Sun*. Retrieved August 29, 2012 from http://www.torontosun.com/news/torontoandgta/2009/05/17/9483606-sun.html.

Part IV
Circulation
Assemblages and Experiences of Mobility

11 Selling the Region as Hub
The Promises, Beliefs, and Contradictions of Economic Development Strategies Attracting Logistics and Flows

Markus Hesse

INTRODUCTION

This chapter deals with the policy and governance dimension of logistics and freight distribution (including services such as trucking, warehousing, freight forwarding, container handling, and the like), related land uses, and circulation modes. It examines how these strategies are being pursued and explores the way in which logistics are discursively framed and thus communicatively constructed. The chapter underlines how contested and contradictory such issues are in the context of urban and regional practices, which leaves a big challenge for the political process. In particular, local economic development initiatives in logistics indicate a certain range of policy shifts that have occurred over the last decades and have transformed political constellations quite significantly. One of these changes includes the shift from traditional infrastructure policy, long predominant in development contexts, toward more operational activities focusing on the stimulation of network building, active company acquisition, and place promotion. The other modification is related to changing perceptions of trucking, warehousing, and freight forwarding by municipalities. During the 1970s and 1980s, freight forwarding and transport companies were not actively promoted at the local level in many countries because of their negative impact on the environment and the quality of life and since their economic productivity per unit area (e.g., in terms of job generation) is generally considered to be much lower than manufacturing (McKinnon 2009). More recently, cities and regions—facing deindustrialization and the lack of traditional development options—have started to engage actively in this particular area, which triggers the demand for something new to follow. To take advantage of growing freight flows, some regions are even marketing themselves as becoming 'logistics regions,' meaning that the corresponding sectors are achieving relatively high shares in added value or employment, or are expected to do so in the near future. Driven by the promise of job creation and tax generation, many municipalities have started to engage here recently.

However, it is by no means clear that cities and regions benefit from competing for logistics acquisitions, and a range of difficulties and problems may arise when pursuing such goals. First, freight and logistics services are essentially concerned with flows, and the related entities seem to be much more mobile, volatile, and less embedded than core industrial activities (for example) have been. Making local sense of global flows is a challenge. Locational dynamics depend upon the demand from shippers or receivers, and they shift according to their respective mobilities. Second, local policy strategies face a dilemma that results from the extension of supply chains and logistics networks: The place where a problem occurs (which often enough is the urban area) tends to be quite remote—in both spatial and institutional terms—from the place where decisions are being made. Third, powerful corporate players have much more steering capacity compared to local planners or economic development managers. Logistics systems imperatives and corporate competition seem to determine the modus operandi of the circulation of goods and commodities, and they leave little space for intervention or strategic response. As a consequence of these difficulties, logistics as a means of regional development is often orchestrated through broader claims of modernity, growth, and prosperity, though concrete proof for such promises may be lacking.

In this context, the aim of this chapter is not to assess the pros and cons of logistics-related development in certain detail, nor draw any general conclusions (see, for example, O'Connor et al. 2012; Hesse forthcoming). Instead, my core argument is that there is a great deal of discursive framing going on that is associated with these policy strategies, whatever the empirical evidence that favors competing views, which thus unfolds in regional development practices. Logistics appears as a spatial imaginary, imbued with meanings of modernity, growth, and prosperity, and makes promises with which local policy endeavors to get the political process going. It is this very dimension of logistics that may correspond further to the aims of this volume, bringing together different ways of approaching transport and mobilities in an urban context (Hesse 2013). Against the more generic background that urban places necessarily intersect with transport networks and flows (see both classical and more recent models and concepts of urban studies; e.g., Harris and Ullman 1945; Jacobs 2012), this chapter is particularly concerned with the question of how urban places and transport networks are being 'co-constructed' through the practices of mobility and circulation (see Prytherch and Cidell in this volume).

In the remaining sections of this chapter, I will explore the diffuse and contested field of local, economic development strategies focusing on logistics and freight transport, precisely by highlighting such issues of co-construction, discursive framing, and imagination. I begin by shedding light on the significance of logistics for regional development, the reasons why this sector appears on the radar of economic development managers, and theoretical concepts for approaching such issues. Next are two empirical case studies from Western Europe, particularly on the Province of Limburg in the Southeast Netherlands and the Wallonia region in Belgium. The former appears rather advanced in this respect as it sets out to diversify from the logistics hub function, while the latter is just starting to engage with logistics. The next section presents selected frames and strategies with which economic development managers and policy practitioners advertise their region as a logistics locale, interpreted as the communicative or discursive construction of both places (clusters) and flows. Finally, the chapter concludes by reflecting on the relevance of this co-construction of economic development and intersecting places and flows for political processes, and why it deserves more recognition in research and practice.

TRANSPORT, MOBILITIES, AND URBAN/REGIONAL ECONOMIC DEVELOPMENT

The Changing Nexus of Places and Flows

It is widely accepted that transport and mobility have long been essential for urban and regional development processes (see, for example, the overview provided by Beyers and Fowler 2013). Traditional understandings of the link between urban places and flows were advanced further by Harris and Ullman (1945), who distinguished between cities as *central places* and those that evolve into *transport places*, with the latter specialization being a consequence of strategic location on transport channels (Harris and Ullman 1945, pp. 8–9). Transport places have thus undermined the natural order within which central places had developed. Some of them grew to become major gateway cities, which were usually situated at the sea-land interface, at major junctions of different transport lines or modes, or at the borders of two states/customs areas (Burghardt 1971).

Meanwhile, many of the old gateways have lost their essential role, compared to logistics hubs that developed as mere interfaces between various nodes within larger logistics and transport networks. The massive provision of transport infrastructure systems, most notably the motorway networks in industrialized countries after World War II, helped accessibility become somehow ubiquitous, further developing the spatial division of labor and thus of transport volumes. The operation of flows—and their spatial imprint—was massively impacted by the shift to logistics and integrated supply chain management, the emergence of complex logistics networks (Aoyama et al. 2006; Christopher 2005), and the globalization of supply chains and rise of the global logistics corporations (Frémont 2007; Bowen 2012). Technological change and information and communication systems allowed an unprecedented level of locational freedom of economic processes to develop, which enabled corporations to orchestrate a huge amount of freight flows quite efficiently, even at a global level.

This new locational freedom—dating at least from the 1990s—coincided with changing patterns of political regulation, most notably the deregulation of major freight and transport industries, and free trade agreements such as the European Single Market or NAFTA. In concert with geopolitical changes (i.e., new consumer markets in Central and Eastern Europe), goods handling and associated territorial manifestations were transformed significantly. The predominant centers of goods handling remain the large metropolitan regions in the urbanized cores and coastal areas of Western Europe, North America, Asia, and Australia (O'Connor 2010). But competitive pressure, land prices, and congestion at the major gateways have accelerated the trend towards regionalized logistics developments. The move of related facilities into the broader hinterland, where massive regional and interregional distribution complexes were newly formed, is indicative (Bowen 2008; Hesse 2008; Cidell 2010, 2011). In these areas, the logistics industry assumes a disproportionate significance. Even peripheral or formerly rural regions, with their substantial industrial land reserves, now enjoy new locational advantages as a consequence of the elimination of state borders and a new geopolitical order.

The Janus Face of Local Governance: Circulation as a Means of Development, Transport as a Source of Problems

Amidst this interplay of economic transformation, a global trade regime, and favorable regulatory conditions, the logistics sector is increasingly perceived as a growth machine and targeted by local economic development and regional policy. While the increasingly circulatory nature of the global economy has contributed to the overall rise of trade, commodity, and transport flows, it is in the interest of local places to participate and become a node within broader networks. Popular strategies of local economic development, first and foremost, focus on providing infrastructure and bundling local and regional assets in order to be recognized as an ideal place for corporate investment (Bowen and Leinbach 2011). As a consequence, cities and regions position themselves as being, or becoming, 'hubs' for managing international flows of goods. In most cases, this trajectory starts with an existing piece of infrastructure (like an airport or sea port), the conversion of a former military airport, or a freight industrial area that had evolved more by chance. While single actors try to develop such sites as a means for acquiring more firms and promoting place, local governments aim at crafting a framework that integrates the site within a bigger picture or development scenario, thus "jockeying for position" (Malecki 2004). In the second instance, once initial activities have been generated, a network of local actors—known as a logistics network or logistics region—has to be established to provide further visibility. Then public or private development interests get together, as in the case of the European freight villages, and combine infrastructure provision, company acquisition, networking, and cluster management.

The visibility and further development of such local initiatives can also depend on policies undertaken at national and even supranational levels. National governments have already been rather active in supporting local and regional logistics clusters, as in the case of the Netherlands. In Germany, the federal government has developed a Master Plan for Freight Transport and Logistics, together with public and private stakeholders and business representatives. In the United States, the Federal Highway Administration (FHWA) runs an Office of Freight Management Operations that provides a framework for related activities at various levels, including cities and metropolitan regions. Intermediary bodies such as OECD or the World Bank are increasingly developing knowledge bases, indicators, and policy initiatives for including freight transport and logistics in broader regional development strategies. The Logistics Performance Indicator (LPI), for example, represents a measurement for assessing the related performance of entire countries (World Bank 2014). This tool is already extensively used by governments for economic development and marketing purposes, not only intensifying the competitive notion of urban-regional development, but also contributing to the further communicative construction and performance of logistics and flows.

In addition to these public and private efforts to attract flows to create economic value, there is also a second, different perspective on the same subject: the critical assessment of material flows and their environmental burden, impact on the condition of infrastructure, and effect on quality of life in neighborhoods and communities (De Lara 2013; Flämig 2013; Hall 2007; Hesse 2008). In the realms of both research and practice, there is an important concern and emphasis placed on such issues, with the aim of trying to make vehicles flow seamlessly and minimize related impacts (TRB 2003, 2013). While environmental regulation targets rather specific issues of reducing air pollution (such as NOx or SO_2) or noise abatement, traffic safety policies are concerned with the particular risks trucks pose for car drivers, cyclists, and pedestrians. In broader terms, heavy-duty vehicles can affect the quality of life in districts or across city levels, so the placement of those facilities that generate traffic will have to be balanced against other urban interests. While environmental policies are quite extensively concerned with freight traffic and the related problems, this does not apply to urban planning, and less so to economic development strategies. At a more abstract level, the transport and logistics industries are indeed subject to various strategies of making them greener (including even ports and airports). Such concepts and measures are certainly also well marketed. However, it seems rather difficult to achieve a comprehensive improvement in the social and environmental impacts of freight transport operations as yet, given the systemic properties that characterize the logistics system and the predominantly economic imperatives driving its operation.

Intersecting Places and Flows—Conceptual Dimensions

Up to now, research has been particularly concerned with the emergence of logistics services, with technological changes that allowed the management of information and communication, or the evolution of specialized functions such as the provision of raw materials, manufacturing, and distribution logistics (as Coe [2014] has recently emphasized in the context of global production networks). In contrast, the mechanisms of how logistics usually *takes place* in territorial terms (and how this has changed in recent times) are less well understood (but see Hall and Hesse 2013). This is even truer for cases where such activities are subject to deliberate concepts of development and planning. While no consistent theoretical models exist that provide an explanation in this respect, there are at least two different ways of asking how the intersection of places and flows could be approached in conceptual terms. One deals with logistics and commodity chains; the other is concerned with clusters and networks.

Analyses of global commodity chains (GCC) or value chains (GVC) aim at tracking the entire transformative life cycle of a given commodity from origin of production to destination, as well as the market and its variegated mobilities (Bair 2009). When considering the chain, this concept distinguishes between the *value chain* as representative for the entire web of inter-firm relations in economic space, and the more frequently used *commodity chain*, which includes institutions and governance, while the logistics or supply chain also comprises material flows and physical distribution (Derudder and Witlox 2010). Regional development comes into play when certain stakeholders—including places—can 'capture' value from being a node within a network, or from becoming 'inserted' into a commodity chain. MacKinnon (2012) offers a renewed perspective on these issues by discussing various forms of 'coupling' (including re- and decoupling) of certain entities with the value-adding process. Moreover, viewed from an evolutionary perspective to regional development processes, strategic coupling can be considered a means of gaining network embeddedness, by linking global production networks (or distribution networks, respectively) and related chains with regional assets like knowledge and expertise, but also services or infrastructure (Ducruet and Lee 2006).

One blueprint idea for logistics as a means of development is Michael Porter's (1998) concept of clusters—based on the assumption that the concentration of economic activity provides a certain degree of interaction of economic actors, generating further added value. Meanwhile, a myriad of cluster concepts has emerged in a variety of industries. Most recently, Yossi Sheffi (2012, 2013) applied the concept of clusters to freight and logistics, considering this as relevant for regional economic development. According to Sheffi:

> Logistics intensive clusters are agglomerations of several types of firms and operations: (1) firms providing logistics services, such as 3PLs, transportation, warehousing and forwarders, (2) the logistics operations of

industrial firms, such as the distribution operations of retailers, manufacturers (in many cases after-market parts) and distributors and (3) the operations of companies for whom logistics is a large part of their business.

(2013, p. 463)

By locating different firms of the same sector close to each other, such 'neo-Marshallian' benefits of agglomeration are expected to further mobilize economic impacts such as local employment, network building, or innovation. The same applies to the clustering of production firms and producer services, such as logistics and supply chain management, warehousing, and transport. O'Connor et al. (2012), Heuvel et al. (2014), and Rivera et al. (2014) presented some evidence that such logistics services do agglomerate and thus can contribute to regional development.

However, it seems difficult to strategically design and implement freight clusters based on industrial linkages provided onsite. McCalla et al. (2001) explored the linkages around freight terminals and found only weak evidence for local interactions; see also the related warning by Gouvernal et al. (2011). In more general terms, critical analyses of the cluster concept were presented by Spencer et al. (2010). The crucial question—and cause for caution—here is to what extent cluster concepts need to follow the business orientation of their individual users, and how a common rationale of networking can provide system-wide benefits. On the one hand, clusters are increasingly considered to contribute to growth: "[M]ost importantly . . . logistics clusters generate a large number of jobs" (Sheffi 2013, p. 499). On the other hand, they were already referred to as a "mesmerising mantra" (Taylor 2010), given the high expectations that clusters would generate endogenous growth, provide network effects, and foster innovation. In an increasing number of cases, logistics cluster or network strategies are no longer confined to one single territory, such as an airport, sea port, or an industrial area, but are extending across various scales. Given the prominence and frequent use of the cluster term and concept, also with regard to logistics, it actually appears that the bundling of local associations and networks to clusters also became subject to discourse and communicative construction. Besides the material evidence that is inherent to the two case studies presented in the next section, these cases may simultaneously illustrate how the regional economic benefit provided by logistics and the spatial clustering of related functions is framed by discourse.

MAKING A DEAL WITH FLOWS: EVIDENCE FROM CASE STUDIES OF THE LOW COUNTRIES

Venlo, Limburg/The Netherlands

The case of the Netherlands illustrates a circulation industry that is prevalent at local, regional, national, and international levels and thus contributes to structuring and articulating urban space (cf. Prytherch and Cidell in this volume). It unfolds as a historically distinct policy trajectory based on the ability of corporations and institutional actors to make local sense of overarching global flows, traditions that reach as far back as the old seafaring nation. Levelt (2010) explored this rich tradition of freight distribution activities in the Netherlands on a large scale and distinguished various types of trade hubs that have played a major role in this particular economic development trajectory for centuries: (1) as a distribution node that physically connects demand for and supply of goods through distribution activities, (2) a marketplace where goods are traded and change their ownership, so supply and demand are linked together onsite, and (3) as a place where scattered demand and supply become connected, mediated through traders and their specialized networks. This sort of trade network node seems to be a more sophisticated interface, where the transformation of value is even more important compared to the mere physical distribution. This is also the defining feature that allows the generation of economic value—the distinction between a physical interface (such as a logistics hub) and a trading place that includes processing value transactions from various activities beyond mere distribution. Whereas the former might be less attractive for regional development policies, most regions wish to become part of the latter, however difficult to organize or provide this may be. Also, the related policy frames, narratives, and discourses tend to be both complex and contested, as Huijs (2011) was able to reveal in the case of Schiphol, Amsterdam.

The national aptitude of the Dutch to organize trade, travel, and commodity shipments unfolds through local and regional places, where distinct platforms for freight transport, logistics, and associated industries have been developed (Levelt 2010, p. 11). These platforms not only organize shipping activity and related services, but are also deeply embedded in the Dutch specialization of reexporting processed commodities, such as produce, cut flowers, or clothing. As a template case, the province of Limburg and particularly the city of Venlo come into play here, since this area represents one of the most important inland hubs in Europe. Limburg and Venlo are already moving towards a more diversified economic portfolio, no longer confined to managing flows but trying to create value from a broad range of activities. The province is situated in the southeast of the Netherlands, covers a territory of roughly 2,200 square kilometers, and has a population of about 1.1 million people. The city of Venlo is located close to the German border at the intersection of several motorways and on the river Meuse; it has about 100,000 inhabitants (as of 2014) and about 58,000 jobs (as of 2013; data after Centraal Bureau voor de Statistiek [CBS]).

At the local level, the city's evolution as a transport and logistics hub was ignited by the decision of the European Container Terminals (ECT) corporation, situated at the port of Rotterdam, to establish its first inland terminal there in 1992 (Rodrigue and Notteboom 2009). This was mainly due to the scarcity of handling capacity within the port, and also to the provision of land and multimodal transport access in Venlo. Following the placement of the ECT, Venlo's strategic location along the main logistics axis from both Rotterdam and Antwerp to Germany and further to Continental Europe, as well as the presence of barge and rail terminals for inland container transport, made it an important transshipment point for logistics. With two industrial areas in the city hosting more than 900,000 square meters of warehousing and distribution space, Venlo follows the two main ports of the Netherlands—Rotterdam and Amsterdam—in terms of cargo volume handled. Meanwhile, the entire province of Limburg has specialized in the logistics industry by attracting corporate investment, establishing required infrastructure, and promoting a sort of network-building economic development approach. Based on related achievements, Venlo and the province present themselves as having "Europe at one's feet" (see Figure 11.1). While this statement could certainly be misunderstood as exaggerated and indicative of policy *hubris*, it can also be considered metaphorical, indicating the self-assessment and self-consciousness of the regional players and their ambition to create something important. Last, but not least, this sort of 'pretended,' apparent, or effective centrality appears as a common frame with which various regions specializing in logistics promote themselves.

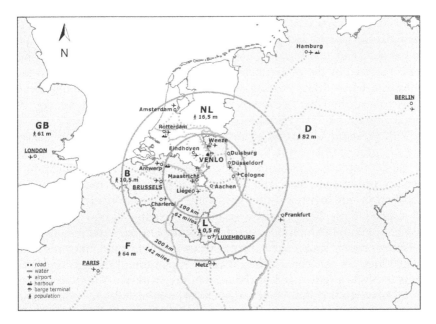

Figure 11.1 The spatial representation of *Venlo central*.

(*Source:* Author, after Limburg Development and Investment Company; cartography: Malte Helfer, University of Luxembourg).

Venlo obviously represents a rather advanced case among European regions trying to become a hot spot in logistics, not only in terms of infrastructure and the built environment, but also a range of economic activities effectively triggered in the process. The massive establishment of distribution centers has had a significant impact on the labor market. In 2013, 5,400 of the total 58,300 jobs in Venlo were in the transport and warehousing sectors (personal communication with the city of Venlo, Ruimteen Economie). By reaching more than 9% of the city's total employment, the relative proportion of logistics is much higher than the usual range of 3–5%. Over the last decade, however, employment in Venlo—including the trade port—has tended to stagnate. This is one reason behind more recent attempts to establish a still more diversified trajectory of economic development.

Beyond its function as one of the main inland hubs hosting large-scale distribution centers and related freight transport flows, the region is pursuing an advanced economic profile, both in terms of trade and associated logistics, due to Limburg's specialization in cultivating produce and flowers (judging from various sources, the northern part of the province is the largest or second largest horticultural area of Europe). And starting in the mid-2000s this area was designated to host one of six 'greenports' in the Netherlands. This concept was initially framed by the government and then further developed by regional agencies, adding the particular notion of greenport to the existing two 'mainports' (Schiphol, Rotterdam) and also to the 'brainport,' as the neighboring region around Eindhoven has been called. The decision to pick Venlo for this particular purpose was based on the idea of combining the strengths of the two sectors—horticultural and agro-industrial production on the one hand, and logistics and freight distribution on the other—which are considered by the government to lead the national economy: agro-food being first, horticultural production second, and logistics fifth among the top 10 sectors of the Dutch economy (Ministry of Economic Affairs, Agriculture & Innovation 2011, p. 8).

Both the national spatial strategy that includes the greenport concept and the strategic vision for Venlo 2030 articulate the need for further economic development, innovation, and adaptive growth. The combination of primary production, food and horticultural processing, and also storage, distribution, and logistics is thus key. While Venlo has already been successful in hosting the 2012 Floriade—a major international flower, gardening, and horticultural exhibition (awarded by competition)—the region is now pursuing projects such as new mixed farming, healthy food innovation, and related logistics concepts. Also, a massive expansion of the agro-industrial areas in Venlo is envisaged in the near future, with the 5,000-hectare site of the Klavertje 4 (K4, or four-leafed clover) that includes the combination of cultivation, processing, and distribution activities, which are intended to be organized according to the cradle-to-cradle principle (Greenport Venlo Development Company 2012; Laurentzen et al. 2009). In accordance with the envisaged growth in the food and flower industries, a nearby 230-hectare

'Trade Port North' logistics area is also under development. Such projects, which are massive in scope and huge generators of commodity circulation (even given the optimization of internal flows as the K4 intends to provide), have become iconic for the business aspirations and ambition of the region.

Wallonia/Belgium

The Belgian region of Wallonia represents the second case to be examined in more detail here. It is located about 150 kilometers south of the city of Venlo and comprises an area of about 3.5 million people over an area of 16,844 square kilometers, thus representing half of the Belgian territory. Wallonia is one of the three regions that represent contemporary Belgium, besides the northern region of Flanders and the capital region of Brussels. While Flanders represents the wealthier part of the country and Brussels has developed quite well as the political and administrative capital of the European Union, Wallonia has always lagged behind the other two. Massively hit by deindustrialization and the demise of coal mining and steel production, the Walloon region came under severe pressure to strengthen regional economic development by investing in assets such as physical infrastructure, education systems, and research capacity in order to attract investment from abroad. As population and settlement densities are much lower in Wallonia compared to the 'Flemish diamond'-like urbanization pattern of the northern part of Belgium, Wallonia appears as a hinterland space serving the congested economic activities that are situated on the North Sea coast, most notably the port of Antwerp, which is Europe's second most important maritime port (in terms of both tonnage and container handling). Another rising location for logistics and freight distribution activities has emerged in the western part of the region, where Liège airport has gained significance as one of the top 10 European air freight hubs. Liège also hosts Europe's third largest inland port and is accessible by inland waterways from both Antwerp and Rotterdam sea ports via the Albert Canal and the Juliana Canal, respectively.

For the purpose of fostering regional development, the Walloon government released a 'Marshall Plan' in 2005, spending about €1 billion for a broad set of measures to be undertaken in that particular context. (The title is borrowed from the recovery plan that the U.S. government had put in place for the post–World War II reconstruction of Germany; it certainly indicates the size of the challenge and the ambition of the actors.) The Marshall Plan focuses on the establishment of various clusters, which were named 'pôles de competitivité' (competitive growth poles), assuming that they may contribute to regional development in particular ways. One cluster is dedicated to supporting logistics activities and supplemented earlier, more experimental measure to attract related investments in the region. According to the Walloon Economic Development Agency (AWEX), the aim of a competitive cluster is "to promote innovation by supporting corporate projects involving strengthened cooperation between innovating companies and research centres or university departments" (Strale 2008, pp. 195–196). Since it was launched, the cluster has been involved in various research and training projects, and it promotes the logistics economy by fostering exchange between the actors involved. As an exemplary case, the press reported on the U.S. pharmaceutical firm Johnson & Johnson, whose consolidation of freight distribution led to the establishment of a European Distribution Center (EDC) close to the city of Charleroi, Wallonia—in order to "build on hub status" (*Mee* 2011). Other global players in the pharmaceutical and health care industries, such as Pfizer or GlaxoSmithKline, are targeted by the Biolog subsidiary of Logistics in Wallonia, a platform specializing in serving life sciences industries and thus strengthening the link between manufacturing, hightech, and logistics.

The material developments that were triggered in the field of logistics are obvious, yet also contentious. Recently, a major emphasis has been focused on investments in the regional infrastructure, particularly to support waterways and railway transport modes. Multimodal terminals connecting barge, rail, and road transport are provided in Mouscron-Lille, Mons, and Charleroi in the west and Central Ardennes and Athus in the south. The port of Liège, Europe's third largest inland port, is currently becoming a 'Trilogiport,' connecting the various transport modes and offering massive space for development. However, the recent closing of the Arcelor-Mittalsteel production facility in Liège has reduced the demand for freight and logistics services quite significantly. In terms of network building, the logistics cluster is considered to be successful, as it comprises about 265 members, mostly from the corporate world. Labor market figures reveal a constant increase in jobs in warehousing (from about 15,000 in 2003 to more than 23,000 in 2012), while core transport employment tends to stagnate. However, the cluster performance as such—that is, the steering process undertaken by the regional agencies—has recently been critically evaluated by a government committee. Without mentioning the six clusters specifically, it was recommended that they operate more strategically, improve decision-making processes, and link their activity with related industries better (IWEPS 2014).

It was not necessarily the predetermination of Wallonia as a logistics space that helped the region to become the subject of further economic development policies. It was a very particular discourse and communication event that created Wallonia as a place worth investing in, particularly a ranking study that measures the performance of places with respect to various criteria, not least in logistics. The case in question here is the 2009 edition of the report delivered by Cushman & Wakefield consultants (Cushman & Wakefield 2009). This was the first report ever that ranked Wallonia highly among other European regions focusing on logistics. According to this study, the scaling up of economic processes due to globalization causes a previously peripheral or disadvantaged place such as Wallonia to become central. Thus, similar to the case of Venlo and the Province of Limburg, centrality seems to be essential (see Figure 11.2). It is argued that a large part of

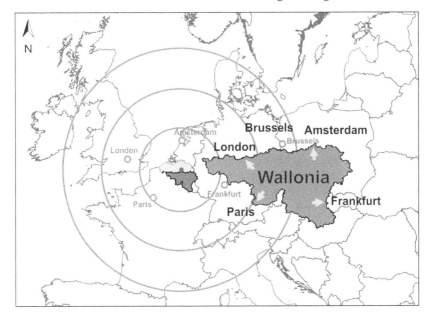

Figure 11.2 The spatial representation of *Wallonia central*.

(*Source:* Author, after Logistics in Wallonia Innovation Network; cartography: Malte Helfer, University of Luxembourg).

European consumption is located in the vicinity of the geographical area of Wallonia; also, the region is situated close to the 'North Range' mainports (Antwerp, Rotterdam, Hamburg, Zeebrugge, Dunkirk, Le Havre) and also to Europe's main passenger and cargo airports. These features obviously contribute to a particular production of space.

However, Strale (2012) has meticulously addressed the shortcomings and difficulties that are associated with the rankings of cities and regions, both in substantive and methodological terms. Thus he deconstructed the particular value and delivery of reports such as the one from Cushman & Wakefield. What was already, and rather euphemistically, called the "bible of worldwide logistical localisation" (Magain 2005, p. 5) can also end up as a myriad of data constraints, methodological questions, and mere regional public relations. Regarding policy recipes that are promised to evolve from such rankings, these are also criticized for fostering unconsidered regional competition. As a consequence, the author concludes by stating that it "seems to us risky and somewhat incredible to use this kind of study as a decision-making orientation tool. One can even wonder about the public authorities' need to sponsor and finance such classifications, which are somewhat unreliable, strongly connoted ideologically and poorly operational" (Strale 2012, p. 12). Notwithstanding the urgent need for old, industrialized regions to recover and generate growth and employment, the question of how to pursue this and how to achieve proper justifications for related policies remains obvious.

FRAMES, DISCOURSES, AND IMAGINARIES: THE POLITICS OF MODERNITY, GROWTH, AND BEING THE CENTER OF ALL/AT ALL

The cases investigated in the previous section reveal that the subject matter without doubt has a communicative dimension, going far beyond material developments. Since logistics has been successfully branded and marketed for the purpose of development in both cases, it seems appropriate to turn to an interpretative perspective and take a critical look at the underlying discourses and imaginaries that have helped to reconfigure logistics in more general terms. This approach is inspired by the emerging literatures on policy discourses, on the ways in which agency and practices are embedded in overarching normative frames or ideologies, and how institutions may provide reasons and justification for decisions they have made. It particularly borrows from Healey's (1999) account of the dissemination of economic development discourses in the context of the English planning system, from Peet's (2000) cultural analysis of regional economic development and the related role of imaginaries, and Bevir and Rhodes' (2006) policy analysis that brings forward an interpretative, institutionalist perspective, as was also applied by Gibbs and Krueger (2012). In their analysis of political processes and institutions, Bevir and Rhodes use three conceptual lenses for tracking the rhetorics and ideologies behind the practice of governance: traditions, beliefs, and dilemmas. The authors argue that traditions do not determine decisions, but prepare the corridors in which they will be made. The concept of beliefs enables the researcher to explore how individuals construct their world, "including the ways they understand their location, the norms that affect them and their interests" (Bevir and Rhodes 2006, p. 6). Dilemmas pose a problem, challenge existing practices, call for an adjustment of strategies, and include related contradictions. This differentiated lens appears rather useful for exploring the 'politics of flows' in regional development contexts, and for deciphering how actors and institutions make sense of the prejudices of their decision-making processes.

My interpretative analysis of how logistics is being 'sold' by institutions as a regional development recipe has detected three different narratives, strongly embedded in the beliefs and traditions of the acting institutions: *modernity, growth*, and *centrality*. First, the production of related imaginaries (Peet 2000, p. 1221ff) has helped a new perception of logistics to emerge, from being viewed as a disturbing land use to becoming subject to economic development strategies. This perception has only a little in common with the mere activity of physical distribution (that is, truck driving, warehousing, etc.), but it is apparently modern and provides a new economic appeal to cities and regions. This was also the case when e-commerce firms started to place their distribution centers (DCs) and sold the idea of high/new technology to those regions who applied to be selected. While the DCs were indeed packed with technological devices, software-driven conveyer belts, and the like, the physical appearance compares to the old era of warehousing: flat concrete spaces that inhibit big box industrial developments and trigger frequent lorry flows. Economic development managers believed they were getting something new, while the spatial imprint remained rather conventional. The transformation of the old world of trucking and warehousing into something now called logistics and supply chain management has been, first and foremost, a semantic one. This is also quite visible in the representation of the industry and its associations at trade fairs, conferences, and in business circles. This pattern of interpretation may be indicative for the very first stage in the lifecycle of a logistics region, a stage that was already left behind in the case of Venlo, yet still seems characteristic for the case of Wallonia. In the latter case, it may also take some time to see a trading place evolving here, instead of a mere transshipment interface.

Second, the underlying promise of logistics in economic development processes is that of growth and job creation. "The cluster aims at boosting innovation in the sector, stimulating networking and support to enterprises and promoting Wallonia abroad to attract new investors and new activities in Wallonia" (Logistics in Wallonia Innovation Network n.d., p. 2) So far, this has been a quite contentious and somehow contradictory narrative: Given the tendency of clusters to concentrate, logistics growth is often confused with the mere delocalization of firms from one place to another. This clearly happens to the benefit of one region, yet often at the cost of another (see Oosterlynck and Swyngedouw 2010, on the case of Brussels v. Leipzig). Also, the apparent growth of logistics value is partly statistical, based on the outsourcing of fleets from retailers and manufacturers to specialized logistics firms, and due to changing classifications of previously public entities that are now being privatized, such as postal and railway service firms, and now appear in the statistics. Finally, there are fundamental questions as to the role goods distribution can still play *locally*, under the conditions of an increasing locational mobility of *global* firms and processes. This dilemma refers to the prevailing complexities of getting local actors to become part of global chains, and it probably sits behind the question of how to link the mere logistical function with trade and manufacturing better, in order to create and locally bind added value.

Third, the way logistics is framed in regional economic development processes is also about the desire of places to become *central*, mainly by redefining centrality. The most popular imaginary in this respect, which is used by almost all regional development agencies and discourses concerned, is the concentric pattern that puts the place of reference in its center (see again Figures 11.1 and 11.2). Formerly peripheral regions present themselves as being the center of all—drawing a map with distance circles around their particular locale. This applies both to powerful cases, such as Limburg, Netherlands, but also to newcomers such as its neighbor, Wallonia, Belgium, whose government is trying to develop a profile in logistics as well. In fact, however, neither will become truly central. Rather, they are preparing to serve as a transport place, which adds in some ways to their economic profile, yet the associated share of logistics employment or added value may usually remain in the one-digit realm. In any case, it is also evident that not all regions can qualify to be equally successful in the same area. Another issue points out the robustness of an urban environment that is required if one is to accept the shadow side of vehicle circulation, industrial activity, and demand for land. Regions with a long-standing engagement in managing flows, such as port cities or traditional gateway cities, may present themselves quite successfully as logistical hot spots; others may not easily adapt to the related requirements and accept the associated burden.

The points made above briefly demonstrate that logistics processes are not only an important measure of economic development and thus targeted at local levels. What has become clear as well is that discourse, and the meaning made by discourses, play an important role in creating these developments, and add to the apparently hard and undisputed facts. In this respect, new investments convey an abstract promise of—belief in—development and well-distributed benefits (Cidell 2006). Planning theorist Patsy Healey (1999) remarked that these narratives bundle "site, jobs and (corporate) portfolios" in order to promise that a place would experience a boost. All this discursive framing eventually leads to emerging interrelations and interdependencies between the various fields of discourse, not least since the semantics of a hub are used to play a role in local development narratives. Moreover, assets such as infrastructure and even apparent 'non-places' such as transport hubs have already been discussed in terms of their potential flagship role for cities and regions (Warnaby 2009). In some cases, self-assigned logistics regions just inhibit symbolic dimensions of place-making and self-ascription of certain attributes in the regional competition—and reveal only limited evidence that these imaginaries refer to objective properties *per se*. Logistics regions are therefore interpreted not only in terms of their constitution, but also as ideal types of a communicative construction of space.

What is sitting squarely behind these apparently rational investments and their discursive co-construction is something that McCann (2013) has called 'policy boosterism.' This includes not only the successful implementation of a freight cluster, a sea port terminal, or an air cargo facility, but the promotion of good and best practices that rely on simplified assumptions of transfer. City governments and mayors, as they manage inner city investments, claim to know what major airport or sea port investments may bring to their region and how to steer them to optimize growth and prosperity. This notion of both policy boosterism and transferability of concepts and recipes from one place to another is also inherent in the increasing number of logistics regions that have emerged recently. They are increasingly seeking to sell their products to a wider audience; see, for example, the European logistics cluster platform (socool@eu; apparently being 'so cool') or the network of 11 major logistics initiatives in Germany (available at www.logistik-initiativen.de). Obviously, and contrary to the beliefs of policy boosters, the ways in which place-based policies foster local economic development are yet to be understood, and this raises questions on the territoriality of these developments and on the assumed linear mobility of policies and ideologies (Barca et al. 2012). Combined with the positivistic boast that such developments are linear processes, this discursive cover is necessary in order to prime the political process to be supportive. Both boosterism and policy boosterism represent discourses that are somewhat disconnected from a realistic assessment of the logistics industry and have rarely been subject to a critical analysis.

THE PROMISES AND LIMITATIONS OF SELLING THE REGION AS A HUB

This chapter gave an overview of recent strategies focusing on logistics for regional development purposes. It also took a particular look at how these regional concepts were being framed, justified, and communicated ('coded'). In the case of Venlo, it was the recent past that helped the place to emerge as a distribution hub, and future ideas for the greenport—that is, the massive expansion of agro-food industrial activity—for which logistics seems an ideal vehicle and catalyst. However, the pressure on economic development remains high, as employment figures indicate saturation or stagnation. In Wallonia, it is not yet certain whether their aim makes sense (cluster, 'extended gateway to port hinterland,' which is borrowed from Flanders). The picture is also somehow mixed here: Employment figures actually indicate a certain growth trajectory which corresponds to the early stages of the lifecycle of regional policies. The recent evaluation of the government's cluster policy, however, was rather critical.

Analyzing the two cases has also emphasized two distinct dimensions of the emerging logistics hubs and regions and their role for local economic development. One of the two—the economies of hubs and gateways—is rather popular, and meanwhile has been met with critical engagement by scholars and practitioners alike. In contrast to the first, the discursive and communicative dimension of logistics as a means of regional development has received rather little attention so far. This needs to be changed, since discourse provides a mental infrastructure that pushes human perception and societal debate in a very particular direction. Only against the background of such an informed approach—revealing the beliefs, the imaginaries, and the hidden promises of related discourses—should one engage with questions of how to manage and design the urban airport interface and related landscapes.

And, for sure, this debate would then also have to deal much more extensively than before with the political frames that address the dark side of circulation, such as the environmental, health-related, and neighborhood impacts, as well as other cases of contention, as, for example, Oosterlynck and Swyngedouw (2010) did in the case of airports and air travel in Brussels. The particular promise of a greenport is certainly the notion of the green, not only referring to primary production but also to the apparent sustainability of the envisaged transformation. Moreover, this discussion also seems essential given what could be called the 'scalar mismatch' or 'paradox' of the economies of transport hubs: While their benefits tend to spread quite broadly, their negative impacts remain concentrated locally. This is also the point where traditional concepts of territory no longer bring the debate forward. We need to better understand the complex, dynamic ways in which flows are going to be spatialized in two ways: the interplay of governance at various spatial scales (vertical) on the one hand, and the broader, increasingly city-regional and international scopes (horizontal) of urban-regional development on the other hand, situated in between the global and local. Judging from a geographical perspective, goods movement is relevant, since it allows the exploration of new territories in order to create further economic value, and conversely it thus shapes these places. However, the underlying dynamics are not simply the outcome of factor combination and economic fortune: They can only be fully understood by including the variegated realms of policy, governance, and regulation (local, overall), and also by approaching the emerging geographies of distribution as being discursively framed and communicatively co-constructed.

WORKS CITED

Aoyama Y., Ratick S., and Schwarz G. (2006) "Organizational dynamics of the US logistics industry: An economic geography perspective," *The Professional Geographer*, 58(3):327–340.

Bair, J. (ed) (2009) *Frontiers of Commodity Chain Research*, Palo Alto, CA: Stanford University Press.

Barca, F., McCann, P., and Rodríguez-Pose, A. (2012) "The case for regional development intervention: Place-based versus place-neutral approaches," *Journal of Regional Science*, 52(1):134–152.

Bevir, M., and Rhodes, R.A.W. (2006) *Governance Stories*, London: Routledge.

Beyers, W.B., and Fowler, C.S. (2013) "Economic structure, technological change and location theory. The evolution of models explaining the link between cities and flows," in P. Hall and M. Hesse (eds) *Cities, Regions and Flows*, Oxford: Routledge, pp. 23–41.

Bowen, J.T. (2008) "Moving places: The geography of warehousing in the US," *Journal of Transport Geography*, 16:379–387.

Bowen, J.T. (2012) "A spatial analysis of FedEx and UPS: Hubs, spokes, and network structure," *Journal of Transport Geography*, 24:419–431.

Bowen, J.T., and Leinbach, T.R. (2011) "Transportation networks, the logistics revolution and regional development," in A. Pike, A. Rodriguez-Pose, and J. Tomaney (eds) *Handbook of Local and Regional Development*, Abingdon, UK: Routledge, pp. 438–448.

Burghardt, A. F. (1971) "A hypothesis about gateway cities," *Annals of the Association of American Geographers*, 61:269–285.
CB Richard Ellis (CBRE) (2011) *Understanding Logistics in the Netherlands*, Amsterdam: CB Richard Ellis.
Christopher, M. (2005) *Logistics and Supply Chain Management: Creating Value-Adding Networks* (3rd ed.), London: Prentice Hall.
Cidell, J. (2006) "Air transportation, airports and the discourses and practices of globalization," *Urban Geography*, 27(7):651–663.
Cidell, J. (2010) "Concentration and decentralization: The new geography of freight distribution in US metropolitan areas," *Journal of Transport Geography*, 18:363–371.
Cidell, J. (2011) "Distribution centers among the rooftops: The global logistics network meets the suburban spatial imaginary," *International Journal of Urban and Regional Research*, 35(4):832–851.
Coe, N. M. (2014) "Missing links: Logistics, governance and upgrading in a shifting global economy," *Review of International Political Economy*, 21(1):224–256.
Cushman & Wakefield, with Logistics in Wallonia and AWEX (2009) "Comparison of prime locations for European distribution and logistics (abridged ed.)." Retrieved March 9, 2014 from www.investinwallonia.be.
De Lara, Juan D. (2013) "Goods movement and metropolitan inequality. Global restructuring, commodity flows, and metropolitan development," in P. Hall and M. Hesse (eds) *Cities, Regions and Flows*, Oxford: Routledge, pp. 75–92.
Derudder, B., and Witlox, F. (2010) "World cities and global commodity chains: An introduction," *Global Networks*, 10(1):1–11.
Ducruet, C., and Lee, S.-W. (2006) "Frontline soldiers of globalisation: Port-city evolution and regional competition," *GeoJournal*, 67:107–122.
Flämig, H. (2013) "Infrastructure and environmental policy on regulating road vehicle emissions. From top-down policy directives to the local level," in P. Hall and M. Hesse (eds) *Cities, Regions and Flows*, Oxford: Routledge, pp. 209–225.
Frémont, A. (2007) "Global maritime networks: The case of Maersk," *Journal of Transport Geography*, 15(6):431–442.
Gibbs, D., and Krueger, R. (2012) "Fractures in meta-narratives of development: An interpretive institutionalist account of land use development in the Boston city-region," *International Journal of Urban and Regional Research*, 36(2):363–380.
Gouvernal, E., Lavaux-Letilleul, V., and Slack, B. (2011) "Transport and logistics hubs: Separating fact from fiction," in P. V. Hall, R. McCalla, C. Comtois, and B. Slack (eds) *Integrating Seaports and Trade Corridors*, Farnham: Ashgate, pp. 65–79.
Greenport Venlo Development Company (GVDC) (2012) *Structuurvisie Klavertje 4 Gebied*, Venlo: GVDC.
Hall, P. V. (2007) "Seaports, urban sustainability, and paradigm shift," *Journal of Urban Technology*, 14(2):87–101.
Hall, P. V., and Hesse, M. (eds) (2013) *Cities, Regions and Flows*, Oxford: Routledge.
Harris, C. D., and Ullman, E. L. (1945) "The nature of cities," *The Annals of the American Academy of Political and Social Science*, 242:7–17.
Healey, P. (1999) "Sites, jobs and portfolios: Economic development discourses in the planning system," *Urban Studies*, 36(1):27–42.
Hesse, M. (2008) *The City as a Terminal. Logistics and Freight Transport in an Urban Context*, Aldershot: Ashgate.
Hesse, M. (2013) "Cities and flows: Re-asserting a relationship as fundamental as it is delicate," *Journal of Transport Geography*, 29:33–42.
Hesse, M. (forthcoming) "International hubs as a factor of local development: Evidence from Luxembourg City, Luxembourg, and Leipzig, Germany," *Urban Research & Practice*.

Heuvel, F.P.V.D., Langen, P.W.D., Donselaar, K.H.V., and Fransoo, J.C. (2014) "Proximity matters: Synergies through co-location of logistics establishments," *International Journal of Logistics Research and Applications*, 17(5):377–395.

Huijs, M. (2011) *Building Castles in the (Dutch) Air. Understanding the Policy Dead lock of Amsterdam Airport Schiphol 1989-2009*, Delft: Delft University of Technology.

IWEPS (2014, February) "Assessment programme of PM2.V (Marshall Plan 2. Green, *Plan Marshall 2.Vert*). Executive Summary."

Jacobs, J.M. (2012) "Urban geographies I: Still thinking cities relationally," *Progress in Human Geography*, 36(3):412–422.

Laurentzen, M., Kranendonk, R., and Regeer, B. (2009) *The Making of the Greenport Venlo. Eindrapportage Streamlining Greenport Venlo*, Wageningen: Wageningen University.

Levelt, M. (2010) *Global Trade & the Dutch Hub. Understanding Variegated Forms of Embeddedness of International Trade in the Netherlands. Clothing, Flowers, and High-Tech Products*, Oisterwijk: Uitgeverij BOX Press.

Logistics in Wallonia Innovation Network (n.d.) "Logistics in Wallonia is the business cluster and the competitiveness cluster for the transport and logistics sector in Wallonia." Retrieved September 12, 2014 from http://www.logisticsinwallonia.be.

MacKinnon, D. (2012) "Beyond strategic coupling: Reassessing the firm-region nexus in global production networks," *Journal of Economic Geography*, 12:227–245.

Magain, M. (2005) "Wallonia, land of logistics," *W+B Wallonia Brussels*, September: 4–6.

Malecki, E. (2004) "Jockeying for position: What it means and why it matters to regional development policy when places compete," *Regional Studies*, 38(9): 1101–1120.

McCalla, R., Slack, B., and Comtois, C. (2001) "Intermodal freight terminals: Locality and industrial linkages," *The Canadian Geographer*, 45(3):404–413.

McCann, E. (2013) "Policy boosterism, policy mobilities, and the extrospectivecity," *Urban Geography*, 34(1):5–29.

McKinnon, A. (2009) "The present and future land requirements of logistical activities," *Land Use Policy*, 26:293–301.

Mee, S. (2011, November 14) "Logistics: Aim is to build on hub status," *Financial Times*. Retrieved March 2, 2015 from http://www.ft.com/intl/cms/s/0/4dc0145e-0542-11e1-a3d1-00144feabdc0.html#axzz3TGY0Zzw4.

Ministry of Economic Affairs, Agriculture & Innovation (2011, February 4) "To the top. Towards a new enterprise policy [official letter]." The Hague.

O'Connor, K. (2010) "Global city regions and the location of logistics activity," *Journal of Transport Geography*, 18(3):354–362.

O'Connor, K., Holly, B., and Clarke, A. (2012) "A case for incorporating logistics services in urban and regional policy: Some insight from US metropolitan areas," *Regional Science, Policy & Practice*, 4(2):165–177.

Oosterlynck, S., and Swyngedouw, E. (2010) "Noise reduction: The postpolitical quandary of night flights at Brussels airport," *Environment and Planning A*, 42(7):1577–1594.

Peet, R. (2000) "Culture, imaginary, and rationality in regional economic development," *Environment and Planning A*, 32(7):1215–1234.

Porter, M.E. (1998) "Clusters and the new economics of competition," *Harvard Business Review*, 76(6):77–90.

Rivera, L., Sheffi, Y., and Welsch, R. (2014) "Logistics agglomeration in the US," *Transportation Research Part A: Policy and Practice*, 59:222–238.

Rodrigue, J.P., and Notteboom, T. (2009) "The terminalization of supply chains: Reassessing the role of terminals in port/hinterland logistical relationships," *Maritime Policy & Management*, 36(2):165–183.

Sheffi, Y. (2012). *Logistics Clusters: Delivering Value and Driving Growth*, Cambridge, MA: MIT Press.
Sheffi, Y. (2013) "Logistics-intensive clusters: Global competitiveness and regional growth," in J. H. Bookbinder (ed) *Handbook of Global Logistics, International Series in Operations Research & Management Science 181*, New York: Springer, pp. 463–500.
Spencer, G.M., Vinodrai, T., Gertler, M.S., and Wolfe, D.A. (2010) "Do clusters make a difference? Defining and assessing their economic performance," *Regional Studies*, 44(6):697–715.
Strale, M. (2008, Juin) "La miseen place d'une politique wallonne de promotion des activités logistiques; quelsenjeux pour le territoire régional? (The introduction of a Walloon policy that promotes logistics activities; Challenges for the regional territory)," *Territoire(s) wallon(s)*: 191–202.
Strale, M. (2012, Decembre 1) "For a critical approach to benchmarking studies: The example of logistics in Wallonia (Pour une approche critique des études de benchmarking: l'Exemple de la logistiqueen Wallonie)," *Territoire(s)*: 1–16.
Taylor, M. (2010) "Clusters: Amesmerising mantra," *Tijdschrift voor Economische en Sociale Geografie*, 101(3):276–286.
Transportation Research Board (TRB) (2003) *Integrating Freight Facilities and Operations with Community Goals. A Synthesis of Highway Practice*, Washington DC: TRB of The National Academies (NCHRP Synthesis 320).
Transportation Research Board (TRB) (2013) *City Logistics Research. A Transatlantic Perspective. Summary of the First EU-U.S. Transportation Research Symposium*, Washington, DC: TRB of The National Academies (Conference Proceedings 50).
Warnaby, G. (2009) "Non-place marketing: Transport hubs as gateways, flagships and symbols?," *Journal of Place Management and Development*, 2(3):211–219.
World Bank (2014) *The Logistics Performance Index and Its Indicators*, Washington, DC: World Bank.

12 The Politics of Public Transit in Postsuburban Toronto

Christian Mettke

INTRODUCTION: TECHNICAL INFRASTRUCTURES AND POSTSUBURBANIZATION

Toronto is one of the 'show rooms' of postsuburban development in North America. The city of Toronto, with 2.8 million inhabitants, is the largest city in Canada. The city itself is part of a city region known as the Greater Toronto Area (GTA) that has a population of over 6.5 million and is predicted to grow to 8.6 million by 2031. Structurally, the city includes both older, relatively dense urban forms and newer, dispersed, auto-oriented suburban forms. Explosive regional population growth and immigration into the inner and outer suburbs of the region have changed the socio-spatial patterns and flows within the city region significantly. Nowadays, Toronto is an archipelago of various sub/urban types in terms of their spatial forms and functions (see also Addie, this volume).

The existing transportation infrastructure has failed to match those socio-spatial dynamics. Two contradictory phenomena currently characterize public transit in Toronto. On the one hand, the spatial and infrastructural needs vary heavily within the city-region but are generally shaped by an immense increase of demand for transit services. On the other hand, the current physical network is shaped by path dependencies, obsolescence and capacity problems, and the fragmented governance structure that failed to readjust itself to the new spatial dynamics (Arthur 1994, North 1990).

Thus, public transit provision in Toronto offers a paradoxical perspective on the impacts and phenomena of a city that is shaped by postsuburbanization (Teaford 1997; Phelps 2010; Phelps and Wu 2011). This chapter suggests an understanding of postsuburbanization as a process of intraregional differentiation that is based on a suburban, socio-spatial, economic, and political increased significance. As cities evolve to globalized city-regions (Scott et al. 2001), the societal demands and requirements for technical infrastructures change. In the process of postsuburbanization, many technical infrastructures fail to fulfill their specific functions due to the territorial limits of their physical networks or due to limited (spatially restricted) governance structures, and therefore do not meet the needs of expanding, increasingly integrated, global urban societies. Manifold postsuburban phenomena (Keil and Young 2008; Soja 2000; Teaford 1997; Lang 2003; Phelps et al. 2006), including economic and political emancipation, functional differentiation, and the socio-spatial marginalization of poverty, as well as the stigmatization, vulnerability, and fragmentation of those spaces, not only challenge monocausal suburban theoretical understandings. They also challenge current institutional structures and actors to readjust traditional mechanisms of infrastructure and urban politics. This chapter highlights this spatial shift and offers a symptomatic perspective on the dialectic of postsuburbanization and urban mobility, as well as on the challenges of the *politics of mobility* within such postsub/urban settings.

In a first step the chapter explores the notion of 'technical infrastructures' and 'postsuburbanization' before it proposes an analytic framework for the analysis of any given technical infrastructure system in an urban context. In the second part of this chapter this framework is used to highlight the development of public transit provision in Toronto before using the case study of 'Transit City' to illustrate the relationship of postsuburban dynamics and public transit and to point out the distinct characteristics of the politics of (transit) mobility in a postsuburban context.

TECHNICAL INFRASTRUCTURES, PUBLIC TRANSIT, AND POSTSUBURBANIZATION

Technical Infrastructures

Modern urban societies rely heavily on highly specialized and integrated technical infrastructure networks. The visible and sometimes invisible sedimentations, the physical artifacts, and the built environments of such interscalar interactions create the local landscape for the 'system builders' of technical infrastructures (Hughes 1989). The state of those infrastructures as materialized places of flow strongly defines the intermediate function of cities and city-regions in the capitalist, globalized system.

Against this background, technical infrastructures have become an essential precondition for the worldwide intensification of economic, social, and political integration, widely known as 'globalization.' At the very same time, those technical infrastructures have to fulfill very local demands within global city-regions like the GTA. Due to that dialectic and the social/urban consequences of political reorganizations and restructurings of infrastructure provision, their vulnerability, inertia, or momentum tend to stay centered within scientific and political arenas, as the case study of public transit in Toronto highlights.

Public Transit

Public transit fulfills societal functions that are not, or only insufficiently, satisfied by other transport systems. They are supposed to meet mobility demands within cities and city-regions, but also in the countryside. Therefore, one of the critical functions of public transit systems is to provide a degree of accessibility and connectivity that serves both individual and collective mobility demands. Accessibility hereby does not only include the physical access to such networks, but also the political and planning dimension of transit or the question of financial restrictions for individuals due to pricing strategies. The significant *function* of public transit is the formation of mobility options for people. In modern societies, (access to) mobility is a basic urban necessity (Bonß et al. 2004). The constitution of 'mobility spaces' through public transit can be understood as the spatial manifestation of human activity and the need for spatial exchange. In this sense, the existence or absence of mobility options within (urban) societies can be directly linked to the 'right to the city' (Lefebvre 1970; Harvey 2008). Public transit provides one significant, sustainable option for individuals, the young and the elderly alike, to be physically mobile within city spaces.

This understanding can conceptually be connected to the discussion about the *politics of mobility* (Cresswell 2010). If "[m]obility is a resource that is differentially accessed" (Cresswell 2010, p. 21) and if postsuburbanization is based on an intraregional differentiation, mobility is a central part of this process, and the politics of mobility must be a focal point to address in order to understand both processes. In this regard, the discussion about the politics of mobility aims to shed light on the re/production of social relations through mobility (and vice versa) and focuses on the topology of power relations within that ratio. Public transit as a multidimensional mobility space with individual and collective accessibility patterns, as well as its historically close linkage to urban spaces, seems to be best placed for a techno-urban analysis in a postsuburban setting.

Postsuburbanization

Postsuburbanization, in the sense of a postmodern urban reality, puts the 'urban fringe' within the center of societal and political arenas and power struggles. The predicted "urban revolution" (Lefebvre 1970) appears much more decentralized today, with spaces of concentration outside traditional city cores. Thus the dominant form of global urbanization is more peripheral than central (Filion 2010). The intraregional emancipation of traditional suburban, peripheral spaces has led to an empirical and theoretical debate within and around these spaces. In response to the observed global increased significance of the urban fringes, some researchers have tried to translate this phenomenon into their own concepts, such as the In-Between City (Sieverts 1997, 2001), the Boomburbs (Lang and LeFurgy 2007), Exopolis (Soja 2000), Postsuburbia (Teaford 1997; Phelps and Wu 2011), Edge Cities (Garreau 1991) or Edgeless Cities (Lang 2003), amongst others.

Nowadays, postsuburbanization is increasingly shaping city-regions (Scott 2002) in manifold ways. Within these regions, more and more original features and functions of traditional urban centers, like corporate headquarters, administrations, industrial parks, or residential areas, are decentralized or rather polycentric. Existential technical infrastructure systems like water treatment plants, airports, transportation hubs, or waste treatment plants are also placed in the peripheries of city-regions (Hesse 2007). That spatial reorganization and transformation reflects the theoretical and empirical displacement of the classic urban system. Boundaries, materialized or discursive, are disappearing or shifting while territorial definitions often remain. The urbanization of the suburbs (Masotti and Hadden 1973) has now led to regional urbanization (Soja 2000). As Soja puts it, "what once could be described as mass regional suburbanization has now turned into mass regional urbanization, with virtually everything traditionally associated with 'the city' now increasingly evident almost everywhere in the Postmetropolis. In the Era of the Postmetropolis, it becomes increasingly difficult to 'escape from the city'" (Soja 2000, p. 242).

The specific case of postsuburban public transit offers a particular view on the emerging spatial needs that have their roots in the changing conditions within affected urban realms and related disruptive moments of postsuburbanization. Because changing environments provide moments in which entities negotiate the current and future places of flows, the existing and emerging dynamics are leading to multidimensional spaces of conflicts and tensions, but also to new spaces of possibilities and enablement. A multidimensional perspective on public transit in the context of postsuburbanization acknowledges the possibility of various existential factors and actors, and it sheds light on normative concepts and embedded values within the local or regional debate about public transit. It particularly recognizes the sociotechnical characteristics of public transit systems and the techno-urban development path.

One of the leading questions of this chapter is, are the 'sociotechnical' characteristics of transit infrastructure systems (more details in the next section) incapable of readjusting their functional geometry accordingly to postsuburban dynamics? Or, can this *techno-political mismatch* best be understood as a distinctly infrastructural manifestation of current postsuburban spatial dynamics?

ANALYTIC FRAMEWORK FOR APPROACHING TECHNO-URBAN DEVELOPMENT PATHS

If one of the main purposes of this chapter is to shed light on the re/production of social relations through a public transit lens in a postsuburban dynamic, then the analysis of the transit system needs to take into account different distinct dimensions in order to highlight the contextual specifics. It is the conviction of this chapter that the below presented analytic framework allows us to discuss a variety of *techno-urban development paths*, independent from the actual technical infrastructure system or the given urban/spatial context.

In this respect I propose four conceptual dimensions (see Figure 12.1) that are suitable to analyze infrastructure developments paths: (1) 'Pattern of Usage' refers to distinct preferences, appropriations, and practices by the users and inhabitants of the infrastructural space, which changes over time and significantly accommodates to the development path of technical infrastructures. (2) 'Technologies' point to the specifics and meanings of technological artifacts and physical networks and their transformation. Local, regional, and, of course, national idiosyncrasies of certain technology styles, as for example the AC/DC case, vary from context to context. (3) 'Industry Structure' highlights the distinct, internal governance structure of technical infrastructures. Due to traditional and cultural practices and experiences, technical infrastructures tend to evolve specific internal regulation structures. Those structures might then be disrupted by external or internal innovations, such as transnational regulations or those of international agreements. Last but not least, (4) 'Governance' refers to the external regulation that is the interplay of actors, institutions, norms, and laws on different political levels, which shape the trajectory of technical infrastructures due to juridical power. At the same time, such external structures often are (intentionally) influenced by the expertise and know-how of a specific technical infrastructure regime. Of course, local *socio-spatial conditions*, such as demographic, economic, or ecological characteristics and processes, shape those four dimensions or infrastructures.

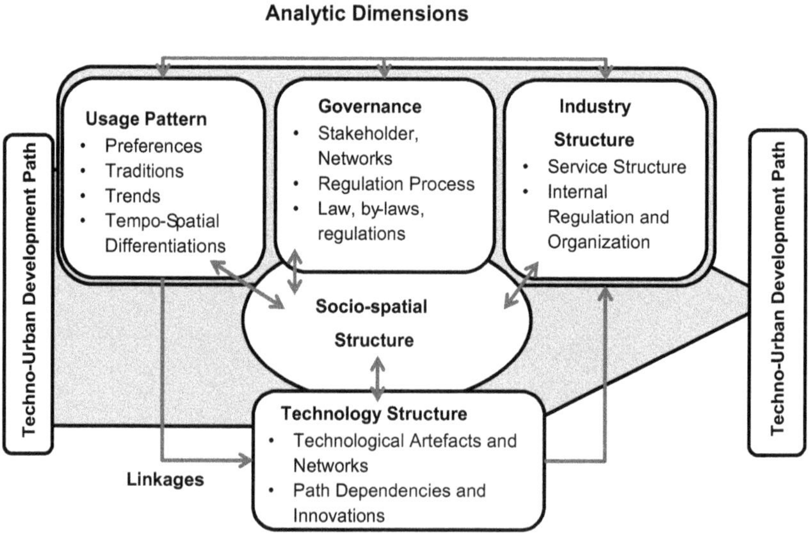

Figure 12.1 Conceptualizing the techno-urban development path.
Source: Author.

The complex and sometimes messy interaction of those factors is an *interscalar phenomenon*, which defines any given *techno-urban development path*. This perspective highlights how tightly technical infrastructures are interwoven with their urban context. *Cities as places of flows* (Prytherch and Cidell, this volume) understands urban spaces as the relation among mobile entities and practices, physical transportation networks and collective or individual places; in other words, as a space of intersecting flows. Technical infrastructures are a precondition and the foundation of every networked flow of people or objects within the 21st-century city.

If urban research aims to analyze the political economy or the politics of infrastructures (McFarlane and Rutherford 2008; Young and Keil 2009), the proposed analytic dimensions help to understand the interplay of relevant structures, stakeholders, and processes. The politics of technical infrastructures within cities can be best understood and studied through this multidimensional perspective, as it emphasizes various arenas of infrastructural and spatial conflicts on constraints and initiatives and on emerging actors, networks, and interplays. In this case study and in the context of postsuburbanization, public transit in a global city like Toronto illustrates a remarkable example of 'places of flows' and the political economy behind them (Brenner 2004).

PUBLIC TRANSIT IN TORONTO. A POSTSUBURBAN CONTEXT

The postsuburban dynamics of the socio-spatial development in Toronto over the last two decades rely on explosive regional population growth and immigration into the inner and outer suburbs of the region. Almost all of the surrounding municipalities of Toronto were among the fastest growing municipalities in Canada. The economic and social importance of the traditional suburbs within the GTA has led to their political emancipation from the traditional core, which affected the governance structure of the public transit system.

In 1998, the provincial government of Toronto merged six municipalities into a single municipality, the city of Toronto. The political belief was that a new, bigger city would be more cost-efficient. This amalgamation brought together two very different urban settings (Filion 2000) and lifestyles. The old city of Toronto has a high-density core and a grid of streets that provide a mix of uses, and the inner suburbs (amalgamated municipalities) were the first ring of post–World War II suburbs built around the old city of Toronto. In a later wave of construction, throughout the 1970s and 1980s, high-rise apartment buildings were added along arterials and at the major intersections that now define the hybrid character of the inner suburbs.

Toronto is challenged by the need to manage various sub/urban types in terms of their spatial forms and functions: the now booming, dense downtown of the old city of Toronto; the in-between spaces of the inner suburbs, accommodating typical suburban bungalows but also hundreds of high-rise towers and infrastructural corridors; the also booming, new downtowns of the outer suburbs, like Markham or Brampton; the seemingly endless strips north- and southbound; or the international airport of Toronto, the central air transportation hub of East Canada, which is located at the fringe of Toronto in the 'suburban' municipality of Mississauga.

Pattern of Usage

The socio-spatial dynamics of the last decades in the GTA have changed the socio-spatial patterns and flows within the city region. In response to those shifts, the existing transportation infrastructure has been confronted with capacity and distribution problems. Within the last 20 years, the GTA and the suburban municipalities in particular have experienced a tremendous transformation, with extremely high growth rates in both population and transportation journeys. Car ownership rates and traffic congestion have been growing more quickly than the population during this time, with daily car trips in the outer suburbs having almost doubled.

The transportation infrastructure has failed to match the population growth (TTS 2014; interview with Eric Miller 2012). In the last 25 years the rapid transit network in the city of Toronto grew only about 10 kilometers and the road network by only around 7%. As a consequence, transportation has been identified as a bottleneck preventing regional and global competitiveness in Toronto (Keil and Young 2008). Today, the GTA is one of the most congested city regions in North America, something that costs the regional economy around $6 billion every year, according to the Toronto Board of Trade (2010). Through the lenses of transportation, public transit usage mirrors the postsuburban trajectory of the GTA as a whole. The overall balance of transit trips is shifting to an increasing significance of the inner and outer suburbs.

Today, poverty in Toronto is increasingly a phenomenon of the inner suburbs and of newer cohorts of immigrants (Hulchansik 2007). Almost all of the 13 'Priority Investment Neighborhoods'[1] are located within the inner suburbs. They have become fragmented spaces that combine different lifestyles, biographies, and socioeconomic statuses. Poverty is now predominantly a 'suburban,' vertical phenomenon in Toronto, as many of the new immigrants and poorer inhabitants live in the thousands of high-rise towers in the city (United Way 2004, 2011). Research has shown that new immigrants are more likely to use public transit to commute; newer cohorts of immigrants have higher rates of transit use than earlier cohorts (Heisz and Schellenberg 2004). The residents of the high-rise towers in the inner suburbs rely more on transit, walking, and cycling than do other residents of the region. Many of these neighborhoods have higher than average rates of these modes and lower than average car ownership (E.R.A. Architects 2010). The mismatch, the gap between urban/demographic development and the spatial adaptability of the transit network, has led to an unsustainable trajectory within the city region. The current situation has not only influenced the daily life of the inhabitants, but has also become a major burden on the local economy. Without significant investments (and especially in public transit), the transport infrastructure will become increasingly dysfunctional over the next years.

Postsuburbanization in Toronto also includes all phenomena of traditional suburbanization, such as car ownership, single-family homes, and the lack of infrastructure provisions. From an operational perspective, the structural heritage (or the architectural landscape) of the car-oriented paradigm from the 1960s and 1970s of the inner suburbs often limits an economic, efficient transit provision (Miller and Soberman 2003; interview with Eric Miller 2011). At the same time, as more and more people are moving to the GTA and as many of the neighborhoods of the old city of Toronto have become desirable places to live, work, and play, the gentrification of downtown and the inner ring has led to a marginalization of the poor, a movement of the immigrants and/or the old towards the inner suburbs and the in-between spaces of the city-region (Hulchanski 2010). The utopian idea of suburbanism and suburban spaces as spaces of individual freedom and dreams now is shadowed by dystopian certainty and fear of spatial injustice (Soja 2011), manifesting in the socio-spatial polarization in postsuburban Toronto. In this socio-spatial context, infrastructures and particularly public transit are seen as a potential solution to tackle this fragmentation and splintering of urban cohesion (Graham and Marvin 2001).

Governance and Industry Structure

In Canada, transit is constitutionally very closely tied to municipal governments, much more than in the United States or many European countries, which have powerful, independent authorities at the regional level. In Toronto, there are two major public transit players that are shaping the trajectory of the system. On the local level, the city of Toronto is responsible for providing local public transit through the municipal transit authority, the Toronto Transit Commission (TTC). It acts as a full service transit agency that provides planning and operation for the city of Toronto. The TTC works as a municipal department with policy and funding decisions made by the city council. Even with a separate commission that has been created to oversee the TTC, the board is composed of municipal politicians with three additional civic representatives. The city of Toronto is responsible for ensuring full funding of the commission's capital program. In accordance with the Municipal Act, any funding for the commission's capital program from other governments flows through the city. The second major stakeholder is the regional transit authority, Metrolinx, which was founded by the province in 2006/2007 and oversees the regional dimension of transit provision within the GTA. It provides planning, funding, coordination, and operation (through the regional commuter service GO Transit) on a regional scale. GO Transit is the operating division of Metrolinx but was already founded in 1967 by the government of Ontario.

Technology Structure

Toronto's current public transit network is the third largest public transit system of North America (behind New York City and Mexico City) in terms of daily ridership. The TTC's weekday ridership exceeds all monthly ridership of the other regional transit providers in the GTA combined. Its physical network consists of a backbone of rapid transit in the form of a subway that runs north–south and east–west from downtown, and a network of feeder bus and streetcar lines. This network is thickest in the inner city, roughly contiguous with the old city of Toronto boundary. Although there have been several plans for extending transit service over the last 20 years, no major coherent network extension has been implemented since the subway system was built in the 1950s, 1960s, and 1970s. The expansion program 'Transit City' (2006) aimed to extend the rapid transit network of Toronto and to improve particularly the connectivity within the inner suburbs.

TRANSIT CITY AND THE POLITICS OF MOBILITY IN TORONTO—AN OVERVIEW

Since the 1990s, the development path of public transit in the GTA has been affected mainly by two phenomena. First, explosive urban growth with a rebalancing of the intraregional hierarchies and the socio-spatial dynamics challenged the trajectory of public transit significantly. Second, the instability of the governance structure limited the ability of the transit system to adapt to these changing urban requirements. The regulative instability has its roots in the horizontal and vertical fragmentation of the planning process, in financial uncertainties, and in the absence of a national transit strategy or framework.

During the last 10 years, the provincial government of Ontario increasingly influenced the governance structure of public transit provision in Toronto and the GTA. In 2003, with a new provincial government in place, public transit returned to the provincial political agenda, after years of shifting responsibilities and cost to the municipalities (Boudreau et al. 2009). After installing a new regional transportation body in 2006, the 'Greater Toronto Transportation Authority,' later to be renamed Metrolinx, the provincial government aimed to tackle congestion by an integrated, regional, 25-year transportation plan called 'The Big Move' (Metrolinx 2008). It represented the effort of the provincial government to upgrade transportation planning to the regional level (1) by expanding and improving the public transit network within the GTA and (2) by establishing a regional transit body that would coordinate the development of the regional transit system. However, even in 2014 the regional transportation plan still lacks reliable and stable funding

Politics of Postsuburban Transit   237

mechanisms that would provide the local actors and the expansion projects in the region with long-term financial certainties. To this day, financial commitments are entirely project based and require a Memorandum of Agreement between the local actors and the provincial level. Nevertheless, 'The Big Move' can be seen as the first relevant achievement of Metrolinx and the 'reentrance' of the provincial government into the public transit sphere.

The introduction of 'The Big Move' itself is important, as it provided the political framework for the city of Toronto to introduce its expansion plan, branded as 'Transit City,' and integrate it within 'The Big Move.' But with the absence of a long-term and reliable federal public transit strategy, regional transit funding mechanisms and the lack of serious, institutionalized, regional coordination and cooperation, Transit City was vulnerable on various frontiers.

Transit City

Transit City can be best understood as a mostly 'postsuburban' transit expansion program that aimed to extend the rapid transit network of Toronto (see Figure 12.2) and improve particularly the connectivity within the inner suburbs, which are increasingly challenged by spatial marginalization and polarization. The plan can be seen as a physical manifestation of

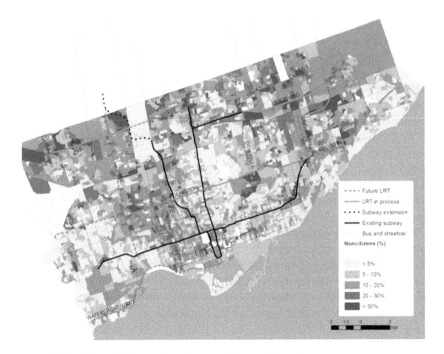

Figure 12.2   Transit City and residents without a Canadian passport.
Source: Kramer and Mettke 2014.

postsuburban challenges within Toronto, as it aimed to connect physically high-density pockets of the (new and old) high-rise cluster along the arterial roads of the inner suburbs (compare to Enright, this volume).

Transit City was introduced and championed by former Mayor David Miller in 2007, but was later incorporated into the 'The Big Move.' Transit City initially called for seven new light rail transit (LRT) lines and several bus rapid transit (BRT) lines, as well as the replacement of the overaged and overcrowded Scarborough Rapid Transit with light rail. In June 2007, the province of Ontario committed $8.4 billion for the first five lines initially, under the 'five in 10' plan for the first 10 years, but later postponed $4 billion of this funding due to budget constraints. The project was scaled back to just three new east–west LRT lines and the SRT replacement (see Figure 12.2). As documented above, Toronto would have significantly expanded its rapid transit network for the first time in decades and therefore overcome inertial path dependencies. Financial reasons alone cannot explain the postponement and scale back by the province, as the city of Toronto offered to prefinance the whole plan until the province would have solved its budget problems. "From a transportation and financial perspective, it [was] not a rational decision" (interview with David Miller 2011).

The political battle over Transit City was a messy, conflicting, and disruptive process that could be split into three phases. The first period can be directly linked to the introduction of Transit City by David Miller and the funding by the provincial government. The second phase had its political start with the election of a new mayor in 2010, Rob Ford, who cancelled Transit City with the words "Transit City is dead, ladies and gentlemen" and "it's time to stop the war on the car." After his victory in December 2010, and as a consequence of his idea of public transit, Ford reached a new Memorandum of Understanding with the province in March 2011, which stated that the new expansion plan—a partially underground cross town line on Eglinton Avenue—needed to be approved by the city council in Toronto. In this scenario, all financial contributions from Metrolinx would contribute to that one single underground line, which would have covered much less territory for the same cost. Until the end of 2011, Ford did not ask the city council for approval of his Memorandum with the provincial government. The third phase of Transit City can be seen in the city council actions in 2012/2013. By the time the vote came to council in early 2012, the council voted in favor of motions to resume work on the reduced Transit City plan, defeating Rob Ford's campaign for subways. After a long and complicated decision-making process, Transit City as an extension plan survived with the intended lines and technology (LRT), but died as a political program. It led to a broad discussion about transit governance, transit funding, and transit technologies within the whole region. The transit stakeholders repositioned themselves, the TTC board was restructured, the TTC management structure was reorganized, the relationship between the local TTC and the regional transit authority Metrolinx was, and still is, under reconsideration, and the whole transit regime of the GTA currently seems to be in a disruptive moment. Even in 2013, and after all the political tensions and conflicts, the TTC chair, Karen Stintz,[2] and some other counselors came up with a new plan. They proposed to convert the Scarborough RT, which was part of the LRT expansion, to a full subway line. Transit City clearly symbolized the political and societal importance of public transit within the highly dynamic and transforming postsuburban context of Toronto.

Conceptually, Transit City aimed to expand the rapid transit network towards the spaces that were underserved and disconnected by rapid transit options. It seemed surprising that it was largely the inner suburban voters that elected Rob Ford. This paradox may be partly explained by the discrepancy between those who (are legally allowed to) vote and those who rely on transit provision, as well as by the needs of current residents compared to future populations in these areas. One explanation of that decision-making can be suspected in the lack of political representation of many inhabitants of the inner suburbs. Some postsuburban demands of inner suburban populations are politically unheard. One in seven Torontonians are not eligible to vote in municipal elections because they are not Canadian citizens. Some of the neighborhoods within the inner suburbs have over 30% of residents who are noncitizens (Siemiatycki 2010). Most of these spaces are located in areas that would have benefited from Transit City, as Figure 12.2 illustrates.

Transit City as a political process highlighted that neighborhoods of immigrants, noncitizens, and the poor are less likely to have political influence and representation. The combination of a higher than average demand for transit services and the lack of appropriate transit provision within the inner suburbs can be understood as a lack of political representation and power, which structurally disadvantages this part of the population in terms of political integration and infrastructural inclusion.

Therefore, this techno-political mismatch can be understood as a distinctly infrastructural manifestation of current postsuburban spatial tensions and conflicts. The politics of Transit City point to the inertial structure of the existing transit regime, to the dysfunctionalities, to the intraregional conflicts, to the politicization of transit planning, and to the techno-political mismatch of mobility within Toronto. But it might also be a turning point towards an *incrementalization of transit planning* and a better *techno-urban* and *techno-political integration*. The transit regime might become more adaptable due to the political battle and societal discussion/awareness about public transit and transit funding in general. Maybe politicization will lead to more participation, to more democratization of transit, and therefore to an increased accessibility of the transit system in Toronto.

POSTSUBURBANIZATION AND PUBLIC TRANSIT IN TORONTO

The economic and social dynamics in Toronto and the significant investment backlogs (CUTA 2010) led to an increasingly dysfunctional situation of the transit system. Financial restrictions due to inadequate funding mechanisms within the governance structure had a direct impact on the performance and adaptability of the current postsuburban public transit regime. At the same time, that cannot be used as the sole explanation for the inertia of public transit in Toronto, as the chronology of Transit City has shown. The "political death" of Transit City symbolizes significantly the "governance failure" (Bakker 2010) of public transit in Toronto. The antiquated governance and industry structure (see Figure 12.3) would need far-reaching organizational restructuring to become more adaptable and regionally integrated.

On the regional level, the internal governance structure of Metrolinx can be characterized as highly vertical or top-down, with little political representation of local municipalities, in combination with the strong influence of the provincial government. Thus, the internal power structure of Metrolinx follows the political tradition of the relationship between the provincial and municipal level in Ontario. However, the regional expansion plan, 'The Big Move,' can be understood as a strategic shift from a monocentric transit network towards a more networked, polycentric public transit logic with multiple transit hubs surrounding the city of Toronto.

This case study verified the technological and institutional path dependencies of infrastructure systems in shifting urban contexts such as post-suburbanization. In Toronto, the transit system faced several disruptive moments and unstable trajectories that highlighted the infrastructural

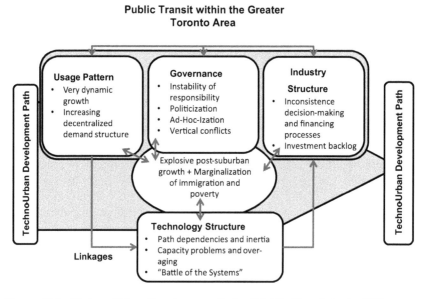

Figure 12.3 Techno-Urban Development Path of Public Transit in the GTA.
Source: Author.

inertia and inability of the internal and external governance structure. In this regard, Transit City is one of the most prominent of many examples of these conflicts and this mismatch over the last two decades.

Postsuburbanization within Toronto has led to hybrid spaces in which the functional, geometric, and architectonic characteristics appear contradictory. The variety of urban realms within Toronto emphasizes the empirical overcoming of theoretical dichotomies like the concept of urban versus suburban. Those hybrid spaces, not to mention the whole topology of the region, symbolize a general transformation towards a globalized, spatial reorganization of the urban society. A postsuburban perspective does not aim to describe urban spaces as something limitable and closed, but rather to emphasize a new geometry of urbanization (Phelps et al. 2006, p. 32) in which conceptual juxtapositions, such as 'core' and 'periphery,' do not fit anymore to describe and analyze today's urban development or mobilities.

CONCLUSION

The analysis of public transit in Toronto has advanced an understanding of the variety and multidimensionality of contemporary urban development in the context of postsuburbanization and the multiplicity of places of flows within modern city spaces. The proposed analytic framework for the analysis of technical infrastructures helped to differentiate the various contested relations and variables of such a complex system. Transit City as a case study illustrated the interwoven and conflictual relationship within the mobility space of postsuburban Toronto.

Postsuburbanization as an infrastructural phenomenon is a moment of change and disruption, of splintering and fusion, a moment of new demands and contradictions (Coutard 2002, 2008). Last, but not least, it is a moment of emerging and suppressed spatial, infrastructural, and political tensions. In this regard, postsuburbanization through the lens of mobility is a contested moment in the modern city (see also Addie or Van Neste, this volume).

For Toronto, the highly dynamic postsuburban development led to multiple inertial symptoms of the public transit system that highlighted the discrepancy between institutionalized practices and structures on the one hand and the current and upcoming socio-spatial demands in Toronto on the other. The overcoming of those symptoms—that is, the capability of being responsive to socio-spatial changes in order to actualize their societal function—might be the biggest challenge of technical infrastructures in the 21st-century city. The proposed analytic framework will help to increase the understanding of how the techno-urban relations and interlinkages really work and how, in this case, the politics of mobility can be understood through the lenses of the infrastructure itself.

On a more abstract level, a sustainable infrastructure development path would combine infrastructural sensitivity and flexibility to the changing urban demand structures *and* a stable momentum of physical and organizational reliability. In Toronto, that complex and idealistic necessity is far from realistic due to the current postsuburban tensions and various mobility conflicts. The dis/connectivity and un/accessibility of the present public transit network in Toronto reflects postsub/urban and societal power relations. Even if public authorities or transit agencies point to cost recovery rates and financial restrictions as a justification for uneven provision, one could argue that an uneven socio-spatial geography of access can be understood as socially/politically constructed and therefore as a materialization of the political economy of the GTA (see also Addie, this volume).

As the trajectory of public transit in Toronto has documented, postsuburban public transit regimes lose their momentum and are challenged to create new institutional and physical arrangements that can fulfill the new societal and spatial requirements. This search for a new, regional infrastructural balance can lead to inertial moments and dysfunctional structures, as the case study of Toronto and Transit City has shown. But it also points to new opportunities and constellations in which the needed adaptations are negotiated and sometimes implemented.

Cities have always been infrastructure spaces. And the flows of mobilities within and between them are constitutive of (urban) places (see Prytherch and Cidell, this volume). Within the context of postsuburbanization this constitution becomes challenged and contested, as other contributions in this volume have also shown (e.g., Van Neste or Addie). But as this case study demonstrated, despite those multiple moments of inertia, crisis, or dysfunctionality, disruptive spatial transformations such as postsuburbanization also always open up new spaces of opportunity and change, creating new networks of flows.

NOTES

1 Priority Neighborhoods are defined by the city of Toronto as areas with a lack of access to social and technical infrastructures, combined with high rates of household poverty.
2 Who then announced to run for mayor in the next election in 2014.

INTERVIEWS

Crowley, David (15.11.2011)—Transit expert and consultant
Kirkpatrick, Jamie (15.11.2011)—Transit expert and former consultant/strategist at Toronto Environmental Alliance
Miller, David (24.11.2011)—Former mayor of Toronto, 2003–2010
Miller, Eric (05.03.2012)—Professor at University of Toronto and lead on an expert panel in the course of Transit City discussion, consulting city council 2012
Perks, Gordon (8.11.2011)—City councilor at the city of Toronto

WORKS CITED

Arthur, W. B. (1994) *Increasing Returns and Path Dependence in the Economy*, Ann Arbor, MI: University of Michigan Press.
Bakker, K. (2010) *Privatizing Water—Governance Failure and the World's Urban Water Crisis*, Ithaca, NY: Cornell University Press.
Bonß, W., Kesselring, S., and Weiß, A. (2004) "Society on the move—Mobilitätspioniere in der Zweiten Moderne," in U. Beck and C. Lau (eds) *Entgrenzung und Entscheidung: Perspektiven reflexiver Modernisierung*, Frankfurt: Suhrkamp, pp. 258–280.
Boudreau, J.-A., Keil, R., and Young, D. (2009) *Changing Toronto—Governing Urban Neoliberalism*, Toronto: University of Toronto Press.
Brenner, N. (2004) "Urban governance and the production of new state spaces in western Europe, 1960-2000," *Review of International Political Economy*, 11(3): 447–488. Canadian Urban Transit Association (CUTA) (2010) *Transit Infrastructure Needs for the Period 2010–2014*, Toronto.
Coutard, O. (2002) "Premium Network Spaces," *International Journal of Urban and Regional Research*, 26:166–174.
Coutard, O. (2008) "Placing splintering urbanism," *Geoforum*, 39:1815–1820.
Cresswell, T. (2010) "Towards a politics of mobility," *Environment and Planning D: Society and Space*, 28:17–31.
E.R.A. Architects (2010) "Tower neighbourhood renewal in the greater golden horseshoe." Retrieved February 19, 2015 from http://www.cugr.ca/tnrggh.
Filion, P. (2000) "Balancing concentration and dispersion? Public policy and urban structure in Toronto," *Environment and Planning C: Government and Policy*, 18:163–189.
Filion, P. (2010) "Reorienting urban development? Structural obstruction to new urban forms," *International Journal of Urban and Regional Research*, 34(1):1–19.
Garreau, J. (1991) *Edge City: Life on the New Frontier*, New York: Doubleday.
Graham, S., and Marvin, S. (2001) *Splintering Urbanism: Networked infrastructures, technological mobilities*, London: Routledge.
Harvey, D. (2008) "Right to the city," *New Left Review*, 53:23–40.
Heisz, A., and Schellenberg, G. (2004) *Public Transit Use Among Immigrants*, Ottawa: Statistics Canada, Analytical Studies Branch.
Hesse, M. (2007) "Mobilität im Zwischenraum," in O. Schöller, W. Canzler, and A. Knie (eds) *Handbuch Verkehrspolitik; 1*, Auflage: Verlag für Sozialwissenschaften, pp. 279–300.
Hughes, T. (1989) "The evolution of large technical systems," in W. Bijker, T. Hughes, and T. Pinch (eds) *The Social Construction of Technological Systems: New Directions in the Sociology and History of Technology*, Cambridge, MA: MIT Press, pp. 51–82.
Hulchanski, J.D. (2010) *The Three Cities Within Toronto: Income Polarization Among Toronto's Neighbourhoods, 1970–2005*, Toronto: Cities Centre, University of Toronto.
Keil, R., and Young, D. (2008) "Transportation: The bottleneck of regional competitiveness in Toronto," *Environment and Planning C: Government and Policy*, 26(4):728–751.
Keil, R., and Young, D. (2009) "Fringe explosions: Risk and vulnerability in Canada's new in-between urban landscape," *The Canadian Geographer/Le Géographe Canadien*, 53(4):488–499.
Kramer, A., and Mettke, C. (forthcoming) "The death and life of 'Transit City'— Searching for sustainable transportation in Toronto's inner suburbs," in R. Thomas (ed) *Planning Canada—A Case Study Approach*, Toronto: Oxford University Press Canada.

Lang, R. (2003) *Edgeless Cities: Exploring the Elusive Metropolis*, Washington, DC: Brookings Institution Press.
Lang, R., and Lefurgey, J. (2007) *Boomburbs: The Rise of America's Accidental Cities* [James E. Johnson Series], Washington, DC: Brookings Institution Press.
Lefebvre, H. (1970/2003) *The Urban Revolution*, Minneapolis: University of Minnesota Press.
Masotti, L.H., and Hadden, J.K. (1973) *The Urbanization of the Suburbs*, Thousand Oaks, CA: SAGE Publications.
McFarlane, C., and Rutherford, J. (2008) "Political infrastructures: Governing and experiencing the fabric of the city," *International Journal of Urban and Regional Research*, 32(2):363–374.
Metrolinx (2008) *The Big Move: Transforming Transportation in the Greater Toronto and Hamilton Area*, Toronto: Metrolinx.
Miller, E., and Soberman, R. (2003) "Travel demand and urban form," Issue Paper No. 9, Toronto: Neptis.
North, D.C. (1990) *Institutions, Institutional Change and Economic Performance*, Cambridge: Cambridge University Press.
Phelps, N.A. (2010) "Suburbs for nations? Some interdisciplinary connections on the suburban economy," *Cities*, 27(2):68–76.
Phelps, N.A., Parson, N., Ballas, D., and Dowling, A. (eds) (2006) *Post-Suburban Europe: Planning and Politics at the Margins of Europe's Capital Cities*, New York: Palgrave Macmillan.
Phelps, N., and Wu, F. (eds) (2011) *International Perspectives on Suburbanization: A Post-Suburban World*, New York: Palgrave Macmillan.
Scott, A., Agnew, J., Soja, E., and Storper, M. (2002) "Global city-regions," in A. Scott (ed) *Global City-Regions. Trends, Theory, Policy*, Oxford, UK: Oxford University Press, pp. 11–32.
Siemiatycki, M. (2010, March 12) "Toronto's lost voters," *The Mark*. Retrieved February 18, 2015 from http://pioneers.themarknews.com/articles/1090-torontos-lost-voters.
Sieverts, T. (2001) *Zwischenstadt: Zwischen Ort und Welt, Raum und Zeit, Stadt und Land* (3rd ed.). Wiesbaden: Vieweg.
Soja, E. (2000) *Postmetropolis-Critical Studies of Cities and Regions*, Oxford: Blackwell Publishers.
Soja, E, (2011) *Seeking Spatial Justice*, Minneapolis, MN: University of Minnesota Press.
Teaford, J. (1997) *Post-Suburbia: Government and Politics in the Edge Cities*, Baltimore: Johns Hopkins University Press.
Toronto Board of Trade (2010) *The Move Ahead: Funding "The Big Move,"* Toronto: Toronto Board of Trade.
Transportation Tomorrow Survey (TTS) (2012–2014) *Internet Data Retrieval System (iDRS)*, Toronto: TTS.
United Way (2004) *Poverty by Postal Code: The Geography of Neighbourhood Poverty, 1981–2001*, Toronto: United Way.
United Way (2011) *Vertical Poverty: Declining Income, Housing Quality and Community Life in Toronto's Inner Suburban High-Rise Apartments*, Toronto: United Way.
Young, D., and Keil, R. (2009) "Reconnecting the disconnected: The politics of infrastructure in the in-between city," *City*, 27(2):87–95.

13 Place-Framing and Regulation of Mobility Flows in Metropolitan 'In-Betweens'

Sophie L. Van Neste

This chapter takes the broader topic of the book, the intersection of place and flows, through its manifestations in public debates and contentious action. Mobility is defined, following Cresswell (2010), as 'socially produced motion.' Mobility flows are debated in relation to meaning given to place, yet they remain little studied as such. The politics of mobility includes a politics of place. Activists deploy 'place' arguments in mobility debates, while state and private authorities also format and selectively represent places through their regulation of mobility flows.

This chapter considers mobility and flows through the politics of place; i.e., how the future of a place is negotiated. It aims at proposing a framework to study these 'place' arguments in mobility debates. The framework is applied at a particular type of place in metropolitan areas, the 'in-between cities' (Young and Keil 2014), spaces for which metropolitan flows are often emphasized, but the sense of place is forgotten or at least heavily contested. These metropolitan 'in-betweens' thus constitute key cases from which to start an exploration of place-framing in relation to the regulation of mobility flows. The empirical illustration is based on what appears an emblematic case of a metropolitan in-between, a zone between Rotterdam and The Hague in the southern part of the Randstad, in the Netherlands.

THE POLITICS OF PLACE, THE CONCEPT OF PLACE-FRAMING

The term *place* has several intertwined meanings. Agnew (1987) distinguishes between three understandings of place. Place is used to speak of a location, a specific point on the earth. Place is also discussed as a locale; that is, a site as 'a setting and scale' at which daily practices and interactions are experienced. Finally, place also refers to the subjective meaning given to the site and/or the location, the 'sense of place.'

The *politics of place* is defined by Amin (2004) as the negotiation of spatial juxtaposition on a site. If place is defined as a site of daily practice, shared by its different users and imbued with meaning, Amin (2004, p. 38) emphasizes how places are "sites of heterogeneity juxtaposed within close spatial proximity, and as sites of multiple geographies of affiliation, linkage and flow." Different flows and connectivities are hence constitutive of places, intersecting in them "within the same turf."

The politics of place consists then in the debates between "the different micro-worlds on the same proximate turf," including those involving flows and connectivities with other places. Amin (2004) proposed to consider spatial juxtaposition as a "field of agonistic engagement."

> This means seeing the local political arena as an arena of claims and counter-claims, agreements and coalitions that are always temporary and fragile, always the product of negotiation and changing intersectional dynamics, always spreading out to wherever a claim on turf or on proximate strangers is made or to where novelty is generated by juxtaposition.
>
> (Amin 2004, p. 39)

How can the politics of place as a field of agonistic engagement be concretely studied? I propose to use the notion of place-framing and to ground it in the theory of articulation of Laclau and Mouffe (2001). To start, let's define generally a place-frame as a discursive representation of a place meant to change its planned future. The place-frame is understood as a discourse in the sense of Laclau and Mouffe.

Laclau and Mouffe (2001) conceptualize the constitution of discourse in a theory of articulation, where dominant and counter-discourses elaborate themselves in relation of opposition to one another. Opposing discursive formations try to redefine the same contested terms, but in articulating them with different elements. The meaning given to a term comes from its articulation to others, in what they call a 'chain of equivalence.' The meaning of the term 'local,' for example, would take a whole different meaning if associated with solidarity and engagement than if associated with particularism and exclusion. In the process of articulation, nodes from dominant discourses are being redefined by being linked to new discursive elements from the broader discursive field.

The term antagonism is to denote that a new discourse constitutes itself in relation to the discursive opponent. Yet, the term 'agonistic engagement,' used by Amin and Mouffe, indicates a situation where actors with different political projects nonetheless "recognize the legitimacy of their opponents" and the goodness of democratic debates with a diversity of points of views (Mouffe 2005, p. 20). This is emphasized to note that even if discourses evolve through opposition and the meaning of place constitutes itself through the intersection, 'on the same turf' of contrasted and different perspectives, it does not mean that open war is upon each place. Debates can take place in an agonistic attitude and allow for cordial discussions and even collaborations, even with oppositions in the discourses; yet, in some cases, the conflict is direct and palpable in concrete interactions.

The term place-framing has particularly been used by Pierce, Martin, and Murphy (2011) and Martin (2013) in studying the place-making at work in claim-making. For Pierce, Martin, and Murphy, places are "bundles" of the different individual experiences of space, but

> place-framing articulates the iterative co-bundling process through which social and political negotiations result in a strategic sharing of place. Place-frames represent only a fraction of any place, the socially negotiated and agreed place/bundle that is rhetorical and politically strategic—not fully a place but a place-frame.
>
> (2011, p. 60)

Place-framing hence means a selective representation of place. If there are numerous, subjective senses of place, to act and be heard, one needs to find allies with which to share a powerful representation of place. Martin (2013) uses the term place-frame to emphasize the strategic joint definitions given by activists to places. This includes for her the elaboration of a joint diagnostic structuring of the action of a collective and of consensual solutions in space. Situating this in the theory of articulation, the problematization and solutions constitute counterframes to the dominant framing of place by authorities.

Yet, for the solutions and the vision to be implemented in place, activists may have to position themselves in a politics of scale or of territories. McCann (2003) focused on how activists argue for a certain distribution of political power, for example to their neighborhood, in order for their ideal organization of a place to come true. Such strategies have also been documented in the mobilization for car alternatives, especially to change the territory and scale through which public authorities had framed mobility needs in concrete sites (Van Neste and Bherer 2014). The politics of place, as a field of agonistic engagement, could include the contestation of territorial boundaries as natural and immutable, especially when activists feel they imply a certain framing of issues in space. More generally, Nicholls (2008, pp. 849–851) speaks of the form of the regulatory state and its institutions as a 'cage,' framing issues in terms of specialized and distinct geopolitical fields, making contestation more difficult. The addition I propose to the term place-frame used by Martin and her colleagues is thus meant to consider explicitly the geography of power involved in the production of place: Certain visions of place will be obstructed by the boundaries set by the political powers in place and their framing of place in relation to its existing position in the wider geography of governance. The institutional arrangements, and their specific geographies, do potentially limit the range of possibilities on what a place can become. If these institutional or governance arrangements may limit the political opportunity for a reframing of place, activists may also strategically denounce them and propose alternative geographies of governance (Gamson and Meyer 1996; Jonas and While 2005).

Table 13.1 Definitions of the key notions used to study place-framing.

Place	Site where daily practices are experienced, to which a subjective and collective meaning is given, where different microworlds meet on the same turf
Politics of place	Negotiation of spatial juxtaposition
Place-frame	Selective representation of a place from a collective to have influence in the politics of place, including 1. A certain problematization of the politics of place 2. A solution in space 3. A geography of governance which would make that solution possible
Articulation	Process through which discourses are elaborated, including place-frames. Articulation implies dynamic interrelations between dominant discourses and counter-discourses, redefining the same contested terms in articulating them with new elements of the discursive field

Place-frames are thus considered discursive chains of equivalence constituted through a process of articulation, in relation to the dominant discourse. Visions of places, with diagnostics and prognostics, are articulated with a particular geography of governance: These different terms of the place-frame are linked in a dependence relation imbued with meaning. Activists articulate the links differently from what the dominant discourse implies, but do so in redefining the same contested terms. This means that if a dominant discourse on place emphasizes a certain framing of mobility flows, mobility flows, to be contested, would need to be given a new meaning by the counter-discourse, through new articulations in relation to place. This new meaning could imply a vision of the place in the future and of its geography of governance, allowing for that vision to come true.

Table 13.1 summarizes the key notions of the framework, which will be illustrated in the special cases of the metropolitan 'in-betweens.'

The case presented to illustrate the framework, a debate concerning a space inbetween Rotterdam and The Hague, was documented with a discourse analysis of the transcriptions of parliamentary, municipal, and participatory debates, with documents produced by civic and public actors, along with 20 interviews and two focus groups with public officers and activists.

FRAMING PARTICULAR PLACES OF FLOWS: METROPOLITAN IN BETWEENS AND THE EMBLEMATIC CASE OF THE SOUTHERN PART OF THE RANDSTAD

Metropolitan areas are contested political spaces in which the landscapes defy easy and traditional categorizations (Young and Keil 2014). Beyond the central cities, the rest of the regional or metropolitan spaces may have no clear *common* image, markers, or identity (Dembski 2013). Spaces have been defined as *Zwischenstadt* (Sieverts 2003), 'in-between' cities not corresponding to the old urban core or new suburbs, but "complex urban landscapes of mixed density, use and urbanity" (Young and Keil 2014, p. 2). They come from an overlap of urban, suburban, and rural legacies, and may be residuals left inbetween infrastructural networks and redevelopment sites. Certain sectors are 'disconnected' from the infrastructural networks, crossed by them with yet no access to the population, affecting conditions of accessibility and well-being, of risks and vulnerabilities (Young and Keil 2010, 2014; see Addie and Mettke in this volume). Other sectors contain landscapes which actors wish to keep intact from the more acute pressure of metropolitan flows (Gilbert 2004). Different stakeholders see in them different future purposes.

Scholars have documented efforts to attribute images and identities to metropolitan, in-between landscapes, through what Dembski (2013) has coined, for example, "symbolic markers," spatial projects emphasizing certain meaning to the in-betweens, or through what Boudreau (2007) coined "spatial imaginaries," linking new imaginaries with existing spatial practices, in using strategic planning tools. Yet these spaces are also characterized by their 'in-betweenness' in terms of governance, making this meaning-giving process complex politically, but also selective in the spatial perspectives represented, certain perspectives *in situ* not necessarily having power of say. In-between cities "find little political representation, hardly any symbolic valuation and often become residual terrain in metropolitan governance" (Young and Keil 2014, p. 2).

The opposition to a new highway segment in-between Rotterdam and The Hague area is an interesting case to discuss this, in a context qualified as "the most fascinating citified landscape in the Netherlands and can be considered as a Zwischenstadt par excellence," with the mix of port facilities with industrial, recreative, rural, and residential functions between the two core cities (Dembski 2013). The negotiation between these different uses and valuations of place are rooted in a historic planning imaginary in the Netherlands of the 'Randstad' and of the 'Green Heart.' The Randstad, which is the name and metaphor given to the more urbanized section of the Netherlands, is composed of the four main cities placed in a ring around the 'Green Heart' (see Figure 13.1). For decades, the Randstad has been characterized by a heavy pressure on land. In the Randstad are concentrated not only urban and socioeconomic built uses, but also agricultural production, industries, and transport-heaviest networks. The Green Heart is meant to provide close green amenities for urban dwellers, strengthen local agriculture, and limit the expansion of the urban (Hajer and Zonneveld 2000). In the Randstad, space is tight and well optimized, and open space is considered to be rare. The Green Heart has, through the years, received some urban satellite growth centers and transport infrastructures, exacerbating

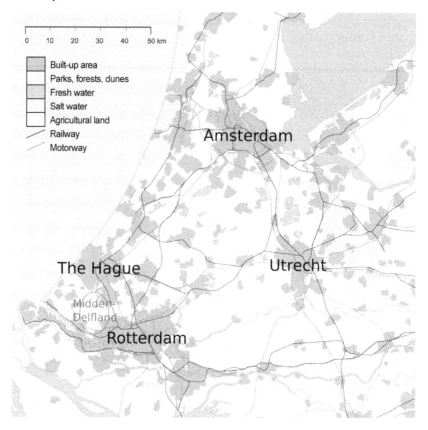

Figure 13.1 The Randstad, with its four main cities
(*Source:* Wikipedia, Creative Commons. Modified by Sophie L. Van Neste).

fragmentation and the mix of functions, but always with debates on compensation to minimize the impact on the open space (for example, locating parts of highways and trains in tunnels) (Eeten and Roe 2000). Rotterdam and The Hague compose the southern part of the Randstad and are also separated by an open space (agricultural and recreative), which previous policies have striven to preserve: Midden-Delfland.

In parallel to this ambition for Midden-Delfland, however, a highway segment linking Rotterdam and The Hague and passing through Midden-Delfland—the A4—was planned and debated for decades, and just decided upon recently, with building starting in 2012. Several environmental organizations ended up agreeing to the highway as long as the segment was tunneled (an agreement which was not respected). Yet, a few months after the approval of the A4, a new highway segment was announced as a priority for Rotterdam and the port, to reduce car congestion. The new highway segment (NWO) would connect the two shores of the Meuse river, on which is located the port of Rotterdam. Two options were considered: a location closer to Rotterdam (Blankenburg option) and going through part of the green area of Midden-Delfland, or one closer to the agglomeration of The Hague (the Oranje option), outside Midden-Delfland. If, in the first case of the Blankenburg option, opponents argued that the consequences on nature and open space would be greater, the segment would, according to the transport analyses from the minister of infrastructure and environment, better respond to the congestion problem in the ring of Rotterdam. A more radical counter-discourse on the protection of Midden-Delfland and its importance for the region was then elaborated, as we will see.

In addition to this hybridity and contested juxtaposition in the uses of the 'in-between,' the area is also an emblematic case of a 'metropolitan in-between,' because of the governance scheme relating to that landscape. Although their mobility flows and periphery overlap, The Hague and Rotterdam (at a distance of 30 kilometers from one another) have separate spatial planning agendas. They have different priorities for future development and are known to be reluctant to cooperate in regard to each of their peripheral zones (Kreukels 2003), which has made the planning of intermetropolitan mobility flows difficult (Salet 2008). Each agglomeration has built separate ties with the minister of infrastructure and environment in regard to transport infrastructures. For Salet (2008), these relations are detrimental to a more polycentric planning of the region as a whole. The planning and assessments for infrastructures would still be made in territorial containers, in the urban agglomerations, with no relations between them. In recent years, this privileged container for the relation to the minister in regard to infrastructure has been further strengthened in Rotterdam by the fact that the Rotterdam municipal and agglomeration councils are led by the same political party as the national cabinet.

In addition to this political structure, the port of Rotterdam, among the most important in the world, is also a key actor in the production of the space inbetween the two agglomerations of Rotterdam and The Hague. Spatially, the port area is now much closer to The Hague, but institutionally, it is part of the municipality of Rotterdam. The port has historically been tied to Rotterdam and participated closely in its development (Kreukels 2003). It has a strong influence on the definition of the zones inbetween, yet has been tempered by regulations from the national government, ensuring up to now some buffer zones between the core of industries and residences, for health and nature purposes.

There are municipalities in the relatively open space between Rotterdam and The Hague—Midden-Delfland, Maasluis, Vlaardigen, and Schiedam—which are remnants of older villages spread in the industrial/rural area and have received some of the agglomerations' sprawl. They are included within the main cities' agglomeration bodies, but as of recently, they cannot oppose decision-making taken at higher scales; they can only voice their concern within the agglomeration arena. In sum, the municipalities 'inbetween' Rotterdam and The Hague, including Midden-Delfland, have little political say. Rotterdam and The Hague comprise an emblematic case of place-framing in metropolitan in-betweens, both because of the characteristic mixed landscape and because of the governance scheme with power differentials in regard to place-framing and the regulation of mobility flows.

Metropolitan in-betweens are hence great cases to consider place-framing and its relation to the regulation of mobility flows because (1) the meaning of the in-betweens as places is far from clear, very much contested, and part of ongoing debates, and (2) the regulation of mobility flows is tied to a contested status of in-betweenness in the governance scheme, with little power to the local authorities in the metropolitan in-betweens.

Let's now consider the place-frame elaborated to affirm a certain meaning to the space inbetween Rotterdam and The Hague in opposition to a highway going through it. The contrast between this place-frame and the spatial articulations of the dominant discourse will then be presented. Finally, the rearticulations by the civic coalition of their meaning of place with alternative regulations of flows are presented.

PLACE-FRAMING: MIDDEN-DELFLAND IN A GREEN METROPOLIS

The new highway segment NWO would go through part of a green open area between Rotterdam and The Hague, Midden-Delfland, if the Blankenburg option was chosen (the Oranje option, in comparison, goes through a more industrial landscape). The landscape of Midden-Delfland is characteristic of the agricultural and (man-made) meadow landscape of the Dutch country. Yet the area had been neglected in the past. Starting in 1977, a 'reconstruction law' was adopted by the national government to valorize this landscape. The surrounding local municipalities, inhabitants, farmers, and nature groups were involved in the 35-year process. Resident associations were also created to ensure the primacy of the 'green open character' of the territory. Part of Midden-Delfland had been designated as a 'state buffer zone,' one of the green open areas to be preserved for the benefit of urban dwellers. This was a particularly important buffer zone, in terms of quality of life and air quality, because of the limit it set to the spreading of industries from the port of Rotterdam.

Yet in the 1990s, a plan of landfill for toxic waste on the site became known to residents; this was the first large breach in the policy for the area's preservation, with justifications pertaining to its position close to the port area. An association was created to block it. On December 12, 1992, almost 8,000 people planted a total of 16,000 trees on the site, which became known as 'the People's Woods.' The action successfully blocked the landfill project. The Vlaardigen-based group, which had mobilized their whole community for the preservation of the green buffer, re-resuscitated

20 years later to oppose the Blankenburgtunnel, which would go through the People's Woods.

The association entered in an opposing coalition with other resident groups and nature protection and environmental organizations, such as Friends of the Earth and Nature monuments. With regional organizations, local actors positioned the loss for Midden-Delfland not only in relation to the local municipality (as had been done in 1992), but as a loss for the whole region:

> Is the Blankenburgtunnel and highway through our last piece of green space really the solution? No. For if you arrive with no traffic to your work, you nevertheless would like to enjoy cycling through the nice Zuuidbburt or recreate at the Krabbe lake. But there you would see asphalt and hear the traffic hurry. Not really a relaxing situation. Then you would have again the feeling that you live in an inhospitable environment.
>
> (Personal translation from Dutch, Groeiend Verzet 2011, p. 1)

In the report *Green Metropolis*, a utopia was formulated of no more highways in the whole region of Rotterdam and The Hague, for the preservation and accessibility to green open spaces. If this utopia was considered radical even for groups in the opposition to the Blankenburg highway segment, a new consensus had been reached that no more highways in Midden-Delfland were compatible with the preservation of landscape. The value of Midden-Delfland, of this area in-between, for the whole region, giving it identity and coherence, was promoted with slogans, demonstrations, tweet campaigns, videos, maps, images, and direct lobbying with political parties in parliament. The campaign built on existing organizational networks in place between municipalities, farmers, and heritage and ecological associations in Midden-Delfland, and on the strong lobby of nature conservation in the Netherlands with the organization Nature monuments.

The scope of the mobilization was associated with the continuous impinging of highways on the landscape. This new highway segment was presented as bringing irreversible damage to Midden-Delfland in a dependency loop with ever more roads. The construction of the highway A4 in Midden-Delfland had not yet begun, and already it was planned to bring further congestion and necessitate a new highway connection. And after the Blankenburg segment would be built, it seemed that yet another section would be necessary to complete a new ring around Rotterdam—the A24—to deal with increased traffic and congestion. This would be the end of Midden-Delfland, crossed then by three highway segments. The civic actors hence diffused a discourse of Midden-Delfland as constitutive of the urbanized region, but incompatible with the logic of highway extension.

THE SPATIAL ARTICULATIONS OF THE DOMINANT DISCOURSE

This place-framing of Midden-Delfland as a green open space constitutive of the region was directly in opposition to the dominant framing of the place by the alliance between the national government and the Rotterdam agglomeration. In the context of the economic crisis, legal provisions for nature conservation were relaxed in the Netherlands, permitting a project of 'national interest' to be built in a natural area if need be (Verschuuren 2010). Transport infrastructures were part of the infrastructures on which public participatory procedures and nature provisions were reduced to ensure their rapid construction, since their building was expected to participate in economic recovery (ibid.; MIM 2011c). The place-framing from the government hence contains a very different starting point than the ones from civic associations wishing to protect Midden-Delfland. The emphasis is on congestion, which materializes itself in two focused categories of mobility flows in the region.

First are the flows related to the growth of the port of Rotterdam. A large growth of the port is planned for 2030—Maasvlakte 2—with an extension of platforms in the sea. New road connections were considered necessary to ensure that the port businesses continue their activities without excessive time lost in congestion (expected to occur in 2030), even with the port authority's objective to increase freight transport by rail and inland navigation (MVW et al. 2009; Port of Rotterdam 2011; Transumo 2009). A new river crossing also appeared necessary for safety reasons, since there is at the moment only one main road to leave the port facilities. If an additional exit in case of calamities was important for the justification of the NWO, the Oranje option (outside of Midden-Delfland) scores better on this criterion than the Blankenburg (MIM 2012).

The second focus around which the dominant discourse is framed is the Benelux tunnel, which ought not to be congested, since it would be part of a new economic corridor, the backbone and accessibility axis of South Holland and of the economic route toward Belgium: "[T]here needs to be sufficient capacity in the Benelux tunnel" (MVW et al. 2009, p. 52; MIM 2012, pp. 8–9). So a second highway crossing of the river, the closest possible to this axis, is needed. The Blankenburg option responds to this priority on the Benelux better than the Oranje option. Yet the choice of this economic route is a selective one which was still under construction at the time, and several actors (independent advisors' council to the state, a university professor, the association of car drivers) argued that other strategic links could be built in relation to the Oranje option (College van Rijksadviseurs 2011; Geerlings 2012; ANWB 2011). But this economic route went through the agglomeration of Rotterdam and meant direct investment inside the territory of Rotterdam. For the national government, the Blankenburg option was also cheaper than the Oranje option, where the water crossing is wider

Place-Framing and Regulation of Mobility Flows 255

and involves a more complex infrastructure. The fight against congestion framing the region in a certain way is thus both linked to a global discourse on the economy of freight flows, and is territorially selective. This fits with the picture depicted by Cidell (2011) that even with a discourse of global freight flows, the spatial imaginaries and territorial priorities from local authorities determine the geography of the transport network. In the case of the NWO, the interests in the growth of the port (to keep attracting its share of freight global flows) were intermingled with the city of Rotterdam's territorial priority for road fluidity and state investments on its territory, through the projected economic transport axis still in process.

As for the value of Midden-Delfland, the actors of the dominant discourse maintained that it could be accounted for through compensation measures. The leading governmental coalition put emphasis on the fact that it is not exactly 'real' nature, but more 'recent' nature. Hence the value and attractiveness it had acquired in the last decades could be remade elsewhere (Tweede Kamer 2011, p. 36:76; Tweede Kamer 2012, p. 39:76). With the civic mobilization and the opposition from local municipalities and some national political parties, the parliamentary and agglomeration debates largely focused on landscape 'integration' to minimize the impacts of the highway. But such measures could only be marginal: Only a quarter of the road after the water crossing could be underground, because of the dam and a junction to another highway above ground. Many discussions hence concerned tunnel technicalities, and little on the other highway route sparing Midden-Delfland, the Oranje option, and even less about no highway at all in the region.

Civic actors had tried to mobilize a "community of fate" (Hajer 2003) around the survival of Midden-Delfland, positioning it in the region's livability, but it had proven insufficient. Their criticism of the logic of highway extension was discussed in strict relation to the threat it posed to Midden-Delfland. The implications for mobility flows were left unspecified, although the governing coalition had a strong discursive articulation linking economic growth with the new highway, particularly the Blankenburg option. At that point, it seemed that the debate about mobility flows could not be avoided.

THREE REARTICULATIONS OF PLACE AND FLOWS TO SAVE MIDDEN-DELFLAND

The civic coalition debated mobility flows and their intersection with Midden-Delfland in three ways: first, in showing alternative mobility modes with less impact on the place; second, in changing the territorial scope of the traffic studies and participatory process; and third, in reconsidering the actual growth in mobility flows, through trends toward its 'dematerialization' in virtual connections and flexible working locations. In considering solutions to the intersection of place and mobility flows, activists also tackled directly the geography of their governance.

Alternative Types of Mobility Flows

From the beginning, the place-framing from the civic coalition directly attacked issues of mobility, but it was put aside by civic leaders because of its 'radical character.' Friends of the Earth Netherlands had emphasized mobility issues in its plan called *Building a Green Metropolis*, submitted to the province as a 'citizen initiative.' The plan illustrated a set of alternatives to avoid all highways planned by the government in the region. It included eight public transport investment projects. To further reduce car traffic and finance the public transit projects, the report *Green Metropolis* advocated the introduction of a pricing system for car use and freight transport. In a context in which the national authorities did not go through with a national congestion charge (this had been put aside in 2010, canceling engagement from the previous government), Friends of the Earth proposed to set up a pricing system at the *metropolitan scale* of Rotterdam and The Hague. What Friends of the Earth proposed was hence a metropolitan project of car regulation and public transit. It received relatively little support. It was considered a real joke by the right-wing party leading the governing coalition in parliament, "like a pie in the sky" (Volkspartij voor Vrijheid en Democratie (VVD) in Tweede Kamer 2011, p. 8). In the agglomeration and the province, the comparison made by Friends of the Earth between the areas, which they characterize as 'a port area,' with the successful congestion charge in London, was really considered too far of a stretch:

> The port area and the Westland are really not to compare to a successful example such as the center of London. The connections in public transit to the port area and the Westland are problematic because of the spreading of destinations. There are also large freight flows from the greenport and mainport which need to move at an exact moment (just in time), making it difficult to conduct by boat or train. A regional congestion charge is hence expected to have little positive effects.
> (PZH 2012, p. 2, translated from Dutch)

Considering Another Territorial Scope to Define Place and Evaluate Mobility Flows

Although this first attempt of reframing mobility flows to protect place received limited support, it evolved into two other ways to critique the regulation of mobility flows in the region. In the report *Building a Green Metropolis* from Friends of the Earth, the term 'metropolis' was used as a scale to plan mobility beyond individual cities' interests, with a congestion charge (Milieudefensie 2011). In keeping aside the more radical proposition of no highway, the other civic actors went on to criticize the territorial selectivity of the infrastructural choice. Nature monuments and the local opponents to the Blankenburg tunnel took the lead and asked for a broader territorial scope to the traffic study justifying the NWO. Their critique was the following: The analysis from the national government focused too much on the ring of Rotterdam, the Benelux tunnel especially, and the access to the port (see above), and did not evaluate the traffic implications of the interrelation with the agglomeration of The Hague. This would have strong implications for the choice of the localization, since a focus on Rotterdam favored the Blankenburg tunnel option in contrast to the other option. The territory of the traffic engineering analyses objectified the focus on specific segments linked with economic priorities, argued the civic actors. They stated that accessibility to economic destinations could not be the sole objective for a region; the scope needed to be broadened and discussed. This territorial reframing was abundantly used in parliament by opposition parties to put in doubt both the traffic study justifying the choice of the Blankenburg and the participatory process, which appeared very narrow. The project went through several steps of debate between parliamentary members, with political parties asking for accurate information, causing delays. For several months, the civic actors' campaign used this territorial argument in their focus on place protection. Yet, after the majority votes for the Blankenburg option in parliament and within the Rotterdam agglomeration, the sense of a need for a highway segment serving economic prosperity seemed to be an inevitable topic to address.

The Dematerialization of Mobility Flows

A new leader from Nature monuments took the leadership of the campaign; he was based within the region of Midden-Delfland and had been involved in highway oppositions before. The perspective on the 'risk' and radical character of contesting mobility flows to protect Midden-Delfland changed with his involvement. With his new leadership, the need for mobility flows *on* highways, and the necessary condition of highway flows for economic growth, was directly contested. The leader also presented the protection of this green open space as a duty toward the region's residents. This contrasted with the previous attitude to place the organization had voiced, which considered legitimate the port needs and had accepted nature compensation measures elsewhere. As explained in the place-frame, the new segment seemed to bring irreversible damage which could not be accepted; other regulations of flows had to be put in place.

Digging into the statistics on mobility flows, the leader pointed toward a change in mobility trends, documented by the state independent mobility agency, but not accounted for in the models justifying the new highway (MIM 2012; Project NWO 2011). Since 2005, the number of kilometers traveled nationally and regionally had reduced, and the growth in kilometers by cars (and, since 2008, road freight transport) had begun to stagnate (KIM 2012). An increasing annual growth in car use was not the reality; hence, the need for the new highway should be reevaluated on these facts. The content of the critique was also closely tied to programs in which the government, the Rotterdam agglomeration, and the port were closely involved, at an unprecedented extent in comparison to other countries and regions of the world: facilitating the entrepreneurial movement for flexible work schedule and flexible work space, made possible by virtual flows of connections (employees could work at home, in coffee working centers, with virtual meeting platforms, etc.) (PSWSR 2012). In other spheres of governance, this 'smart working = smart traveling' was proudly presented by these same authorities as proof of their innovation in the tackling of congestion, through a deterritorialization of mobility flows at peak hours. The growing numbers of employees working at home or in a flexible time schedule significantly reduced peak hour congestion, *hence* reducing the need for new highways, the civic actors argued. But the above-mentioned trio still considered that this trend should not yet affect large transport investments decisions, the risks of delaying large transport investment in context of congestion being too high for businesses (Int SR, Int PR, MIM 2011b).

Despite the documented changes in mobility, after two years of delays and debates on the NWO and national elections in 2012, the leading party, VVD, regained leadership of the governing coalition and explicitly included in the coalition agreement the implementation of the Blankenburg option for the new highway (ensuring a majority and no further discussions in parliament). The arenas of dialogue had been closed, and the last critique left unanswered.[1]

In sum, the place-frame of 'green metropolis' in relation to Midden-Delfand evolved in its articulation of place and flows. The desire to preserve the green open character of Midden-Delfland remained throughout as a defining characteristic for the regional space. The new government, however, proclaimed the right to construct a (second) highway segment through the place. The reason given was the economic urgency of fighting congestion. This discourse coproduced by Rotterdam, the port, and the national government presented Midden-Delfland as a sum of characteristics that could either be preserved by mitigating the impacts of the road or compensated by reinvestments for nature in other sites. The civic coalition engaged in the discursive work of showing the incompatibility of the highway with the place. Midden-Delfland had been reduced to a space inbetween cities, characterized by the 'missing highway segments.' Much of the debate against the NWO was structured around these lines: 'Can Midden-Delfland survive the highway?' 'Can the region survive the highway?' But also, in the dominant discourse, 'can the region survive *without* the highway?' The opponents hence had to directly tackle the regulation of mobility flows, which they did in three ways, building on existing trends and discourses available in the field: first, in proposing alternative types of mobility flows; second, in changing the territorial scope for the assessments of mobility flows; and third, in emphasizing the dematerializing of mobility flows through virtual connections and flexible work locations.

CONCLUSION

In this chapter, the intersection of mobility flows and place was considered through a concrete public debate. This concrete public debate provided an opportunity to reach two objectives. First, to consider the process of place-framing in debates on mobility flows; i.e., how selective representations of a place are elaborated dynamically by collectives as counter-discourses to dominant ones. Second, to consider the specificity of the 'metropolitan in-betweens' as places of flows, through their discursive and regulatory construction.

We saw that the meaning-giving to Midden-Delfland, and through it to the fuzzy metropolitan region around it, was articulated in relation to the conflict with the proposed highway. This meaning-giving process corresponds to place-framing. One of the questions posed by studying place-framing in a politics of place defined as a negotiation of spatial juxtaposition (Amin 2004) was whether actors could, with place-frames, tackle the issue of the regulation of mobility flows (and not *per se* exclude them from their representation of a 'natural' or 'authentic' sense of place). In this Dutch case, actors opposing the highway had direct attachments to place and defined their mission in relation to that place (and little in relation to mobility flows, the majority being nature- or resident-based organizations), but they still tackled mobility flows and their geography of governance to react to the framing of public authorities. The adaptable place-frame constituted a tool of collective action, working on the level of both representations of a site and its embeddedness in a wider territory and web of mobility flows, each raising specific issues in the geography of governance. The issues raised went beyond place and touched upon the politics of mobilizing for car alternatives in urban and metropolitan areas (Henderson 2013; Van Neste and Bherer 2014), as well as on the politics of infrastructure in city-regions, tied up in specific territorial alliances among public authorities and private ones (Jonas et al. 2013). If these issues went beyond Midden-Delfland, they were tackled in a tentative redefinition of the whole region of Rotterdam and The Hague and of its regulation of mobility flows, as a 'green metropolis.'

I chose the case of the 'in-betweens' to particularly focus on the articulation of place and flows in a co-constructing process. Metropolitan in-betweens are not urban cores nor exurbia suburbs, but between them, in the spaces where infrastructures hosting the flows of the agglomerations tend to concentrate. 'In-between' metropolitan spaces are constituted by metropolitan flows and other hybrid uses *in situ*, implying constant negotiations. The Dutch case I presented could have been analyzed as a case of a social mobilization against the destruction of green open space. Yet, such a perspective would be truncated without considering the positionality of that open space in the metropolitan region and how this positionality has meant, through time, as a certain sense of place, even if a negotiated and constantly contested one. The contrast between an ideal of a past inherited from a true sense of place separate from flows, and the perspective of a co-construction with flows (Massey 1994) appears particularly obvious in the politics of place for the metropolitan in-betweens. Yet, to many extents, this contrast could also be made for any busy street corner of a city, or within any suburban quarter faced with new transport corridors (Cidell, this volume). Cities and metropolitan areas are full of places of flows, even though they may not all be framed as such.

NOTE

1 This last critique, however, continued to be raised in the media and caused conflicts in the second major political party in the Netherlands. The Labor Party had accepted the coalition agreement to be in power, even if it had in the past strongly opposed the Blankenburg. Local factions of the Labor Party brought the issue angrily in the national council, hoping to invalidate the coalition agreement on the Blankenburg which would have been based on inaccurate statistics on the growth of mobility (an attempt which did not succeed).

WORKS CITED

Agnew, J. (1987) *Place and Politics: The Geographical Mediation of State and Society*, Winchester, MA: Allen and Unwin.
Amin, A. (2004) "Regions unbound: Towards a new politics of place," *Geografiska Annaler*, 86(1):33-44.
ANWB (2011) "ANWB Voorstander Oranjetunnel." Retrieved December 6, 2011 from http://www.anwb.nl/verkeer/nieuws-en-tips/archief,/nederland/2011/december/Oververbinding-Nieuwe-Waterweg.html.
Boudreau, J.-A. (2007) "Making new political spaces: Mobilizing spatial imaginaries, instrumentalizing spatial practices, and strategically using spatial tools," *Environment and Planning A*, 39(11):2593–2611.
Cidell, J. (2011) "Distribution centers among the rooftops: The global logistics network meets the suburban spatial imaginary," *International Journal of Urban and Regional Research*, 35(4):832–851.
College van Rijksadviseurs (2011) *Rondetafelgesprek over de Nieuwe Westelijke Oeververbinding*, The Hague: Atelier Rijksbouwmeester.
Cresswell, T. (2010) "Towards a politics of mobility," *Environment and Planning D*, 28(1):17–31.
Dembski, S. (2013) "In search of symbolic markers: Transforming the urbanized landscape of the Rotterdam Rijnmond," *International Journal of Urban and Regional Research*, 37(6):2014–2034.
Eeten, M. V., and Roe, E. (2000) "When fiction conveys truth and authority," *Journal of the American Planning Association*, 66(1):58-67.

Gamson, W., and Meyer, D. (1996) "Framing political opportunity," in D. McAdam, J. McCarthy, and M. Zald *Comparative Perspective on Social Movements*, Cambridge: Cambridge University Press, pp. 275–290.
Geerlings, H. (2012) *Nu Verplaatsen, in de Tijd van Morgen. Uitdaginden Voor Duurzaamheid, Mobiliteit En Governance in Een Dynamische Wereld*, Rotterdam, NL: Erasmus Universiteit Rotterdam.
Gilbert, L. (2004) "At the core and on the edge: Justice discourses in metropolitan Toronto," *Space and Polity*, 8(2):245–260.
Groeiend Verzet (2011) "Files, dat is natuurlijk narigheid! | Snelweg langs het Volksbos? Nee!." Retrieved December 1, 2011 from http://www.volksbos.nl/nieuws/files-dat-is-natuurlijk-narigheid/.
Hajer, M. (2003) "A frame in the fields: Policymaking and the reinvention of politics," in M. Hajer and H. Wagenaar (eds) *Deliberative Policy Analysis: Understanding Governance in the Network Society*, Cambridge: Cambridge University Press, pp. 88–110.
Hajer, M., and Zonneveld, W. (2000) "Spatial planning in the network society—Rethinking the principles of planning in the Netherlands," *European Planning Studies*, 8(3):337–355.
Henderson, J. (2013) *Street Fight: The Politics of Mobility in San Francisco*, Amherst, MA: University of Massachusetts Press.
Jonas, A., and While, A. (2005) "Governance," in D. Atkinson, P. Jackson, D. Sibley, and N. Washbourne (eds) *Cultural Geography. A Critical Dictionary of Key Concepts*, London: I.B. Tauris, pp. 72–79.
Jonas, A.E.G., Goetz, A. R., and Bhattacharjee, S. (2013) "City-regionalism as a politics of collective provision: Regional transport infrastructure in Denver, USA," *Urban Studies*, 51(11):2444–2465.
KiM (2012) *Mobiliteitsbalans*. Den Haag: Kennisinstituut voor Mobiliteitsbeleid.
Kreukels, A. (2003) "Rotterdam and the south wing of the Randstad," in W.G.M. Salet, A. Thornley, and A. Kreukels (eds) *Metropolitan Governance and Spatial Planning: Comparative Case Studies of European City-Regions*, London: Spon Press, pp. 189–201.
Laclau, E., and Mouffe, C. (2001) *Hegemony and Socialist Strategy: Towards a Radical Democratic Politics* (2nd ed.), London: Verso.
Martin, D. (2013) "Place frames: Analysing practice and production of place in contentious politics," in W. Nicholls, B. Miller, and J. Beaumont (eds) *Spaces of Contention: Spatialities and Social Movements*, London: Ashgate, pp. 85–102.
Massey, D. (1994) "Double articulation. A place in the world," in A. Bammer (ed) *Displacements: Cultural Identities in Question*, Bloomington, IN: Indiana University Press, pp. 110–124.
McCann, E.J. (2003) "Framing space and time in the city: Urban policy and the politics of spatial and temporal scale," *Journal of Urban Affairs*, 25(2):159–178.
Midden-Delfland Vereniging (2012) "Midden-Delfland Reconstructieproces." Retrieved June 5, 2012 from http://www.middendelflandsite.nl/gebiedsontwikkelingen/reconstructie.
Milieudefensie (2011) *Bouwen Aan Een Groene Metropool: Een Duurzaam Mobiliteitsplan Voor de Regio Rotterdam-Den Haag*, Amsterdam: Milieudefensie.
Ministerie van Infrastructuur en Milieu (MIM) (2011a) *Notitie Reikwijdte En Detailniveau Nieuwe Westelijke Oeververbinding*, Den Haag: Ministerie van Infrastructuur en Milieu.
Ministerie van Infrastructuur en Milieu (MIM) (2011b) *Brief Programma Beter Benutten*, Minister van Infrastructuur en Milieu.
Ministerie van Infrastructuur en Milieu (MIM) (2011c) *Minister Schultz van Haegen: "Ruimte voor Nederland"—Ontwerp-Structuurvisie Infrastructuur en Ruimte*, Ministerie van Infrastructuur en Milieu. Retrieved June 14, 2011 from

http://www.rijksoverheid.nl/nieuws/2011/06/14/minister-schultz-van-haegen-ruimte-voor-nederland.html.
Ministerie van Infrastructuur en Milieu (MIM) (2012) *Concept-Ontwerp-Rijksstruc tuurvisie Bereikbaarheid Regio Rotterdam En Nieuwe Westelijke Oeververbinding*, Ministerie van Infrastructuur en Milieu.
Mouffe, C. (2005) *On the Political*, London: Routledge.
MVW, VROM, Stadsregio Rotterdam, Gemeente Rotterdam, and Provincie Zuid-Holland (2009) *Masterplan Rotterdam Vooruit. Mirt-Verkenning Regio Rotterdam En Haven: Duurzaam Bereikbaar*, Delft: CE Delft.
Nicholls, W. (2008) "The urban question revisited: The importance of cities for social movements," *International Journal of Urban and Regional Research*, 32(4):841–859.
NWO (2011) *Samenvattende Notitie Conceptresultaten Onderzoeken NWO*, Amsterdam: Nederlandse Organisatie voor Wetenschappelijk Onderzoek.
Pierce, J., Martin, D., and Murphy, J. (2011) "Relational place-making: The networked politics of place," *Transactions of the Institute of British Geographers*, 36(1):54–70.
Platform Slim Werken Slim Reizen (PSWSR) (2012) "Nederland loopt voorop met Het Nieuwe Werken." Retrieved July 9, 2012 from http://slimwerkenslimreizen.nl/index.html?artikel_ID=1111&loc1=2&soc1=2&art=2&lv=1111&naam=Nieuws.
Port of Rotterdam (2011) *The Port Vision 2030*, Rotterdam: Port of Rotterdam.
NWO (2012) *Deelrapport A: Nota van Antwoord, Plan-MER Nieuwe Westelijke Oeververbinding*, Amsterdam: Nederlandse Organisatie voor Wetenschappelijk Onderzoek
Provincie Zuid-Holland (PZH) (2012) *Reactie op burgerinitiatief Milieudefensie*, The Hague: Gedeputeerd Staten Provincie Zuid-Holland.
Salet, W.G.M., Salet, A., Thornley, A., and Kreukels, A. (eds) *Metropolitan Governance and Spatial Planning: Comparative Case Studies of European City-Regions*, London: Spon Press.
Sieverts, T. (2003) *Cities Without Cities: An Interpretation of the Zwischenstadt*, London: Spon Press.
Transumo (2009) *Van Maasvlakte Naar Achterland; Duurzaam Vervoer Als Uitdaging*, Rotterdam: Consortium Transumo A15 project.
Tweede Kamer (2011) *Vaststelling van de Begrotingsstaat van Het Infrastructuurfonds Voor Het Jaar 2011* (MIRT), Tweede Kamer der Staten-Generaal 32500A Nr. 102.
Tweede Kamer (2012) *Structuurvisie Bereikbaarheid Regio Rotterdam En Nieuwe Westelijke Oeververbinding*, Tweede Kamer der Staten-Generaal. 32 598 N. 12.
Van Neste, S.L., and Bherer, L. (2014) "The spatial puzzle of mobilising for car alternatives in the Montreal city-region," *Urban Studies*, 51(11):2406–2425.
Verschuuren, J. (2010) "The Dutch crisis and recovery act: Economic recovery and legal crisis?," *PER*, 13(5):5–25.
Young, D., and Keil, R. (2010) "Reconnecting the disconnected: The politics of infrastructure in the in-between city," *Cities*, 27(2):87–95.
Young, D., and Keil, R. (2014) "Locating the urban in-between: Tracking the urban politics of infrastructure in Toronto," *International Journal of Urban and Regional Research*, 38(5):1589–1608.

14 'Peace, Love, and Fun'
An Aerial Cable Car and the Traveling Favela

Bianca Freire-Medeiros and Leonardo Name

INTRODUCTION

The increase in mobility for many individuals at all scales of the urban experience generates new subjectivities and intersubjectivities, while clearly posing objective challenges, especially in the planning and management of mass public transportation. These challenges are related not only to engineering, but also to political, social, cultural, and financial issues (Sheller and Urry 2006; Elliot and Urry 2010).

The absence or precariousness of citizen mobility and a territory's isolation in relation to others must be understood as factors of social inequity and spatial injustice (Lefebvre 2001; Leibler and Musset 2010; Soja 2014; Boitempo and Carta Maior 2014). In Latin America, where various forms of inequality persist, new urban mobility policies and programs, especially in public transportation, have been assuming central stage since the early 2000s (cf. Paquette 2011). Thus, urban mobility and accessibility policies have put experts to the test in terms of their capacities to deal with increasing dispersion, intraurban discontinuity, and sprawl, sharply present in the largest Latin American cities. Transportation is no longer viewed as an individual problem, but as something to be provided as part of the rights to the city, as much as sanitation or dwelling. In this chapter we discuss the linkages between 'new' transportation technologies and city branding within the contexts of segregation and inequality that so deeply characterize Brazilian metropolises. In dialogue with the 'new mobilities paradigm' (see Sheller and Urry 2006; Urry 2007; Elliot and Urry 2010), we examine how an aerial cable car in a favela of Rio de Janeiro, a city deeply marked by the production of landscape as a commodity, connects to transnational arrangements that invent, market, and sell territories associated with poverty and violence as tourist attractions.

We invite the reader in a journey through the Complexo de Favelas do Alemão, a large cluster of 15 different neighborhoods with a population of 65,000 residents,[1] where an aerial cable car, the so-called Teleférico do Alemão, was inaugurated in 2011 as part of an urban development agenda shared by other Latin American cities. As a matter of fact, a few years before, the Colombian capital had installed three aerial cable car lines—K, J, and L, opened in 2004, 2008, and 2010, respectively, the last a tourist line—that inspired similar investments in other cities in the region. In Caracas, two lines (San Augustín and Mariche, opened, respectively, in 2010 and 2012) now connect to the subway system and enable movements between the city's plains and hillside barrios. These experiences inspired Brazilian public authorities to import the gondola transport system to precarious neighborhoods in Rio de Janeiro.

A vulnerable area located in Rio de Janeiro's North Zone, Alemão, as it is locally known, has a history marked by severe conflict and lack of many basic rights. Part of the so-called Leopoldina suburb (referring to the branch of the railway line around which the different working-class districts developed), it was officially recognized as a neighborhood in 1993. In commonsense perception, Complexo do Alemão is a 'favela' and, as elsewhere, those who live at Alemão do not own a property title and have limited access to formal employment and public services. In the Brazilian imagination (as a vast bibliography demonstrates [Valladares 2005]), favelas have become a central discursive and material reality upon which major issues—inequality, violence, citizenship—are projected, debated, and dealt with in different arenas and by various social actors. In the international imagination, favelas have become part of the stereotypical image of Brazil, along with Carnival, the tropical rainforests, football, and sensuous women. In both cases, favelas are imprisoned in an ambiguous semantic logic, which associates them with solidarity and joy ('cradle of samba, funk, and capoeira'), but also with poverty, moral degradation, and violent criminality ('cradle of marginality'). Despite the installation of a Peacemaking Police Unity (UPP, according to the Portuguese acronym) in the territory since 2011, for example, residents of Alemão still face a routine of harassment by state forces and illegal (especially heavily armed bandits) apparatuses of power.

In this chapter we are particularly inspired by the notion of *traveling favela*, which refers to a kind of poverty tourism, a space of imagination, and a mobile entity that is traveled to while travelling around the world (Freire-Medeiros 2013). A trademark and a touristic destination, it is also an effect and condition of possibility of different but interconnected flows. While legal and illegal capital pours in and out of it, we witness international tourists and worldwide celebrities, always with their cameras, turning the favela into a mediatic landscape which accommodates precarious houses and an amazing view of the ocean within one photographic frame. In the process, the favela is commodified many times in unpredictable configurations, adding market value to fancy restaurants and clubs (the Favela Chic chain being the obvious example), pieces of design furniture, and smart cars. Travel guides, movies, fictional accounts, photologs, and souvenirs are part and parcel of the traveling favela, and so are academic books, articles, theses, and dissertations (see Freire-Medeiros 2013, especially chapter 3).

If tourism destinations are made through ways of touristically seeing, doing, and connecting places, both symbolically and materially (Sheller and Urry 2006), how may a transportation technology participate in the definition of a territory as 'worth visiting?' This paper attempts to address this question, first from a sociohistorical perspective by briefly examining how aerial cable cars have participated in the invention of Rio de Janeiro as a tourist destination. We then zoom in to ask: How may the aerial cable car affect the ways favelas in general and Complexo de Favelas do Alemão in particular—their landscapes and populations—are seen and commodified as a mobile entity? We conclude with some general remarks on the role played by the aerial cable car as a mobility device marking a new chapter on the biography of the favela as an urban form and tourist attraction.

AERIAL CABLE CARS, MOUNTAINS, AND THE SPECTACULAR LANDSCAPE COMMODITY

Aerial cable cars started becoming popular for civilian transportation in 1908, when the Swiss Alps gained a connection between Wetterhorn and Gridewald. If at first the purpose of these installations was to access beautiful and inaccessible sites, contributing to the establishment of landscape gazing as a form of leisure, with time aerial cable cars came to meet the demands for mobility connected to the popularization of Alpine skiing. After all, downhill skiers needed to reach the highest peaks. The gondola connection between the French cities of Megéve and Rocheburne, built in 1933, was the first constructed for that end (Balseinte 1958, 1959; Avocat 1979; Arcay et al. 2003). In the first half of the 20th century, several European mountains received aerial cable cars. Amidst the theoretical-ideological formulation and popularization of urbanist ideas endorsed by the International Congress of Modern Architecture (CIAM), the occupation of, and leisure in, the mountains was a way of turning them into spaces for the 'modern movement's exercise of creativity and rationality and, even more so, into spaces well suited for modernity (Micheletto and Novarina 2002). Installing aerial cables in the continent's mountainous areas thus produced scenic iconographies and tourist imaginaries that associated that form of transportation with abundant snow, the Alpine setting, and winter sports. As Panizza (2001) and Reynard (2005) demonstrate, due to the progressive urbanization and popularization of tourism over the 20th century, mountain areas became economic resources, serving the practice of sports or making their scenic beauty available for contemplation.

The aerial cable car transport technology did not take long to reach Brazil, popularized not as a device for sport but for contemplative leisure within a natural setting. Rio de Janeiro's internationally known Sugar Loaf cable car, an icon of picture postcards, has been in service since 1913. The so-called Bondinho connects the Sugar Loaf and Morro da Urca mountains, and was a result of an urbanistic and symbolic appropriation of these geological formations following the nearby 1908 Commemorative Exhibition of the Centennial of the Opening of the Portsin the Praia Vermelha. That same year, a military academy was built at the foot of Sugar Loaf and, through successive land reclamations lasting until 1913, the Urca neighborhood appeared around it (Chiavari 2000). Unlike other aerial cable cars later installed in Brazil, located in cold weather landscapes connected to hydrotherapy circuits, the Sugar Loaf Bondinho was dedicated to the contemplation of a sunny and mountainous waterfront, a tropical landscape.

The Sugar Loaf long ago became one of Rio's iconic destinations and today transports an average of 2,500 visitors daily. The progressive incorporation of aerial cable car technology into leisure and landscape gazing practices may be seen, therefore, as part and parcel of a process which assigned aesthetic, ecological, economic, and cultural value to mountain landscapes, turning them into spectacles viewed from above and at once. As they incorporated mobile technologies such as suspended gondolas, these landscapes became witness to both the planet's geological history and the technological progress capable of triumphing over their verticality. Often a symbol of the cities in which they are located, mountainous landscapes with transportation through aerial cable cars became a landscape merchandise and are reproduced in various representation devices—photographs, postcards, films, and travel websites, among others—that circulate globally.

Nonetheless, such landscapes—or their transport technologies—are not mere objective facts of reality nor space itself. As understood in the West, the landscape is a modern construction of the manner of apprehending and representing, both visually and subjectively, the variety of natural forms and human interventions to which they are subjected: The landscape is both a part of nature and a cultural object (see Holzer 1999; Claval 2004; Simmel 2007; Name 2010). It is a creation whose origin in large part is connected both to Renaissance techniques of painting and perspective and to the broadening of large-scale movements and travels (Berque 1994). However, in order for the space to be converted into what we are naming here a *landscape commodity*, there needs to be a simultaneous investment in representation, spectacularization, and reproduction technologies that encourage the consumption of these spaces reinvented as suitable for leisure and entertainment. The landscape commodity, therefore, has been the result of actions valuing projects concerning urban fragments to the detriment of planning for the whole, which value beautifying and spectacularization in favor of technical reproducibility as an image which naturalizes, through a touch of aesthetics and harmony, the pattern of spatial occupation and distribution, including its contradictions and inequities (Ronai 1976, 1977; Lacoste 1977).

In a context of peripheral modernization and globalization, equipping cities with urban infrastructure—including technologies for increasing mobility and accessibility—means creating certain visual marks. These landmarks seek not only to increase the perception of meeting fundamental rights, but also to neutralize opposing narratives. As a vast bibliography demonstrates, this is the case of several discourses on management and urban interventions that the public power held throughout history in different Latin American cities (Barbosa 1999; Arantes 2000; Name and Bueno 2013; Pires do Rio and Name 2013). Recently, one of the most praised interventions includes precisely the use of ski slope technology as a means of public transportation aiming to integrate low-income, 'no-go' areas into the fabric of the city, as we discuss in the next section.

AERIAL CABLE CARS AND THE AESTHETICS OF UNEVEN MOBILITIES

Individual, motorized, and nonmotorized transportation devices have been increasingly consolidated in several Latin American cities, whether automobiles or bikes. Still, in terms of public policy, mass transport is the solution most widely implemented in Rio de Janeiro, Brasília, Curitiba, São Paulo, Mexico City, Lima, and Buenos Aires, among others. Different technologies have been recently introduced, including the particularly 'successful' case of the so-called BRT (bus rapid transit) system—large-capacity buses that travel in exclusive lanes—exported to and adopted by several Latin American metropolises.[2] The aerial cable car is another mobile technology that has become progressively widespread as a symbol of not only technological but also (perhaps mainly) social achievement. Suspended gondola systems had been used for landscape contemplation tourism before, as we discussed previously, but the inauguration of Medellín's 'metrocables' in 2004 began to associate such mobility systems equally with public transportation in high-density urban areas, particularly landscapes of precarious and informal neighborhoods struggling to overcome violence and poverty. Central to a 'makeover' conducted by the municipal office in Medellín,[3] the metrocable has been exported to other urban contexts as an inventive strategy for overcoming mobility problems in low-income settlements (Hylton 2007; Fukuyama and Colby 2011).

Investigating the Colombian case, sociologists Bocarejo and Rivadulla argue that, "for politicians and planners, cable cars offer new ways of governing urban poverty at the same time that they bring the promise of marketing their third world cities" (2012). Aerial cable car systems have indeed been advantageous to governments, whether or not in association with the private sector, for several reasons. First, as a widely tested technology, aerial cable cars have a relatively short construction period (Brand and Dávila 2011; Dávila and Daste 2011; Dávila and Brand 2012). The implementation is also relatively cheap, since their aerial nature reduces the need to acquire land (despite a large number of forced removals in places like Rio de Janeiro). Due to their strongly controlled systems, aerial cable cars are also now understood as safe by noninhabitants of these previously inaccessible, unknown, or perceived to be highly dangerous areas, potentiating promises of economic development and tourism growth. Lastly, they also enable linking mobility and local accessibility policies to discourses that believe in ecological modernity and sustainable urban development: It is a transportation system powered by electricity, which Colombia, Venezuela, and Brazil mostly generate through largely carbon-free hydropower.

It is important to keep in mind that the Teleférico is part of a larger strategic investment package, the so-called Growth Acceleration Program (Programa de Aceleração do Crescimento [PAC]), which aims at improving the quality of life in different regions of Brazil. Combining federal and state resources, with contracts given to both private and public companies, PAC was launched by then President Lula da Silva in 2007 and expanded by President Dilma Rousseff in 2011. As a whole, the initiative called for investments of US$349 billion (R$638 billion) in the areas of logistics, energy, and social development in poor areas.[4] In the case of Rio de Janeiro, PAC funded upgrades in several favelas commonly criticized for being monumental rather than utilitarian: an Oscar Niemeyer-designed bridge at the entrance to Rocinha, a panoramic elevator in Cantagalo/Pavão-Pavãozinho, and aerial cable cars in Morro da Providência and Complexo de Favelas do Alemão.

"We are experiencing a very important moment for Brazil," said President Dilma Rousseff at the festive opening ceremony of the Teleférico do Alemão in July 2011. Rousseff also insisted on some 'technical' facts: The cable car system produces less pollution, it shortens travel time between its two extreme stations from 50 minutes by foot, up and down an unpaved hillside, to 16minutes, and all residents of the Complexo do Alemão have the right to two free rides per day. Indeed, the Teleférico is integrated with the suburban rail network via the Bonsucesso/TIM Station and takes a 3.5-kilometer ride from the entrance to the top of the mountain in 16 minutes. The Teleférico thus carries the promise of fostering physical, social, and economic integration. Rousseff and her political supporters are much aware, however, that the Teleférico's value exceeds its impact on the mobility of the underprivileged: The ski slope technology represents to Rio de Janeiro, as a whole, a much valued ticket for riding on the magical wheel of the global tourism cities. 'Rebranded,' the city presents itself—favelas included—as a welcoming site for various mega events (Name 2012), including the Pan American Games (2007), World Military Games (2011), RIO+20 Conference (2012), World Youth Day (2013), FIFA Confederations Cup (2013), FIFA World Cup (2014), Summer Olympic Games and Paralympics (2016), as well as Copa America (2019).

Tourism is a privileged form of converting spaces into landscapes and landscapes into commodities, even encompassing sites that would, at first, be considered not to meet standards for tourism. Such is the case of poverty-stricken and segregated territories that, while capable of instilling both fear and repulsion, are transformed around the world into attractions highly regarded by the international tourist. Whether in the barrios of Medellín, in the townships of South Africa, or in the slums of Kenya and India, tourists are promised authentic encounters with 'exotic,' 'craftsman-like' communities, which are presumably alien to modern temporality and signs of a deprived but happier life. Within the tourism framework, urban poverty and its resulting spaces are advertised and consumed as an authentic urbanity cutting through the pasteurized landscape of globalization (Frenzel et al. 2012; Freire-Medeiros 2009, 2013). In the specific case of Brazil, public power initially ignored—and often openly reproached—the existence of growing flows of tourism toward these areas that it had always sought to hide. Presently, the government in its three levels (municipality, state, and federal) is nevertheless reinventing favela tourism under the principles of city marketing and urban entrepreneurialism. The installation of a spectacular device, such as the aerial cable car at the Complexo do Alemão, a locality far from Copacabana Beach, the Sugar Loaf, and other Rio de Janeiro tourist attractions, should be understood precisely as another example of how favelas are being transformed into official attractions of the 'city as a commodity.'

But not only public actors are investing in the favelas as the new frontier for tourism activities in the city of Rio de Janeiro. At Complexo do Alemão, various social actors and institutions from the private and third sectors are heading different initiatives which have both domestic visitors as well as international tourists as their clientele. In the next section we present a composite of different favela tours promoted by tourism entrepreneurs, which are complemented by accounts of other tourists who shared with us their personal experiences. We obviously do not summarize the modus operandi of every single company and/or travel agent developing tourism activities at Complexo do Alemão, but such participant observation can provide additional insight into the experience of the favela 'seen from above' (extending, for example, Freire-Medeiros 2013).

'PEACE, LOVE, AND FUN:' THE TRAVELING FAVELA SEEN FROM ABOVE

Rio de Janeiro, Summer of 2012. Brazil's biggest group buying site, Peixe Urbano,[5] advertised the "Peace, Love, and Fun: Cable Car Tour at Complexo do Alemão," a guided visit for about US$20. The advertisement promised a "Rio de Janeiro from another angle" experience and "breathtaking views of the whole city." Although three pictures of the favela illustrate the ad, there was no mention of its landscape, landmarks, or tourist attractions. There was no mention either of the fact that the same round trip on the same cable car ordinarily cost, at that time, no more than US$2 (residents have the right to two free journeys, and there is a promotional ticket price of 50 cents of a US dollar). Nevertheless, 29 tours were sold, including ours.

It is a Sunday morning with blue skies. When we get to the station, Washington, a dark-skinned man in his early 30s, is already waiting for us. He introduces himself as "your guide for the day" and asks where we are coming from. He can't help but to be surprised when he realizes that we live not so far from Alemão. "On weekends, I usually do two or three tours, but it's usually for Brazilians from outside Rio, who've heard of the cable car and seen the telenovela" ['Salve, Jorge,' a Brazilian primetime TV show at that time had the Complexo do Alemão as a scenario for its plot]. If other touristic favelas, such as Rocinha and Santa Marta, have been attracting thousands of almost exclusively international visitors for several years now (Freire-Medeiros et al. 2013), Alemão is consolidating itself as a tourist destination for Brazilians themselves. Washington lives in a working-class neighborhood about seven kilometers from the Complexo do Alemão. Later on, our guide will explain that he knows "the complex bottom-up," because for many years he used to be the middle man between a congressperson and local bandits: "It was up to me to get the drug lord's permission to bring in a dentist [sponsored by the afore mentioned congressman] who provided basic dental service to the community. After the pacification, I saw the opportunity to become a tourist guide here."

The 'pacification' that Washington refers to started in November 2011, when Complexo do Alemão witnessed one of the largest military operations in the city's history. In theory, the pacification draws on the ideal of community policing and intends to foster a closer relationship between the police and population, and also to strengthen social policies inside favelas. In this sense, the UPP is considered 'a new model of public security,' in which the state reassumes the monopoly of physical force, restoring control over areas that for decades were occupied by factions connected to drug traffickers. It began in 2008 with the implementation of its first unit in favela Santa Marta, and presently there are 42 UPPs in the city (for the transformation of Santa Marta into a tourism destination, see Freire-Medeiros et al. 2013). In practice, however, the public security project works like this: The BOPE—a special unity of the military police trained to kill—initially occupies the territory and undertakes special operations of tactical dismantling of the drug trade networks, thus clearing the path for policemen who are almost all fresh out of training and have taken courses on 'community policing' and 'human rights.' Those policemen occupy the strategic areas of the favela and stand guard for an undetermined time. In the case of the Complexo do Alemão, due to the high level of confrontation, the 'pacification' counted also with segments of the army and the navy.

Since then, Washington and other tourist guides, as well as cable TV, telephone companies, and other businesses, flocked to the area. A good example is Natura, the biggest, Avon-style, door-to-door sales company in Brazil, which has set up stands in different parts of the community offering free makeup and massage in order to attract not only customers but also potential sales representatives. Natura and other major companies, mainstream media, and some NGOs are all highly encouraging favela residents to redirect their entrepreneurial skills from informal sector activities to 'legal microbusiness.' The flows of people and capital that pour into Alemão are seen, thus, as part of a much celebrated 'social occupation' that 'logically' follows military occupation. After the 'pacification,' and with the supermodern Teleférico constructed to allow for touring the favela *from above* (see Figure 14.1), Alemão is ready to be incorporated as a symbol of 'peace, love, and fun,' as advertised by Washington's tourism agency.

If the Teleférico can carry 3,000 passengers per hour, today the platform is almost empty and impressively clean. Only Washington and the two of us enter the cabin that fits eight passengers. Ahead of us, there are three kids and an elderly lady whom they call 'grandmother.' While they make fun of her being terrified, she prays and nervously laughs. We are not, it seems, the only ones there for the first time to appreciate the scale and complexity of the favela from an aerial perspective.

As we move away, our guide warns us that we will be impressed with the 'shackless favela.' We look through the thick glass at an ocean of *lajes*, the roof tops which play so many different roles in almost every favela, from leisure spaces to backyards for hanging the laundry or raising domestic animals. Blue water tanks, narrow streets, walking men and women, and playful children—all pass underneath us. The immensity of the favela seems to

Figure 14.1 The favela from above: Being inside the cable car is a way to see the spectacular landscape of Complexo do Alemão and Rio de Janeiro mountains in the background.

Source: Leonardo Name 2012.

overlap with the green mountains not so much further down, providing a unique aesthetic trip.

Washington points out to us three buildings—or tourist sites, if you will. They are three physical structures that, within his narrative, embody the past, present, and future of the Complexo do Alemão. First comes an icon of the oppressive past: the ex-drug lord's house ("which even had a swimming pool") that the "pacifying forces took back" rather dramatically by putting down a Brazilian flag on top of its roof—a scene shown repeatedly on national TV. Then follows an icon of the peaceful present: a sports center that may now be enjoyed by all residents who are no longer under the bandits' oppressive gaze. Last, but not least, there are the colorful panels created by world famous Brazilian artist, Romero Britto, strategically placed so they can be seen from inside the cabin in each station. In this sense, the Teleférico stations (see Figure 14.2), with their eye-catching decorations, are celebrations of the future that awaits the favela as a territory to be reshaped according to the logics of the neoliberal model of market-based economic and social organization.

Outside, a huge, blue building with a mirrored glass façade bears the inscription 'UPP.' Nearby are stands selling drinks and snacks, but neither Washington nor the vendors offer us anything. We walk up to a belvedere to find a spectacular view of the entire complex, encompassing the Teleférico, the ocean, and mountains. A young couple, which we assume to be

Figure 14.2 Palmeiras, the last cable car station at Complexo do Alemão, offers to the touristic gaze a large panorama of mountains and precarious buildings.
Source: Leonardo Name 2012.

Brazilians, kiss rather passionately, and two teenage girls have fun posing and taking photos in front of the UPP building.

If green jeeps have become iconic to Rocinha tourism, inspiring comparisons between favela tourism and a 'safari among the poor,' what does this filmic, almost idyllic landscape of a favela seen from far above the Teléferico offer us? The Alemão landscape, now resignified by the new mobility system, is also the spatial representation of the current power regime. As authors such as Lacoste (1977, 1990) and Virilio (2005) have demonstrated, there is an undeniable relation between militarism and tourism, between war strategies and the capture and production of images and landscapes. War and tourism are both activities mediated by optical instruments such as binoculars and cameras. They equally depend on establishing tall and distant points of view through which grandiose field perspectives are revealed.

We head back to the cable car station to enter the cabin. The scope of the Rio de Janeiro landscape one sees from the Teléferico is impressive, a view very similar to others that have been traveling the world since the 1940s through Hollywood movies, postcards, and travel guides. Near the end of the descent, we face an angle which reveals the favela, the 'Christ the Redeemer' statue, the ocean, and the massive mountains behind. The breathtaking view gives one the impression, once again, that the favela has been placed in front of a traditional postcard shot of the city. Washington muses, "Rio de Janeiro is really gorgeous, isn't it? It's just gorgeous . . ." We are silent, and he sighs.

CONCLUDING REMARKS

Lack of mobility is certainly an important part of the set of disadvantages and deprivations connected to social and territorial exclusion. The issue that deserves attention, therefore, is how increases in mobility options in marginalized areas can lead (or not) to a reduction in the conditions of poverty and an increase in well-being; i.e., to social improvements and to reducing inequalities. In this chapter we attempted to discuss, through an account of our ride in the Teléferico do Alemão, how mobility technologies, tourism, and the conversion of the favela landscape into a commodity are intertwined.

The huge popularity of the Teléferico as a tourist attraction rather than a transportation device *per se* inspires us to think that we are facing a new stage in the biography of the traveling favela and its consolidation as a tourist attraction, the logic of which cannot be dissociated from how Rio de Janeiro's tourist landscape in a broader sense has been constructed over the years: a celebration of human intervention upon a majestic natural scenery. The Sugar Loaf is an excellent example: Its singular, geomorphological form was not enough, and a cable system had to be installed. The same can be said about the Corcovado Hill, cut by a sinuous road and crowned by the statue of Christ. In the other hills of the city, tracks, belvederes, and free flight spots keep the observed landscape at a perennially high perspective, which leans towards totality and uniformization. Upon its massive touristic appropriation, certain favelas of Rio de Janeiro—i.e., the 'pacified' ones[6]—also present themselves as a result of violent interventions which can be captured not only through the gaze, but also allow one to *travel* or *pass* through it. Second, in times of 'pacification' it should not come as a surprise that the position of the three levels of government towards the practice of favela tourism has shifted from an initial posture of opposition, followed by indifference to open support. In the specific case of Complexo do Alemão, by guiding the tourists to experience the favela from above, it is possible to redirect the gaze to the achievements of security programs and spectacular infrastructural projects. In the process, the tourist gaze is also diverted from the numerous 'forgotten favelas' at the periphery, as well as from the demolition and eviction measures which are also part of the present rebranding of Rio de Janeiro.

For those who are *inside* Alemão, the Teleférico not only increases and facilitates the mobility of locals and tourists, but also signifies a powerful intervention in the territory, which allows this group of favelas to be labeled as a landscape, both figuratively and literally. Upon entering one of the cabins, one's gaze faces a deep, long-distance perspective from a high point of view, embracing the space ahead. This maximizes, in some measure, the already common situation of foreigners' gazes, which savor the view of the precarious houses piled up underneath them, mingled with the luxurious tropical vegetation.

For those who gaze at the Complexo do Alemão from the *outside*, the Teleférico redraws the favela skyline, aligning it with the tradition of interventions into the city's landscape: their unusual aspects—wires, cables, and cabins—attract the gaze and, consequently, the camera, interested in taking pictures or filming. In both cases one gets the panoramic view, with portions of the ocean and other hills in sight as a backdrop, wherein the Complexo do Alemão is inserted, in constant dialogue with the remaining Rio de Janeiro tourism landmarks.

Even if they do not overcome urban inequality and only serve a small part of the population where they are installed, aerial cable cars have an enormous symbolic efficacy. As we attempted to demonstrate here, they produce perceptions of development, inclusion, and articulation with the formal city. This comes as no surprise if one realizes that they are a stark intervention that makes possible a new, moving, aesthetic experience of the landscape: A terrestrial route and a bus system might be able to serve more passengers, but definitely lack the visual and aesthetic impact of aerial cables within the cityscape. A 'pacified' territory in tune with city branding instructions, the 'traveling favela' that the Teleférico provides is part and parcel of a regeneration project that aims at the "regulating fiction" (Robinson 2003) of the global city. In this sense, implementing mobility through aerial cable cars is part of the trajectory of spectacle cities announced by Débord long ago, or, as Clarke (2011) more recently termed it, "the city as an entertainment machine."

ACKNOWLEDGMENTS

A previous version of this chapter was presented at the Conference Differential Mobilities: Movement and Mediation in Networked Societies, hosted by the PanAmerican Mobilities Network in 2013 at Concordia University (Montreal, CA). We are grateful to participants for their comments and suggestions.

NOTES

1 Although Complexo de Favelas do Alemão's official population, according to the Brazilian Institute of Geography and Statistics, does not surpass 65,000 (IBGE 2011), many estimate it might be as high as 300,000 (Perlman 2010; Sluis 2011).
2 Before becoming a major option in different cities, the BRT system was introduced in Curitiba, Brazil in the 1970s and in the Ecuadorian capital in the late 1990s. According to Paquette (2011, p. 189), the choice for the BRT over other mass transportation options is explained in part by its large capacity, close to that of a subway, while costing 10 to 20 times less. Nonetheless, it is not always the most adequate solution for all contexts, and often its implementation does not foresee transportation systems' intermodality, which limits it in terms of access to a greater variety of locations. The same author notes that Brazil alone has more BRT systems than the entire Asian continent (31 cities in Brazil, 30 in all of Asia). In Mexico City, the implementation of the so-called Metrobús in its historical center was highly criticized by the population due to the area's urban fabric vulnerability to this type of intervention. The Peruvian capital's Metropolitano, in turn, is an example of a BRT system not connected to other transportation systems, unlike what happens in Rio de Janeiro, for example, where a single ticket lasting two hours gives access to several interconnected transportation systems: buses, trains, subways, ferries, and cable cars.
3 Colombia's second largest city, Medellín, made the world's headlines in the 1980s and 1990s as a city of widespread violence 'owned' by the almost mythical figure of drug lord Pablo Escobar.
4 The whole program was organized under six major initiatives: Better Cities (urban infrastructure), Bringing Citizenship to the Community (safety and social inclusion), My House, My Life (housing), Water and Light for All (sanitation and access to electricity), Energy (renewable energy, oil, and gas), and Transportation (highways, railways, and airports).
5 Also known as 'collective buying,' such a shopping strategy, which promises businesses high-volume sales in return for high-percentage discounts (often a 50% to 95% discount rate) for groups that agree to purchase the same item, has been popular in Brazil since 2010, but comparatively is far from being as big as it is in the United States, for instance.

6 The UPP program is by no means a citywide measure; it is confined to selected favelas, and they are not necessarily those with the highest crime rates, but those located in city areas, which are strategically relevant to the events.

WORKS CITED

Arantes, O. (2000) "Uma estratégia fatal: A cultura nas novas gestões urbanas," in O. Arantes, C. Vainer, and E. Maricato *A Cidade do Pensamento Único*, Petrópolis: Vozes, pp. 11–74.
Arcay, A.O., Ordax, M.N., and Rodriguez Bugarín, M. (2003) *Transporte por Cable*, Corunha: Ed. Tórculo Artes Gráficas.
Avocat, C. (1979) *Montagnes de Lumière: Briançonnais, Embrunais, Queyras, Ubaye. Essai sur l'Évolution Humaine et Économique de la Haute Montagne Intra-Alpine*, Villeurbanne: Impr. A. Fayolles.
Balseinte, R. (1958) "Les stations de sports d'hiver en France," *Revue de Géographie Alpine*, 47:129–180.
Balseinte, R. (1959) "Megève ou la transformation d'une agglomération montagnarde par les sports d'hiver," *Revue de Géographie Alpine*, 47:131–224.
Barbosa, J.L. (1999) "O caos como imago urbis: Ensaio crítico a respeito de uma fábula hiper-real," *GEOgraphia*, 1:59–69.
Berque, A. (1994) "Paysage, milieu, histoire," in A. Berque (ed) *Cinq Propositions pour une Théorie du Paysage*, Seyssel: Champ Vallon, pp. 11–29.
Bocarejo, D., and Alvarez Rivadulla, M.J. (2012) "La esperanza de ser vistos. Percepciones de los habitantes de la Comuna 4 frente a la posible construcción de un cable aéreo," in J.D. D'Avila (ed) *Movilidad Urbana y Pobreza: Aprendizajes de Medellín y Soacha*, Bogotá, CO: The Development Planning Unit, pp. 143–148.
Boitempo and Carta Maior (2014) *Cidades Rebeldes: Passe Livre e as Manifestações que Tomaram as Ruas do Brasil*, São Paulo: Boitempo/Carta Maior.
Brand, P., and Dávila, J.D. (2011) "Mobility innovation at the urban margins: Medellin's Metrocable," *City*, 15:647–661.
Chiavari, M.P. (2000) "Os ícones da paisagem do Rio de Janeiro: Um reencontro com a própria identidade," in C. Martins (ed) *A Paisagem Carioca*, Rio de Janeiro: Museu de Arte Moderna, pp. 56–75.
Clarke, T.N. (ed) (2011) *The City as an Entertainment Machine*, Chicago: University of Chicago/Elsevier.
Claval, P. (2004) "A paisagem dos geógrafos," in Z. Rosendahl and R.L. Corrêa (eds) *Paisagens, Textos e Identidade*, Rio de Janeiro: EdUERJ, pp. 13–74.
Comitê Olímpico Brasileiro (2009) *Dossiê de Candidatura do Rio de Janeiro à Sede dos Jogos Olímpicos e Paraolímpicos de 2016*, Rio de Janeiro: COB.
Dávila, J.D., and Brand, P. (2012) "La gobernanza del transporte público urbano: Indagaciones alrededor de los Metrocables de Medellín," *Bitacora*, 21(2):85–96.
Dávila, J.D., and Daste, D. (2011) "Pobreza, participación y Metrocable. Estudio del caso de Medellín," *Boletín CF+S*, 54:121–131.
Denis, R.C. (2000) "O Rio de Janeiro que se vê e que se tem: Encontro da imagem com a matéria," in C. Martins (ed) *A Paisagem Carioca*, Rio de Janeiro: Museu de Arte Moderna, pp. 82–97.
Elliot, A., and Urry, J. (2010) *Mobile Lives*, London: Polity Press.
Freire-Medeiros, B. (2009) "The favela and its touristic transits," *Geoforum*, 40: 580–588.
Freire-Medeiros, B. (2013) *Touring Poverty*, London: Routledge.
Freire-Medeiros, B., Grijó Vilarouca, M., and Menezes, P. (2013) "International tourists in a 'pacified' favela: Profiles and attitudes. The case of Santa Marta,

Rio de Janeiro," *Die Erde—Journal of the Geographical Society of Berlin*, 144:147–159.
Frenzel, F., Koens, K., and Steinbrinck, M. (eds) (2012) *Power, Ethic and Politics in Global Slum-Tourism*, Abington, Oxon: Routledge.
Fukuyama, F., and Colby, S. (2011) "Half a miracle," *Foreign Policy*, 25. Retrieved February 2015 from http://foreignpolicy.com/2011/04/25/half-a-miracle/.
Holzer, W. (1999) "Paisagem, imaginário, identidade: Alternativas para o estudo geográfico," in Z. Rosendahl and R. Lobato Corrêa (eds) *Manifestações da Cultura no Espaço*, Rio de Janeiro: EdUERJ, pp. 149–169.
Hylton, F. (2007) "Medellin's makeover," *New Left Review*, 44:70–89.
Lacoste, Y. (1977) "A quoi sert le paysage? Qu'est-ce un beau paysage?," *Hérodote*, 7:3–41.
Lacoste, Y. (1990) *Paysages Politiques*, Paris: Le Livre de Poche.
Lefebvre, H. (2001) *O Direito à Cidade*, São Paulo: Centauro.
Leibler, L., and Musset, A. (2010) "¿Un transporte hacia la justicia espacial? El caso del Metrocable y de la Comuna Nororiental de Medellín, Colombia," *Scripta Nova*, 331. Retrieved February 2015 from http://www.ub.edu/geocrit/sn/sn-331/sn-331-48.htm.
Micheletto, M., and Novarina, G. (2002) "La montagne vue par les urbanistes (les années 1930–1940)," *Revue de Géographie Alpine*, 90:33–47.
Name, L. (2010) "O conceito de paisagem na geografia e sua relação com o conceito de cultura," *GeoTextos*, 6:163–186.
Name, L. (2012) "Jogos de imagens: Notas sobre o Dossiê de Candidatura do Rio de Janeiro à Sede dos Jogos Olímpicos e Paraolímpicos de 2016," *Revista Intratextos*, 4:277–297.
Name, L., and Machado de Mello Bueno, L. (2013) "Contradição nas cidades brasileiras: 'Ambientalização' do discurso do planejamento com permanência dos riscos," in Departamento de Geografia *Riscos Naturais, Antrópicos e Mistos*, Coimbra: Universidade de Coimbra, pp. 727–739.
Panizza, M. (2011) "Geomorphosites: Concepts, methods and examples of geomorphological survey," *Chinese Science Bulletin*, 46:4–5.
Paquette, C. (2011) "Retos del desarollo urbano," in Institut des Amériques *Los Desafíos del Desarrollo en América Latina: Dinámicas Socioeconómicas y Políticas Públicas*, Paris: AFD, pp. 195–211.
Pires do Rio, G. A., and Name, L. (2013) "O novo plano diretor do Rio de Janeiro e a reinvenção da paisagem como patrimônio," in Encontro Nacional da Associação de Pesquisa e Pós-Graduação em Planejamento Urbano e Regional, 14 Anais..., Recife, BR: ANPUR.
Reynard, E. (2005) "Geomorphosites et paysages," *Géomorphologie: Relief, Processus, Environnement*, 3:181–188.
Roger, A. (1994) "Histoire d'une passion théorique ou comment on devient un Robaliot du paysage," in A. Berque (ed) *Cinq Propositions pour une Théorie du Paysage*, Seyssel: Champ Vallon, pp. 109–123.
Ronai, M. (1976) "Paysages," *Hérodote*, 1:125–159.
Ronai, M. (1977) "Paysages II," *Hérodote*, 7:71–91.
Sheller, M., and Urry, J. (2006) "New mobilities paradigm," *Environment and Planning A*, 38:207–226.
Simmel, G. (2007) "The philosophy of landscape," *Theory, Culture & Society*, 24:20–29.
Soja, E. W. (2014) *En Busca de la Justicia Espacial*, Valencia: Tirant Humanidades.
Urry, J. (2007) *Mobilities*, London: Polity Press.
Valladares, L. do Prado (2005) *A Invenção da Favela: do Mito de Origem a Favela.com*, Rio de Janeiro: Ed. FGV.
Virilio, P. (2005) *Guerra e Cinema*, São Paulo: Boitempo.

Moving Forward

15 Rethinking Mobility at the Urban-Transportation-Geography Nexus

Andrew E. G. Jonas

INTRODUCTION

The "new mobilities paradigm" (Sheller and Urry 2006) in social and cultural studies is transforming the ways in which scholars think about space—especially urban space (Amin and Thrift 2002). It comes on the back of wider discussions about the spatiality of social life in cities, discussions often inspired by the writings of critical geographers and sociologists, such as Doreen Massey (1991) and Manuel Castells (2000), who place emphasis on understanding how urban processes are constituted through relationships, flows, and networks extending far beyond the boundaries of the city. The status of world cities like London, for example, depends upon not just the spatial concentration of global financial institutions within city boundaries but also the nature of global connections shaping the social characteristics of its diverse boroughs (Massey 2007). Relational thinking about cities disrupts an overly containerized view of urban space and opens up new vistas for examining cities and their wider social relationships, connections, and flows.[1]

This book uses urban transportation as a lens through which to rethink mobility. Whilst there have been previous dialogues between urban scholars and transportation geographers, relational scholars of the urban have been a little slow to recognize quite how profoundly new modalities of transportation shape the conduct and governance of mobility in the city today (Cresswell 2010; Shaw and Hesse 2010; Hall and Hesse 2013). At the same time, and paradoxically, certain tropes familiar to transportation geographers, such as mobility, flow, and movement, have increasingly been deployed by critical urban scholars as metaphorical devices for thinking about urban-environmental relations and social injustices in the capitalist city (Heynen, Kaika, and Swyngedouw 2006; Kaika 2005). Yet, for the most part, this has not led to efforts to rethink mobility through the lens of urban transportation *per se*. By adding transportation to the urban mobility equation, this book significantly deepens and extends the empirical scope of relational thinking about the production of urban space.

In their introduction, the editors identify the epistemological challenge as follows: "how best to comprehend and theorize the city as both space and circulatory system. No topic presents this challenge more clearly than transportation, the most explicitly motive force in urban life" (Prytherch and Cidell, this volume, p. 19–20). This is not a straightforward task, however, because the relational and the territorial are always co-constituted in urban space (McCann and Ward 2010). The flows and networks connecting cities and the spaces therein still require the production of physical transportation systems, a fact not lost on the likes of David Harvey, who has consistently emphasized that the production of urban space creates tensions between mobility and fixity (Harvey 1982, 1985a). The kind of fixity Harvey has in mind is not one that examines the urban as fixed territorial container; instead, it recognizes how the tensions between mobility and spatial fixity are bound up in wider social relationships—how capital and labor power are brought together in the city, how urbanization underpins the accumulation of capital, how various urban-based political coalitions organize to channel the circulation of capital through the built environment, and how devaluation constantly poses a threat to fixed capital invested in the built environment (Harvey 1978). Sheller and Urry (2006, p. 210) echo Harvey when suggesting that different forms of mobility require place-specific investments in immobile infrastructure, giving rise to different social and political constructions of locality. So when thinking about geographies of mobility, we should not forget the continuing importance of spatial fixity in its various social and physical forms.

Transportation geography has a long-standing interest in the relationship between transportation, mobility, and urban development (Harris and Ullman 1945; Taaffe, Morrill, and Gould 1963; Berry 1964; Taaffe and Gauthier 1973; Hanson and Giuliano 2004). There is no need to rehearse this history here. What I will say, however, is that the study of urban transportation geography has in some ways contributed to the kind of containerized view of the city that has become the bane of relational urban scholars. It is manifested, for instance, in the division of labor between those who study intraurban transport systems, such as light rail and bus transit, and those interested in interurban systems, such as the airline industry, freight traffic, and high-speed rail. Yet, most urban transportation networks and their systems of governance do not stop at the jurisdictional limits of the city, even if, as the residents of Detroit would probably attest, one's personal experiences of mobility are often profoundly shaped by the presence of urban political boundaries. Similarly, the development of interurban transport systems can have important implications for territorial politics. Thus, for instance, recent proposals in the United Kingdom for a new high-speed rail network linking London and the regions have highlighted a growing territorial-political divide between London and those cities located outside the core urban growth region of the southeast of England (Tomaney and Marques 2013). Approaching urban transportation from the perspective of mobility not only renders the academic division of labor between inter- and intraurban transportation geography anachronistic; it also promises to unsettle the territorial/relational divide in urban theory.

What further sets this book apart from previous encounters between transportation and urban geography is its *approach* to questions of mobility. The emphasis is not on drawing fixed boundaries around the urban; nor is it attempting to treat transportation as an independent, spatial variable—a locational cost surface—shaping urban spatial form. Instead, the chapters animate relational understandings of the city by means of crosscutting themes, such as the changing role of urban spaces and places within wider transportation networks and flows, the governance of urban mobility and transportation systems, the manner in which different modes of transportation are experienced and imagined, and how cities and transportation networks are coproduced through new patterns of circulation. In doing so, the contributors reveal that transport systems are more than engineered structures which physically constrain and limit urban spatial form; rather, they increasingly involve complex social, technical, and political systems and relationships which connect, define, and delimit urban space. I shall refer to this as *the urban-transportation-geography nexus*.

In the remainder of this concluding chapter, I wish to build on each of the sections in the book and propose four themes for further work on the urban-transportation-geography nexus, and how work in this volume contributes to them. First, and elaborating on the discussion of intersections, I refer to the literature on ordinary urbanism and suggest that patterns of mobility in the city are not just the result of decisions by urban planners, public authorities, and growth coalitions; they also reflect how normative rules and regulations governing flows of traffic and people are interpreted, enacted, and performed by ordinary urbanites (whether commuters, truck drivers, pedestrians, or cyclists). Second, I consider how the conjuncture of the politics of mobility, sustainability, and climate change is engendering all sorts of new and unexpected political alignments and coalitions around the urban living place. In section three, I examine how work on the provision and consumption of transportation infrastructures provides an opportunity to explore changes in state territoriality. Investments in urban infrastructure not only create new networks and flows across state territory; they also lend legitimacy to new discourses of territorial competition and city-regional growth. Several of the chapters speak to the evolving relationship between the production of urban mobility, the investment strategies of the competition state, and the governance of city-regions. The fourth section makes a link between reconnecting urban and regional spaces and reimaging urban worlds. It examines how the urban governance challenges of mobility inform how powerful interest groups discursively represent, imagine, and market urban regions, cities, and the spaces within and between them. I conclude by reflecting on how the above themes might point us in the direction of several promising routes for exploring questions of mobility, relations, flows, territories, and the production of urban space.

ORDINARY URBANISM: THE EXPERIENCE OF TRANSPORT AND MOBILITY IN THE CITY

As explored in the first section of the book, fixed investments in transportation infrastructure not only connect urban places to the wider global economy, but also change everyday movement patterns and flows across the city and its jurisdictional limits. In so doing, urban mobility profoundly, yet at the same time surreptitiously, shapes and reshapes what Giddens (1984) has called the "structuration of everyday life" in the city and its constituent locales. Mobility is essential to how cities work in both a literal and figurative sense: how labor is performed (and exploited) in the city, how commuters get to work, and how people make a living, right down to the level of the street (Jonas, McCann, and Thomas 2015). If past urban models expressed a bird's eye view of the city, the editors offer instead a street-level perspective of Chicago:

> Standing at major intersections, like where Michigan Avenue and Wacker Drive join in Chicago's Loop, one is struck less by surrounding skyscrapers than the incessant, negotiated circulation of cars and bike and buses on the streets, pedestrians along sidewalks and up staircases and elevators, elevated trains rumbling above and the subway below, passenger and cargo planes criss-crossing the sky, and boats traveling along the Chicago River whose portage between the Great Lakes and Mississippi watersheds made the city possible . . .
> (Prytherch and Cidell, this volume, p. 21)

Prytherch and Cidell invite us to engage with such spaces with empirical sensitivity and theoretical rigor so that we can better appreciate how cities function as places where ordinary urban lives are experienced, negotiated, and contested.

Several chapters use mobility to explore what Jennifer Robinson calls the geography of "ordinary cities" (Robinson 2006). David Prytherch (this volume) examines street intersections as locales where particular 'rules of the road,' like statutory law and traffic control, choreograph everyday geographies of mobility. Drawing upon Merriman (2012), he interprets mobility as variously comprised of practices, technologies, discourses, and bodily movement in space. Focusing on particular urban spaces—such as an intersection in Hamilton, Ohio—reveals how federal, state, and local rules structure the spatiality of social life and mobility in the city. Meanwhile, Gregg Culver (this volume) invites us to reflect on how traffic engineering tools like levels of service (LOS) reinforce normative spatial visions of mobility. LOS is a way of classifying operating conditions for a given stretch of highway into six different levels, ranging from 'best' (A) to 'worst' (F). Culver attempts to make sense of two interrelated processes: first, how traffic engineering informs normative spatial visions of mobility, and second, how values and assumptions embedded in such tools become naturalized as state-of-the-art science deployed—discursively and materially—in major infrastructure projects, prioritizing fluid automobility over alternative modes.

If traffic codes, rules, regulations, and design standards (like LOS) govern mobility in a formal sense, urban scholars are also interested in how the experience of ordinary urbanism can subvert such received rationalities. Bascom Guffin's chapter (this volume) considers how formal traffic regulations are negotiated and contested in the cities of the Global South and what this says about urban theories based on observations in the cities of the Global North. In Hyderabad, India, drivers use horns rather than brakes to negotiate through traffic. They tend to see abstract traffic regulations as obstacles to be avoided, at best, or, at worst, completely disregarded. Instead, a bodily politics of mobility takes shape, where traffic is channeled more by concrete than compliance with rules. The fact that everyday mobility in Hyderabad diverges from the norm does not make it exceptional or resistant to explanation. Instead, urban theory must fully embrace such differences in the ordinary experiences of mobility.

TRANSPORTATION, MOBILITY, AND THE ENVIRONMENT

Following the Rio Earth Summit Conference of 1992, many urban authorities responded to the call for sustainable development by signing up to initiatives such as Local Agenda 21. Promoting sustainability has since become a concerted focus of activity in the part of urban growth coalitions (While, Jonas, and Gibbs 2004), leading to a new raft of interventions around the built environment, transportation, and urban living place. Interventions—such as redensification, smart growth, and transit-oriented development—attempt to generate socially equitable and environmentally sustainable geographies of urban flow, movement, and encounter. Here the state, via urban planning, attempts to influence the geography of land values, thus altering the relationship between transportation, the urban land nexus, and the process of capital accumulation (Scott 1980). At the same time, new demands are being put on urban leaders to invest in physical infrastructures and urban forms that rely less on the burning of fossil fuels. The discursive landscape of urban development is rapidly becoming colonized by references to sustainability, climate adaptation, and the low-carbon economy, each in its turn a sign of a 'new environmental politics of urban development' (NEPUD) (Jonas, Gibbs and, While 2011).

The spaces where these new politics play out are battlegrounds between conflicting visions and political rationalities underpinning discourses of mobility, sustainability, and economic growth. Focusing on one such space in Vancouver, Canada, Peter V. Hall (this volume) examines the ways particular transportation corridors are planned and accepted (or contested) as routes for truck movement, thereby differentiating and dividing metropolitan communities and places. This results, in part, from the combined effects of the accumulated residue of prior fixed investments, sunk costs, planning laws, and environmental regulations. However, such effects are reinforced by political struggles around mobility on the part of property owners, who are sensitive to differences in land values associated with proximity to known truck routes, and neighborhood organizations concerned about noise, pollution, and health. It seems to Hall that truck routes display rather more resilience than might be expected given the increasingly dynamic and complex patterns of flow they must channel.

As Jason Henderson (2006, 2013) has pointed out elsewhere, there are all sorts of possibilities for unusual political alignments to occur around the politics of mobility. In San Francisco, for instance, a form of these politics has been built on the legacy of prior inner city, urban protests. In the 1950s, the city's regional transportation plans triggered a 'freeway revolt' in lower-income and working-class neighborhoods scheduled for clearance and urban renewal. Henderson (this volume) takes the San Francisco story forward and argues that wider-scale discussions about transport and climate change must be downscaled to the level at which street intersections and bus stops are planned and governed. Localized interventions like prioritizing transit, pedestrians, and bicycles seek to produce measurable impacts, such as reducing greenhouse gas emissions and vehicle miles traveled (VMT), but these techniques are fiercely contested. Likewise, interventions conducted in the name of smart growth, such as transit-oriented development, rezoning, and densification, can achieve certain sustainability benefits, yet can also further exacerbate issues of housing affordability, exclusion, and displacement. This prompts Henderson to conclude that the struggle around climate change is "not just a street fight, but also a struggle over the right to the city."

Urban struggles around mobility sharpen awareness of distributional inequalities across the city. It is well known that suburban development privileges white, middle-class households and exposes low-income groups and people of color to toxic environmental facilities in their communities and neighborhoods (Pulido 2000). In cities like Chicago, people have fled to the suburbs in order to avoid environmental facilities known to be potentially damaging to their health and safety. However, as Julie Cidell's chapter (this volume) shows, new patterns of mobility—including the rerouting of freight traffic from the urban core to suburbs—have sharpened awareness of environmental inequality. And suburbanites' encounters with freight rail crossing provoke new anxieties and fears of the 'uncanny.' What sets this conflict apart from more typical struggles located within the suburban living place is mobility—as material practice and signification—*across* local jurisdictional boundaries.

CONNECTED CITIES/COMPETITIVE STATES: TRANSPORTATION, MOBILITY, AND THE 'GEOPOLITICS OF CAPITALISM'

Relational thinking forces us to rethink how mobility shapes and reshapes urban territory. Instead of restricting urban analysis to economic and political structures neatly contained within fixed jurisdictional boundaries—an approach aptly captured by the title of Paul Peterson's (1981) book *City Limits* – relational urbanism focuses on urban flows, connections, and extraterritorial relations. However, there is no need to throw the baby out with the bathwater; relational approaches are not a substitute for territorial representations of the urban. Instead, mobility allows us to explore the relations of urban space to wider territorial structures of the state.

For example, urban transportation often features in tensions, debates, and political struggles around the state and its territorial structure. On the one hand, mobility shapes how cities are connected across state territory, so decisions about where to build interurban transport infrastructure tend to feed into wider political discourses about the state's internal functionality, as well as its extraterritorial competitiveness (Ward and Jonas 2004). On the other hand, new investments in fixed physical infrastructures can profoundly reconfigure internal state territory in ways that can reinforce existing urban territorial divisions and social exclusions, thereby defining the scope and reach of new urban alliances and coalitions. In each of these respects, investigating the relationship between mobility and the production of urban space can fruitfully provide insights into the reconfiguration of capitalist territoriality, or what Harvey (1985b) has called the "geopolitics of capitalism."

Inspired by the writings of Henri Lefebvre, which emphasize the role of state spatial policy in delivering those physical and social infrastructures essential for sustaining international capital (Lefebvre 1996), a new generation of urban theorists is interested in exploring the changing role of the state in the production of urban space (Brenner 2000; Brenner and Elden 2009). Under Fordism-Keynesianism, the state's role in enabling accumulation often involved building national transportation networks (freeways, motorways, rail systems, etc.) and organizing state territory into a single, functional, and cost-efficient economic space. The rise of the competition state, however, threatens this seemingly fixed state territoriality, raising questions about the scalar division of powers and resources between the national and the urban, particularly with respect to the delivery and consumption of infrastructure. Such questions throw into new perspective struggles around geographies of collective provision at the metropolitan and local scales (Cox and Jonas 1993), which in turn feed into wider geopolitical discourses and practices associated with the rise of the competition state and its internal territorial configurations (Addie 2013; Jonas 2013).

If, as the editors mention in the introduction, technological changes have allowed for an unprecedented level of locational freedom on the part of corporations and capital, new geographies of mobility necessitate new infrastructural connections between urban places and hence also new configurations of state territoriality. Several chapters in this book are able to shed light on how the infrastructural and mobility needs of the competition state engender tensions and conflicts more or less around the scale of the city-region. For example, Theresa Enright (this volume) examines how conflicts over the Grand Paris Express (GPE) rapid transit system are at the same time struggles to define the territorial limits and the extraterritorial reach of the metropolis. The most outspoken voices in the Grand Paris transit debates assume that state powers and resources must be mobilized, both to attract investments to the city-region and enhance its extraterritorial competitiveness. She provocatively describes this new geopolitics of accumulation in terms of a 'regime of metromobility,' which extends well beyond the territorial limits of the city-region. Likewise, Jean-Paul D. Addie (this volume) seeks to bridge the relational/territorial divide and demonstrate how transportation mobility shapes the governance of the Toronto city-region. His analysis reveals the contradictory tension between mobility's rendering of metropolitan space as punctuated by flows and rhythms, on the one hand, and how managing mobility requires regional governance structures that are simultaneously territorialized and containerized, on the other. Both chapters demonstrate how the metropolitan provision of transportation infrastructures underpins the discursive production and active governance of state territoriality through the modern metropolis.

Such changes in state territoriality at the city-regional scale could be indicative of causal connections between accumulation and the growth of regionally extensive urban forms (Soja 2000, 2011). Most of our received models of urban form, such as those generated by the Chicago School of Social Ecology in the 1920s, are based on city-centric processes and patterns of urban development. Whether the competition between business and consumers for central city locations, or the sorting of retail and service activities across the wider settlement system, scholars have imaged the city developing outwards from the center, forming neat concentric zones, sectors, nodes, and spatial hierarchies. To the extent that 21st-century urbanization is amenable to such forms of rationalization, urban scholars now recognize that processes of urban growth have inverted so that, if anything, growth at the territorial periphery drives change at the center. Cities are far more decentralized and spread out than ever before, prompting scholars to develop a new vocabulary to describe peripheral settlement forms, such as "edge cities" (Garreau 1991), "post-suburbia" (Phelps and Wu 2011), and "boomburbs" (Lang and LeFurgy 2007), as well as the inner suburbs and older urban districts seemingly left behind by sprawl: *Zwischenstadt* or the 'in-between spaces' of the metropolitan region (Sieverts 2003; Keil 2011).

As entrenched city-suburban political divisions give way to a plurality of geopolitical possibilities, regional urbanization challenges how we think relationally about territorial politics. What was once fought around the politics of urban versus suburban is now more likely to be manifested as new territorial discourses, such as 'global city,' 'suburban regionalism,' 'new regionalism,' or 'regional collaboration' (Jonas 2011; Jonas, Goetz, and Bhattacharjee 2014). Another possibility is further balkanization around self-governing territorial entities, such as special purpose districts, gated communities, and other privatized forms of suburban development, many demanding premium access to regional utilities and transportation infrastructure (Graham and Marvin 2001). Those inner urban spaces lacking in the corresponding powers and governance capacities, or suffering from the fiscal effects of urban austerity, might be deprived of essential investments in social and physical infrastructures.

Whereas the chapter by Addie focuses on the divergent experiences of Toronto's in-between spaces, yet other possibilities are revealed in Christian Mettke's chapter (this volume) on mass transit in the Greater Toronto Area (GTA). As the GTA expands and attracts global investment, mass transit is caught up in a dilemma of satisfying new urban growth demands at the same time as it needs to serve deprived areas suffering from limited mobility options. Such spatial deficits in collective provision for mobility bring into the public arena new political voices and understandings about city-regional development. Nowhere is this more important than when we examine mobility and transportation in postcolonial urban contexts and how, in turn, such contexts inform our theories and imaginaries of urban spaces once at the colonial core.

REWORLDING, RECONNECTING, AND REIMAGINING THE CITY

Roy and Ong (2011) consider how concepts of territory and political identity are being challenged by the emergence of new urban forms in postcolonial societies. The urbanization of the Global South and the formation of new connections to former colonial centers in the Global North have profoundly influenced the imagination and representation of urban space. That postcolonial urban forms challenge many of the territorial rationalities underpinning the development of the metropolis is highlighted in Ananya Roy's analysis of Malaysia's Multimedia Super Corridor (MSC), which connects Kuala Lumpur to the nation's capital in the new suburb of Putrajaya and the nearby high-tech city of Cyberjaya (Roy 2009). Roy argues that the MSC is emblematic of new imaginaries of nation, state, and territory in a hyperconnected world. Similarly, Anru Lee's study (this volume) of the symbolic and cultural meanings of mass transit in the city of Kaohsiung, Taiwan reminds us of the need to consider how investments in physical infrastructure can become a vehicle of not only physical movement but also of change, breaking away, and becoming. The Kaohsiung Mass Rapid Transit System has enabled not only the possibility of flow for the city's population but also, as Lee suggests, "the flow of the city into a brighter and more prosperous future."

An alternative perspective on mobility and urban representation is offered in the chapter by Bianca Freire-Medeiros and Leonardo Name (this volume). They examine the Complexo do Alemão, a recently pacified favela (shanty town) in Rio de Janeiro, Brazil. A multimillion-dollar cable car system known as Teleférico do Alemão takes tourists on a 16-minute ride above the favela to the top of a mountain. Deploying the notion of the 'traveling favela,' they show how international tourists and celebrities journeying on the cable car have turned the favela into a set of visually appealing photographic images and brochures. The huge popularity of the Teleférico inspires the authors to script a new stage in the biography of the traveling favela, and its consolidation as a global tourist attraction through the production of new transportation mobilities.

Moving beyond the central city into the wider region, one enters into zones, jurisdictions, and areas often lacking in clear territorial markers and political identities. If mobility shapes the relational character of places, the planning and regulation of these in-between spaces nevertheless often requires the imposition of new territorial meanings on the landscape. As Cox and Mair (1988) argue, often it is local actors dependent on growth who strive to fill the void in meaning and signification created by capital mobility. In this manner, the local becomes a space to be colonized by ideologies of community, territory, or place. In his chapter, Markus Hesse (this volume) distinguishes between economic development strategies that promote progressive ideologies and visions of regional growth based around logistics, and other, arguably more restrictive, attempts that draw on alternative meanings and ideologies, such as sustainability and community empowerment. In the former category are the place promotion efforts of regions and states involved in selling and promoting regional hubs or national logistic centers. In the latter are local land use plans and restrictions that seek to defend local use values. Each strategy in its turn entails different and conflicting imaginaries of region and place.

Further evidence of conflict around the meaning of territory is provided in the chapter by Sophie L. Van Neste (this volume), who considers how new infrastructure projects in the in-between spaces of the Randstad region of the Netherlands are strategically framed and counterframed by regional authorities and local residents. At issue is the meaning of a place known as Midden-Delfland in South Holland. In the past, Midden-Delfland had specific qualities and environmental amenities reminiscent of an older Dutch landscape, and which have been promoted by local nature groups, resident associations, and municipalities. However, a new national government, working in partnership with the city and port of Rotterdam, seeks to develop a highway project across the region, giving a new meaning to Midden-Delfland. It has been reduced, in effect, to a space in-between cities: one characterized in terms of missing highway segments rather than having a coherent sense of place. Van Neste refers to this meaning-giving process as one of 'place-framing' and, in so doing, demonstrates that the territorial signifiers used to describe metropolitan areas and the spaces between them are fluid, contested, and inherently political.

FINAL THOUGHTS

This book has much to say on the matter of relational versus territorial representations of urban space. Unlike previous dialogues between transportation and urban geography, however, it does not start out by drawing strict boundaries between intra- and interurban transportation systems, flows, and spaces. Rather, it uses mobility to think about systems, spaces, and flows both within, and stretched across, the jurisdictional limits of urban territory. Investigating new modalities of flow and movement across the city and wider region holds out the promise of changing the way we think relationally about territory—not least urban territory, in all of its various physical and social manifestations. As Mimi Sheller suggests in her preface (this volume), we need "to understand spatiality in more relational ways, and to understand the relations enabled by transport in more mobile ways." This book points us in the right direction for pursuing this task.

I have identified four themes that I think could usefully inform future research on transportation, mobility, and the production of urban space. First, there is further scope to examine how rules and regulations governing flows of traffic and people are interpreted and enacted by residents in 'ordinary cities' and to use these insights to advance critical urban theory. Second, the convergence of the politics of mobility and that of climate change seems to throw into sharp perspective new political struggles and social movements around planning and social provision in the city. Third, transportation mobility provides an opportunity to investigate tensions and struggles around the territorial structures of the competition state. Finally, the governance of mobility feeds into new representations and imaginaries of city-regions and the spaces in between. Whilst I am sure that readers will have their own suggestions, I believe that each of these themes opens up all sorts of new avenues for exploring questions of mobility and the social production of space at the urban-transportation-geography nexus.

ACKNOWLEDGMENTS

Thanks to the editors for their helpful comments and to Andy Goetz and Eric Boschmann for reminding me of the importance of transportation for how we explore and experience processes of urban development.

NOTE

1 Relational approaches to urbanism are not to be confused with Relational Urbanism, a group of professional architects and urban planners who are involved in developing 3D computer graphics models of urban form (see www.relationalurbanism.com, accessed August 2014).

WORKS CITED

Addie, J-P. (2013) "Metropolitics in motion: The dynamics of transportation and state reterritorialization in the Chicago and Toronto city-regions," *Urban Geography*, 34(2):188–217.

Amin, A., and Thrift, N. (2002) *Cities: Reimagining the Urban*, Cambridge: Polity.
Berry, B. (1964) "Cities as systems within systems of cities," *Papers in Regional Science*, 13:147–164.
Brenner, N. (2000) "The urban question as a scale question: Reflections on Henri Lefebvre, urban theory, and the politics of scale," *International Journal of Urban and Regional Research*, 24:360–377.
Brenner, N., and Elden, S. (eds) (2009) *State, Space, World: Selected Essays, Henri Lefebvre*, Minneapolis: University of Minnesota Press.
Castells, M. (2000) *The Network Society* (2nd ed.), Cambridge, MA: Blackwell.
Cox, K. R., and Jonas, A.E.G. (1993) "Urban development, collective consumption and the politics of metropolitan fragmentation," *Political Geography*, 12:8–37.
Cox, K. R., and Mair, A. J. (1988) "Locality and community in the politics of local economic development," *Annals of the Association of American Geographers*, 78:307–325.
Cresswell, T. (2010) "Towards a politics of mobility," *Environment and Planning D*, 28(1):17–31.
Garreau, J. (1991) *Edge City: Life on the New Frontier*, New York: Doubleday.
Giddens, A. (1984) *The Constitution of Society: Outline of the Theory of Structuration*, Cambridge: Polity Press.
Graham, S., and Marvin, S. (2001) *Splintering Urbanism: Networked Infrastructures, Technological Mobilities and the Urban Condition*, London: Routledge.
Hall, P., and Hesse, M. (eds) (2013) *Cities, Regions and Flows*, Abington, Oxon: Routledge.
Hanson, S., and Giuliano, G. (eds) (2004) *The Geography of Urban Transportation* (3rd ed.), New York: Guilford.
Harris, C. D. and Ullman, E. L. (1945) "The nature of cities," *The Annals of the American Academy of Political and Social Science*, 7-17.
Harvey, D. (1978) "The urban process under capitalism: A framework for analysis," *International Journal of Urban and Regional Research*, 2:101–131.
Harvey, D. (1982) *The Limits to Capital*, Oxford: Blackwell.
Harvey, D. (1985a) *The Urbanization of Capital*, Baltimore: The Johns Hopkins University Press.
Harvey, D. (1985b) "The geopolitics of capitalism," in D. Gregory and J. Urry (eds) *Social Relations and Spatial Structures*, London: MacMillan, pp. 128–163.
Henderson, J. (2006) "Secessionist automobility: Racism, anti-urbanism, and the politics of automobility in Atlanta, Georgia," *International Journal of Urban and Regional Research*, 30(2):293–307.
Henderson, J. (2013) *Street Fight: The Politics of Mobility in San Francisco*, Amherst: University of Massachusetts Press.
Heynen, N., Kaika, M., and Swyngedouw, E. (2006) *In the Nature of Cities: Urban Political Ecology and the Politics of Urban Metabolism*, Abingdon: Routledge.
Jonas, A.E.G. (2011) "Post-suburban regionalism: From local politics of exclusion to regional politics of economic development," in N. A. Phelps and F. Wu (eds) *International Perspectives on Suburbanization: A Post-Suburban World?*, Basingstoke: Palgrave Macmillan, pp. 81–100.
Jonas, A.E.G. (2013) "City-regionalism as a contingent 'geopolitics of capitalism,'" *Geopolitics*, 18(2):284–298.
Jonas, A.E.G., Gibbs, D., and While, A. (2011) "The new urban politics as a politics of carbon control," *Urban Studies*, 48:2537–2544.
Jonas, A.E.G., Gibbs, D. C., and While, A. H. (2004) "State modernisation and local strategic selectivity *after* Local Agenda 21: Evidence from three northern English localities," *Policy and Politics*, 32:151–168.
Jonas, A.E.G., Goetz, A. R. and Battarcharjee, S. (2014) "City-regionalism and the politics of collective provision: Regional transportation infrastructure in Denver, USA," *Urban Studies*, 51(11):2444–2465.

Jonas, A.E.G., McCann, E., and Thomas, M. (2015) *Urban Geography: A Critical Introduction*, Oxford: Wiley-Blackwell.
Kaika, M. (2005) *City of Flows: Nature, Modernity and the City*, New York: Routledge.
Keil, R. (2011) "The global city comes home: Internalised globalisation in Frankfurt Rhine-Main," *Urban Studies*, 48:2495–2518.
Lang, R.E., and LeFurgy, J. (2007) *Boomburbs: The Rise of America's Accidental Cities*, New York: Brookings Institution.
Lefebvre, H. (1996) *Writings on Cities* (E. Kofman and E. Lebas, Eds. and Trans.), Cambridge, MA: Blackwell.
Massey, D. (1991) "A global sense of place," *Marxism Today*, 35(6):24–29.
Massey, D. (2007) *World City*, Cambridge: Polity Press.
McCann, E., and Ward, K.G. (2010) "Relationality/territoriality: Toward a conceptualization of cities in the world," *Geoforum*, 41(2):175–184.
Merriman, P. (2012) "Human geography without time-space," *Transactions of the Institute of British Geographers*, NS37:13–27.
Peterson, P. (1981) *City Limits*, Chicago: University of Chicago Press.
Phelps, N.A., and Wu, F. (eds) (2011) *International Perspectives on Suburbanization: A Post-Suburban World?*, Basingstoke: Palgrave Macmillan.
Pulido, L. (2000) "Rethinking environmental racism: White privilege and urban development in Southern California," *Annals of the Association of American Geographers*, 90(1):12–40.
Robinson, J. (2006) *Ordinary Cities: Between Modernity and Development*, New York: Routledge.
Roy, A. (2009) "The 21st century metropolis: New geographies of theory," *Regional Studies*, 43:819–830.
Roy, A., and Ong, A. (eds) (2011) *Worlding Cities: Asian Experiments and the Art of Being Global*, Oxford: Wiley-Blackwell.
Scott, A.J. (1980) *The Urban Land Nexus and the State*, London: Pion.
Shaw, J., and Hesse, M. (2010) "Transport, geography, and the 'new' mobilities," *Transactions of the Institute of British Geographers*, 35(3):305–312.
Sheller, M., and Urry, J. (2006) "The new mobilities paradigm," *Environment and Planning A*, 38(2):207–226.
Sieverts, T. (2003) *Cities Without Cities: An Interpretation of the Zwischenstadt*, London: Spon Press.
Soja, E. (2000) *Postmetropolis: Critical Studies of Cities and Regions*, Oxford: Blackwell.
Soja, E. (2011) "Regional urbanization and the end of the metropolis era," in G. Bridge and S. Watson (eds) *The New Blackwell Companion to the City*, Oxford: Wiley-Blackwell, pp. 679–689.
Taaffe, E., and Gauthier, H. (1973) *Geography of Transportation*, New York: Prentice-Hall.
Taaffe, E., Morrill, R., and Gould, P. (1963) "Transport expansion in underdeveloped countries: A comparative analysis," *Geographical Review*, 53(4):503–529.
Tomaney, J., and Marques, P. (2013) "Evidence, policy, and the politics of regional development: The case of high-speed rail in the United Kingdom," *Environment and Planning C*, 31(3):414–427.
Ward, K., and Jonas, A.E.G. (2004) "Competitive city-regionalism as a politics of space: A critical reinterpretation of the new regionalism," *Environment and Planning A*, 36:2112–2139.
While, A., Jonas, A.E.G., and Gibbs, D.C. (2004) "The environment and the entrepreneurial city: Searching for a 'sustainability fix' in Leeds and Manchester," *International Journal of Urban and Regional Research*, 28:549–569.

Contributors

Jean-Paul D. Addie is provost fellow in urban knowledge and infrastructure in the Department of Science, Technology, Engineering, and Public Policy at University College London. His research engages issues of mobility, governance, and social justice in an era of globalized urbanization. He is currently researching the relationship between city-regional urbanization and the geography of global higher education.

Julie Cidell is an associate professor of geography and GIS at the University of Illinois. She has written on the political economy of transportation and mobilities with regards to airports, container shipping, and railroads, focusing on the relationship of large infrastructure to local governments and territories. She has also studied urban sustainability regarding green buildings and local government regulation.

Gregg Culver (Ph.D., University of Wisconsin-Milwaukee) is a research associate in the research cluster geography of North America in the Institute of Geography at Heidelberg University. His current research interests lie in the intersection of critical urban, mobilities, and transportation geographies, particularly regarding discourses revolving around urban mobility and space, and how these come to matter in the politics of mobility.

Theresa Enright is an assistant professor of political science at the University of Toronto. Her research spans the fields of urban studies, critical theory, and comparative political economy with a focus on global cities, political cultures of mobility, megaprojects, and urban-suburban relations. She is currently completing a book on Grand Paris.

Bianca Freire-Medeiros is senior lecturer of sociology at Getulio Vargas Foundation (Rio de Janeiro, Brazil). She is the author of *Touring Poverty* (Advances in Sociology Series, Routledge, 2013).

Bascom Guffin is a Ph.D. candidate in the Department of Anthropology at the University of California, Davis. His research examines how a rising

296 *Contributors*

new middle class in Hyderabad inhabits city planners' visions of the future as expressed in the newly urbanized periphery called Cyberabad.

Peter V. Hall is associate professor of urban studies at Simon Fraser University in Vancouver, Canada, where he teaches economic development, transportation geography, and research methods. His publications include the coedited *Integrating Seaports and Trade Corridors* (Ashgate, 2011) and *Cities, Regions and Flow* (Routledge, 2013). He is an associate editor of the Journal of Transport Geography.

Susan Hanson is a professor of geography (emerita) at Clark University. An urban geographer, her interests center on gender and economy, transportation, local labor markets, and sustainability. Her research has examined the relationship between the urban built environment and people's everyday mobility within cities; within this context, questions of access to opportunity, and how gender affects access, have been paramount.

Markus Hesse is a professor of urban studies at the University of Luxembourg, Faculty of Humanities. His research focuses on urban and regional development; mobilities, logistics, and global flows; metropolitan policy and governance; and spatial discourses and identities. A particular emphasis in recent research is placed on the relationship between science and practice in geography and spatial planning.

Jason Henderson is professor of geography and environment at San Francisco State University and author of *Street Fight: The Politics of Mobility in San Francisco*, published in 2013 by University of Massachusetts Press. His research includes how culture, economics, ideology, and politics shape urban transportation policy and the geography of cities.

Andrew E. G. Jonas is professor of human geography at Hull University in the United Kingdom. His Ph.D. is from The Ohio State University, U.S. He has published widely on urban development politics and the provision of infrastructure in U.S. and European cities and regions. His latest book, coauthored with Eugene McCann and Mary Thomas, is *Urban Geography; A Critical Introduction* (Wiley-Blackwell). His coedited books include *The Urban Growth Machine: Critical Perspectives Two Decades Later* (SUNY Press, 1999), *Interrogating Alterity* (Ashgate, 2010), and *Territory, State and Urban Politics* (Ashgate, 2012).

Anru Lee is a faculty member of the Anthropology Department at John Jay College of Criminal Justice, the City University of New York. Her research focuses on issues of capitalism, modernity, gender and labor, and urban anthropology. She is the author of *In the Name of Harmony and Prosperity: Labor and Gender Politics in Taiwan's Economic*

Restructuring (SUNY Press, 2004) and is coeditor of *Women in the New Taiwan: Gender Roles and Gender Consciousness in a Changing Society* (ME Sharpe, 2004). Her current project investigates mass rapid transit systems as related to issues of technology, governance, and citizenship. Her most recent fieldwork looks at the newly built mass rapid transit systems in Taiwan in the context of the country's struggle for cultural and national identity.

Christian Mettke successfully defended his Ph.D. thesis at the Technische Universität Darmstadt, in 2014. In his thesis he is comparing the public transit systems of Toronto and Frankfurt. His research examines the contested development of public transit in a postsuburban context. Currently he is working for the GIZ in the field of environmentally sustainable transport.

Leonardo Name is lecturer of architecture and urbanism at Federal University of Latin American Integration—UNILA (Foz do Iguaçu, Brazil).

David Prytherch is an associate professor of geography at Miami University, Oxford, Ohio. He is an urban and cultural geographer interested in the politics and practices of urban planning. His recent research focuses on the regulation and design of the public street, interpreted as a question of socio-spatial rights and social justice.

Mimi Sheller is professor of sociology and founding director of the Center for Mobilities Research and Policy at Drexel University. She is president of the International Association for the History of Transport, Traffic, and Mobility; founding coeditor of the journal *Mobilities*; and associate editor of *Transfers: Interdisciplinary Journal of Mobility Studies*. She is coeditor, with John Urry, of *Mobile Technologies of the City* (Routledge, 2006), *Tourism Mobilities* (Routledge, 2004), and several key articles. Her recent books are *Aluminum Dreams: The Making of Light Modernity* (MIT Press, 2014), coedited *Routledge Handbook of Mobilities* (Routledge, 2013), and coedited book *Mobility and Locative Media* (Routledge, 2014).

Sophie L. Van Neste is a postdoctoral researcher at Université de Montréal and Clark University. She is interested in the spatialities of contentious action, in counter-discourses in the politics of mobility and the politics of infrastructure, and in the dynamic construction of discourse coalitions. Her most recent publications include 'The Spatial Puzzle of Mobilizing for Car Alternatives in the Montreal City-Region' in *Urban Studies*, 2014, and 'Claiming Rights to Mobility Through the Right to Inhabitance' in *International Journal of Urban and Regional Research*, forthcoming.

Index

access 3, 7–10, 33, 138, 182, 230, 275; to city spaces 14, 110; to employment 7; inequality 135, 190, 196, 242; to infrastructure networks 25, 124, 182, 215, 288; to public services 32, 143, 149; roadway 53, 118
accessibility 24–5, 209, 230, 239, 242, 257
activism 7, 9, 110, 112–13, 247–8
Advocating for Container Traffic Off Residential Streets (ACTORS) 126
aerial cable cars 264, 265–8, 269–74
American Association of State Highway and Transportation Engineers (AASHTO) 52–4
anxiety 144, 148
automobiles 105, 139, 143, 197; engineering bias towards 82, 89, 91, 93, 107–9; environmental impacts 103; and flow 5, 8, 82, 86, 108; restrictions on 104
automobility 13, 14, 94–5, 102, 175, 194; and engineering/design 53, 56, 75, 78, 82–6, 92, 106–8; logics of 197; naturalization and culture of 31, 61, 81, 107
auto trip generation (ATG) 108, 111

Bay Area, California (USA) 104–6, 109
Bay Area Rapid Transit (BART) 110–12
behavioral geography 5–6
Belgium 217–23
bicycles: infrastructure and planning 86–9; lanes 104
bodies-in-movement-in-spaces 46, 48, 60
boosterism 222–3

Brazil 263–4, 266–74
Burnaby, British Columbia (Canada) 124
Bus Rapid Transit (BRT) 194, 238, 267, 275

California (USA) 104–14
California Sustainable Communities and Climate Protection Act (SB 375) 109
Caltrain 111–12
Canadian National (CN) 135, 145, 150
Castells, Manuel 28–9, 281
Chicago, Illinois (USA) 5, 21, 31, 119, 134–7, 139–40, 146–7
Chicago School 21
choreography 45, 46, 48–9, 60–1
circulation: urban 20–2, 30, 45, 172, 283; automobile 108–10; of capital 22, 109, 282; of goods 124, 208; kinaesthetics of 48–9, 60
city-as-circulatory-system 19
city-regions 118, 188, 195, 199, 229, 230
civil engineering 5–6, 20, 24, 60
climate change 10, 101, 102–3, 113, 286, 291
cluster strategies 212, 217, 221
commodity chains 212–13
commuter rail 111–12, 143–4, 157–8, 161–8
commuting 16, 110–12, 139, 143, 163, 176–7, 179
Complexo do Alemão 264, 268–75, 289
concrete politics 69, 71, 78
connectivity 4, 6, 173, 187, 192–3, 230, 236–7

300 *Index*

corridors 31, 82–3, 118, 137, 128, 220, 285, *see also* links
Cresswell, Tim 70, 188, 192, 230, 245
critical mobilities studies 14–15, 20, 24–8, *see also* mobilities studies
crosswalks 50, 51, 54, 57, 59
culture 8–9, 15, 23, 162, 164
cultures of mobility 15, 47, 49, 56–7, 60, 61
Cyberabad neighborhood, Hyderabad (India) 64, 66, 71, 76

deindustrialization 160, 208, 217
delay 106–9, 134, 141–2, 147–8
dialectical materialism 189, 191–200
discourse: economic development 220–3; of order and control 45–7; public 131, 156–7, 159–60, 173, 267–8; political 285–7; role in shaping mobilities 8, 194, 197, 220, 222, 246–8; of scientific objectivity 31, 81–4
distribution centers 140, 210, 216, 221
Dome of Light, The 165–6
driving 16, 58, 64–5, 75, 76–8, 104–5, 112
duty of care or duties 51, 58–9

economic development: and logistics 207–10, 223; in the Netherlands 214, 216, 217–23; in Paris 178; in Taiwan 160
economic geography 4–5
economic restructuring: in Taiwan 160; urban 173
economic revitalization 168
Elgin, Joliet, and Eastern (EJ&E) 135, 140, 141, 144–5
Environmental Impact Statement (EIS) 135, 140, 146
environmental justice 140
environmental review 107–8
ethnography 65, 157
exclusion 14, 16, 110–11, 135, 138, 144–5, 190, 273, 287

favela pacification 270, 276
favelas 264, 268–74
flows: of capital 175, 196; and cities 13, 143, 156–7, 175, 187–8, 198, 241–2, 245–6, 281–91; freight and logistics 120–1, 125, 135–7, 146–8, 207–12, 220–1, 224; metropolitan 251–2, 254–60; places of 19–23, 28–9, 45–6, 49, 60–1, 233; politics of 220; as traffic 52, 54, 56, 71–2; in transportation planning/engineering 5, 106–8, 82, 84–6, 108; urban mass transit 167–8
Formosa Boulevard, Taipei (Taiwan) 165, 166
freeway revolts 7
freight 27, 255; distribution 120, 134, 198, 207, 214, 216–8; rail 131, 140–1, 143, 145, 148

gateways 118, 130, 209–10, 223
gentrification 110–11, 113, 235
geometric design 52
geopolitics of capitalism 286–8
global city 178, 183, 198–9, 229, 233, 274, 288
globalization 25, 209, 218, 229, 266, 269; neoliberal 148, 157, 187–9; 199
global warming *see* climate change
Google Bus 102, 110–12, *see also* private commuter shuttle
Gottlieb, Mark 94
governance: and logistics and freight distribution 207, 212, 220, 224; and place framing 247–9, 251–2, 258–9; transit system 168, 173–4, 180, 194, 235–6, 240; urban 13, 121, 190, 228, 232–3, 282–3, 291
Grand Paris 172, 174, 176, 180
Grand Paris Express (GPE) 32, 179, 180, 181, 287–8
Greater Toronto Area (GTA) 32
Greater Vancouver Gateway Council 123–4, 128
Green Heart 249
greenhouse gas emissions (GHG) 101, 102, 104–5, 109, 114
Growth Acceleration Program or Programa de Aceleração do Crescimento 268, 275
growth poles 217

Hamilton, Ohio (USA) 46, 56, 61
hazardous materials 135, 143, 148–9
Highway Capacity Manual 53, 84–5, 88, 93
Hoan Bridge 86–91

Index 301

Hoan Bridge Bicycle/Pedestrian Feasibility Study 87
humanistic geography 22
Hyberabad (India) 64–5
Hyderabad Information Technology and Engineering Consultancy (HITEC) City 67, 76
Hyberabad Traffic Police 66, 77

identity 25, 107–8, 161, 168, 249, 253; national 156
ideology 86, 95, 102, 105, 107, 114; 114; conservative 102, 105, 107; of mobility 105, progressive 102, 109
Ile-de-France 176, 179
imaginaries: and logistics 208, 220–3; spatial 173, 198, 249, 255; touristic 265; and transit 157, 164; urban 19, 130, 189, 289–91
immediate controls 65, 73, 77, 78
immobility 8, 15, 30, 65, 155, 191, 282
in-between city *see* Zwischenstadt
India 64–79
inequality 136, 145, 156, 178, 180–1, 263, 273–4, 286
inequity 7, 110, 113
infill development 104, 111
infrastructure: analytic metrics for 108; and city regions 188–90, 194, 198–200, 210–11; investment decisions 118–19, 123–4, 130–1, 134–5, 156, 283, 284; logistics 217–18; networked urban systems 13–16, 56, 155–6, 173–4, 181–2, 209, 232; politics and policies 251, 254, 259, 287, 289–90; and postsuburbanization 228–9, 234–5; rail 137, 145–8; roadway 66, 69–71, 75–7, 85; sociotechnical 188–9; technical 228–9, 232, 241; transit 160, 167–8, 172, 176–9, 180; transportation 3, 5–6, 10, 26–7, 45–6, 56, 144–5, 153, 188; urban political 173–5, 183; uncanny 149
intercity competition 164, 208
intersections 31, 33, 45–7, 65, 71, 73, 175,
Interstate Highway System 87, 134

jaywalking 57, 59
judicial: case law 58–9; adjudication 59–60; sanctions 57

Kaohsiung (Taiwan) 153, 159–60, 163, 166, 289
Kaohsiung MRT 154–8, 160, 163–4, 167–8
kinaesthetics 48, 56
Kukatpally (India) 6

landscape: urban 19–20, 28, 30, 61, 117, 119–20; in-between 192–3, 195, 248–9; of inequality 136; mass transportation 176; Midden-Delfland 251–3; and mobility 65, 70, 145; as touristic commodity 263, 265–6, 268–9, 273–4, 275; traffic 71
Latin America 263, 267
Lefebvre, Henri 15, 105, 190, 195, 199, 230, 287
Level of Service (LOS): application 89–92; definition 81, 84–6, 106; ideology behind 94–5, 107–8, 284–5; alternatives to 109–19, 111, 113
links, *see also* corridors
livability 102, 103–4, 111, 113
locational analysis: network perspective 22, 32; for regional development 212, 221; in transportation 119, 124, 207–9
logistics hubs 209, 214, 223

materiality 3, 15; and mobility practices 25, 78; transport systems 30, 54, 69, 155, 167, 211
Medellín (Colombia) 267, 275
mediate controls 65–6, 73, 77
megaprojects 156, 173, 182
Metrolinx 193, 197, 235, 236, 240
metromobility 173, 175, 182, 183
Midden-Delfland (Netherlands) 250, 251–5, 258, 290
Milwaukee, Wisconsin (USA) 86–8, 92
Milwaukee County (USA) 86–7
mitigation fees 109, 110–12
mobilities: critical 4, 12, 15, 26, 191–2, 200; new paradigm 12, 21, 105, 174–5, 263, 281; studies 8–9, 12–13, 24, 26–8, 47, 105, 174, 175

mobility 5–9, 46, 49, 52–3, 88–9, 94, 167, 273–4, 281–91; and accessibility 24–5, 267–8; constellations 130–1, 188, 192–4, 200; and immobility 12, 15, 20, 30, 154–5, 191; and justice 14; politics of 14, 27, 29, 45, 81, 102, 135, 192, 229–30, 241, 245; practices 13, 47, 54, 57–9, 121; regimes of 66, 78, 173, 181–3; systems 13, 15,174–5
modernity 137, 159, 163, 173, 208, 220, 265, 268
moral geography 153
motorcycles 75, 154, 163
Muni Forward 105, 108
Muni *see* San Francisco Municipal Railway

National Committee on Uniform Traffic Control Devises (MUTCD) 54
National Committee on Uniform Traffic Laws and Ordinances (NCUTLO) 50–1
National Environmental Policy Act (NEPA) 7
negligence 57
neoliberalism 25, 102, 109–10, 181, 187, 272; policy 64, 110; and urbanization 173–5, 188–90, 199
Netherlands, The 214, 249
networks 21–6, 190, 208–10, 214, 282–3; flows of 28–30, 157, 233, 242; and infrastructure 137–8, 173, 175, 181, 198, 229; rail 134, 139–40; road 53, 64, 108, 117–19; socio-technical 138, 188–9; *see also* places, networked; splintered urbanism
New York City Truck Route Network 117
NIMBY (not in my backyard) 148

Ohio (USA) 46, 61
Ohio Revised Code 50, 57
ordering 45, 49, 51, 60–1
ordinary urbanism 284–5

Pacific Gateway or 'The Gateway' 118, 123
Paris 172, 173
Paris Métro 173, 179
parking 75–6, 104, 107

place 13, 19, 61, 145, 147–8, 209–10, 221, 245–6, 290; of flows 21–2, 29–30, 45, 49, 60–1, 157, 167; making 155, 168; networked 21, 30; place-frames 246, 248, 258, 259; politics of 135, 245–8, 259–60
police and policing 56–8, 66–8
policy 9–10
politics of mobility 14, 20, 23, 27, 81, 252
politics of place 245–6, 259
Port of Milwaukee 87–8
Port of Rotterdam 214, 250–1
ports 123–4, 125, 127
post-suburbanization 228, 230, 235, 240–1
power 50, 61, 240, 247; fleshy 65–7; embodied 66, 77; relations 52, 94, 193, 196, 230, 242; state 46, 67, 78, 198
private commuter shuttle 102
public art 165
public investment 268
public participation 7, 140, 177
public streets 45

qualitative methods 8–9, 29, 174
quantitative revolution 4, 23–4

railroad 134–5, 146; grade crossings 141, 146; North America 134, 139, 140
Randstad (The Netherlands) 32, 249
regime of metromobility 175, 180
regional development 168, 176, 208, 212–14, 217, 223
regional disparity 156–7, 159, 160
regulatory regimes 56
relational geographies 47, 191, 281, 285–6, 290
relational/territorial dialectic 282–3, 287–8, 290–1
representation 86, 94–5, 221, 246–7, 259, 266, 289
RER 176
rhythms 15, 48–9, 138–9, 288
right-of-way 50, 51, 54, 59, 70
rights 51, 58
Rio de Janeiro (Brazil) 32, 263–73
roads and roadways 50, 51, 64, 70, 75–6, 117, 119, 122; classification 53; engineering 52–4

Rotterdam 249, 250, 251, 252, 254
rule systems 47, 60

San Francisco (USA) 101, 102, 104–7, 108, 112
San Francisco Municipal Railway (MUNI) 106, 110–11
San Francisco Municipal Transportation Agency (SFMTA) 105
Sarkozy, Nicolas 172, 176
scientific objectivity 86, 88, 91, 94
scoring 46, 55
signage 54
social science 3, 5, 7
space-time geographies 27, 46
spatiality 23, 29, 130, 155–6, 167, 188, 199, 281
speed bumps 73
splintered urbanism 28, 190, 194, 235
streets 104–5, *see also* roads and roadways
structuration 22, 122, 284
subjectivity 46, 263
suburbs 135–6, 139, 144–8, 196–8, 229; population growth in 147, 148, 233, 234, 236; suburbanization 198; vs. urban 135, 136, 145, 146, 147, 148
Sugar Loaf cable car 265–6, 273
supply chains 209, 212
Surface Transportation Board (STB) 135, 140, 147
Surry, British Columbia (Canada) 128
sustainability 4, 8, 9, 12, 15, 111, 194–5, 283, 285–6
symbolism 73, 168, 174, 181, 222

Taipei (Taiwan) 154, 159–63
Taiwan 32
Teleférico do Alemão 263, 268–73, 289
territoriality 187–8, 287–8
territory 178, 195–6, 198, 224, 247, 272–4, 286–91; and relationality 187, 189–91, 282–3; *see also* relational/territorial dialectic
The Hague (The Netherlands) 249, 251–2
time-geography 6, 22
Toronto 32, 188, 193, 196–7, 228, 233–5, 238–9, 288
traffic 57, 135, 141, 211, 256; control 54–6; engineering 81–3, 94–5,

119; logic 82–93; markings 55; medians 71, 77; planning 66, 69–71; science 82–3, 94–5; signals 56, 73; statutes 50–2
transit 55–6, 106–7, 156, 167–8, 174–6, 181, 229–30; funding 112, 161, 180, 236–7, 238; ridership 105, 112, 154, 164, 236
transit-oriented development 104, 181–2
TransLink (legally South Coast British Columbia Transportation Authority) 112, 128, 130
transportation 181, 188, 209; and environment 211, 224; infrastructure 20, 24, 284; and land use 119–21, 285–6; networks 23, 25, 28; planning 4–6, 25
transportation geography 3–4, 9, 16, 23, 25, 157
transportation investment 119, *see also* public transit funding
truck routes 117–19, 121, 285–6; designated 122, 123, 125–7; driven 129–30
trucks and trucking 117–20, 123, 129

uncanny 135, 137, 147, 148
Uniform Vehicle Code 50
urban anthropology 74, 77
urban geography 4, 6, 13
Urban Geography (journal) 33
urbanization 174, 187–9, 192–4, 196–7, 199, 231, 241, 288; neoliberal 57; post-colonial 289
urban places of flows 21, 28–9
urban planning history 83
urban-transportation-geography nexus 283

Vancouver, British Columbia (Canada) 31, 118, 122, 124
Vehicle Miles Traveled (VMT) 16, 101, 104, 108, 111–13

Wisconsin (USA) 86
Wisconsin Department of Transportation (WisDOT) 87–91, 93–4

Zwischenstadt (in-between city) 189, 192–3, 245, 249, 251, 259, 288

For Product Safety Concerns and Information please contact our EU representative GPSR@taylorandfrancis.com
Taylor & Francis Verlag GmbH, Kaufingerstraße 24, 80331 München, Germany

www.ingramcontent.com/pod-product-compliance
Ingram Content Group UK Ltd.
Pitfield, Milton Keynes, MK11 3LW, UK
UKHW021450080625
459435UK00012B/435